Fascinating Life Sciences

This interdisciplinary series brings together the most essential and captivating topics in the life sciences. They range from the plant sciences to zoology, from the microbiome to macrobiome, and from basic biology to biotechnology. The series not only highlights fascinating research; it also discusses major challenges associated with the life sciences and related disciplines and outlines future research directions. Individual volumes provide in-depth information, are richly illustrated with photographs, illustrations, and maps, and feature suggestions for further reading or glossaries where appropriate.

Interested researchers in all areas of the life sciences, as well as biology enthusiasts, will find the series' interdisciplinary focus and highly readable volumes especially appealing.

More information about this series at http://www.springer.com/series/15408

Jules Janick • Arthur O. Tucker

Unraveling the Voynich Codex

Jules Janick
Department of Horticulture & Landscape Architecture
Purdue University
West Lafayette, IN, USA

Arthur O. Tucker
Department of Agriculture & Natural Resources
Delaware State University
Dover, DE, USA

Contributors
Fernando A. Moreira
Independent Researcher
Calgary, AB, Canada

Elizabeth A. Flaherty
Department of Forestry and Natural Resources
Purdue University
West Lafayette, IN, USA

ISSN 2509-6745 ISSN 2509-6753 (electronic)
Fascinating Life Sciences
ISBN 978-3-030-08421-9 ISBN 978-3-319-77294-3 (eBook)
https://doi.org/10.1007/978-3-319-77294-3

Printed on acid-free paper

This Springer imprint is published by the registered company Springer Nature Switzerland AG.
The registered company address is: Gewerbestrasse 11, 6330 Cham, Switzerland

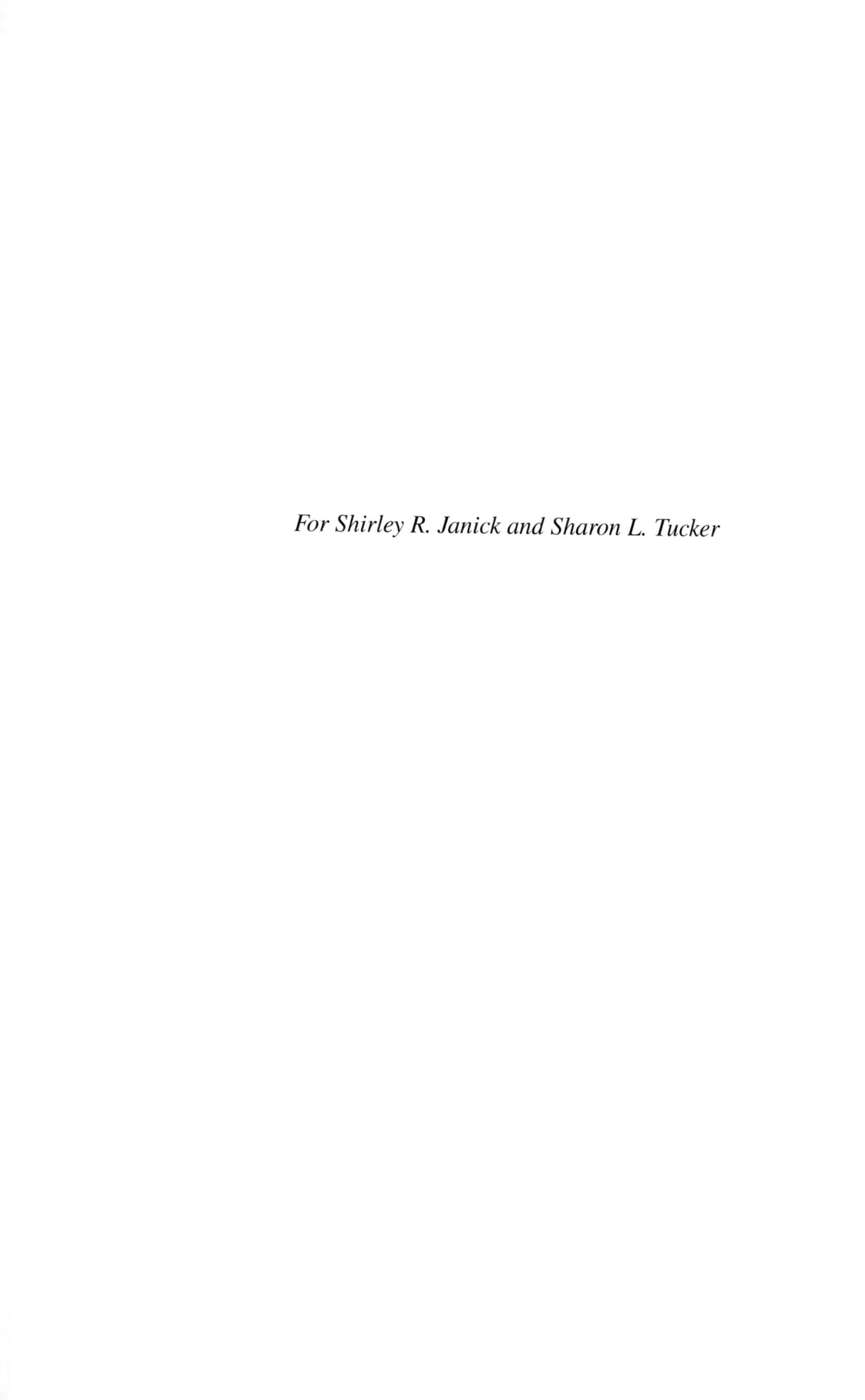

For Shirley R. Janick and Sharon L. Tucker

The cover illustration of three sets of images each of New World plants, animals, and Mexican volcanos provide evidence that the Voynich Codex is a post-Columbian Mexican manuscript.

	jaguarundi	La Malinche volcano
prickly pear cactus + agave	ocelot	Pico de Orizaba cauldron
sunflower	armadillo	Popocatèpetl volcano

Foreword

The *Voynich Codex* is a mysterious, bizarre hand-written manuscript discovered by the book dealer Wilfrid Voynich in 1912. Its unique symbols and text have defied translation attempts by world-eminent cryptologists. The *Codex* is encyclopedic in scope and includes approximately 359 images of plants or plant parts, making it primarily an illustrated herbal, a book that combines traditional plant lore and medicinal properties. But it is much more than that. The *Voynich Codex* also depicts more than 500 nymphs, mostly nude, cavorting in pools with weird plumbing. There are strange magic circles, including ones with zodiac, astronomical, and cosmological depictions. The codex includes a large foldout section with kabbalah-like images that may be interpreted as a map. Many of its pages appear to be medical recipes, poetry, or incantations. The *Voynich Codex* has captured the imaginations of many, but all have failed to make sense of it.

This volume summarizes the collaborative attempts of a botanist and emeritus herbarium director at Delaware State University, Arthur O. Tucker, and a horticulturist at Purdue University, Jules Janick, to unravel the *Codex* from a new perspective. We believe that previous attempts to get to grips with the *Voynich Codex* have taken a wrong approach because they have erred on its origins in time and place, relying upon interpretations rather than the hard evidence. Furthermore, no one previously has been able to make sense of its many parts. No one has been successful in deciphering the codex, which holds its secrets. Although we have not fully succeeded, progress has been made.

The collaboration led to an invited seminar by Tucker at Purdue University in 2014 and a coauthored presentation by Janick at the annual meeting of the American Society for Horticultural Science in 2015. A coauthored joint paper expanding plant identifications appeared in 2016. Finally, a symposium entitled *Mysteries of the Voynich Codex: A Meso-American Herbal,* organized by Janick and Tucker, was held in Atlanta in 2016. The symposium abstract caught the attention of Kenneth Teng, a Springer editor, and this volume is the result of those encounters.

The origins of our collaboration are revealing. We first met in 1990. Later in 2007, Janick invited Tucker to speak at a horticultural congress in Indianapolis concerning herbs, for which Tucker is a recognized expert. Tucker became interested in the *Voynich Codex* in June 2012, when he located a reference to it that coincided with a long interest in Latin American herbs and sixteenth century codices from New Spain. He was amazed at the large number of New World species in the *Voynich Codex* and incorrect identifications by nonbotanists. He sought out collaboration with Rexford H. Talbert, another herb expert and information technologist, formerly at NASA. This resulted in a manuscript entitled *A Preliminary Analysis of the Botany, Zoology, and Mineralogy of the Voynich Manuscript,* based on the identification of 37 plants, seven animals, and the mineral boleite, all indigenous to the New World. The manuscript was submitted in December 2012 to *HerbalGram*, a refereed journal of the American Botanical Council, and was published in 2013. It confirmed a 1944 paper by botanist Hugh O'Neill, which noted that the *Voynich Codex* contained New World plants and must have been written post-Columbus. Furthermore, the Voynichese symbols were decoded into an alphabet based on names attached to some of the plants in the Pharmaceutical section, providing the Rosetta Stone of the elusive codex.

The paper was generally treated with hostility by many members of the Voynich internet community, but received congratulations from academics. It proved a revelation to Janick, who had had minor contact with the *Voynich Codex*, first from a graduate student, Angela Catalina Ghionea, who was seeking advice for her doctoral thesis on magic and science, and later by Professor Lincoln Taiz, who submitted a manuscript on Voynich to Janick, who served as science editor for *Chronica Horticulturae*. Tucker's *HerbalGram* paper was immediately grasped by Janick as a breakthrough and a collaboration was formed that later included Fernando Moreira, a Canadian linguist, and Elizabeth A. Flaherty, a wildlife zoologist at Purdue University. The present book is based on this collaboration.

Janick worked on iconographic analysis of the *Voynich Codex*, and a key finding was achieved when a single foldout page (folio 86v), made up of six sheets, was identified as inspired by kabbalah. It was brought to the attention of Thomas Ryba, a philosophy professor at Purdue University and a theologian at Purdue's St. Thomas Aquinas Catholic Center, who confirmed the kabbalah similarity and suggested that it might be a map. Tucker made the inspired leap that the map was associated with the Celestial City of Jerusalem (Angelopolis), in the present-day state of Puebla, Mexico. The city was founded in 1530 by Toribio de Benavente (known as Motolinía), one of the famous Twelve Apostles, Franciscan missionaries who arrived in Mexico in 1524 upon the recommendation of Hernán Cortés after the Aztec conquest. This information put the *Codex* into context in time and space. It gave a potential earliest date to the *Codex* (1530) and suggested that it must have been written before 1571, the date the Inquisition was formally introduced into New Spain, when it would have been dangerous to put any kabbalah images in a manuscript. This also confirmed that there was a Franciscan connection to the *Voynich Codex,* as follows:

1. The Franciscan order had created El Colegio de Santa Cruz de Tlatelolco (in present-day Mexico City; also called Colegio de Santa Cruz) for the sons of

Nahua (Aztec) nobility, and we believe that the author of the *Voynich Codex* must have been associated with that institution.
2. The Franciscans had been supporters of the kabbalah in Spain and could explain this allusion in the codex.
3. Motolinía was a Franciscan friar.
4. Finally, Bernardino de Sahagún, the great chronicler of Aztec culture, was a Franciscan and at one time a professor and dean at the Colegio de Santa Cruz.

With this discovery, a general hypothesis could be created. The author/artist of the *Voynich Codex* must have been associated with the Colegio de Santa Cruz and brought up with Spanish, Western, and Aztec sensibilities (both transcultural and syncretic). As a result, we re-examined the codex through the lens of this hypothesis. Parts of the *Codex* began to confirm and reinforce our assumptions. For example, the decipherment of the Voynichese symbols (or alphabet) allowed the decipherment of a variant of the name of the city Huejotzingo, written on a drawing of the convent-fortress in circle 2 of folio 86v. The zodiac clearly showed an amalgamation of Aztec and Western sensibility, with the replacement of many of the traditional signs with animals native to the New World. The nymphs could be explained as part of the bathing ceremonies of the Aztecs. The many medicinal plants and bathing facilities for concubines were clearly compatible with the gardens at the palace of Nezahualcoyotl as well. Furthermore, folio 86v distinctly showed three volcanoes, which were major landmarks of central Mexico. The presence of the star cluster universally recognized as the Pleiades fitted our hypothesis, as this star cluster was a vital part of Aztec cosmology, and its 52-year cycle was an essential component of Aztec culture and theology associated with the New Fire ceremony.

We found our hypothesis to be reinforced by events that had occurred at the Colegio de Santa Cruz. Many students and staff, such as Bernardino de Sahagún, had become renowned, including Martinus (or Martin) de la Cruz, a Nahua staff physician, and Juannes Badianus (or Juan Badiano), a Nahua teacher of Latin, who collaborated on an illustrated Aztec herbal (now known as the *Codex Cruz-Badianus*), written in Latin in 1552. Others included the indigenous writer Pedro de San Buenaventura, the historian Juan Bautista de Pomar, great grandson of Nezahualcoyotl, and Gaspar de Torres, master of students at Colegio de Santa Cruz from 1568 to 1572, highly educated as a physician and lawyer, a supporter of Indian rights, and governor of Cuba in 1580. In addition, Torres' name and the ligated initials of Juan Gerson, an indigenous artist (*tlacuilo*), were embedded in the first botanical image (folio 1v) of the codex, which suggested that Torres might have been its author and Gerson its illustrator.

Aside from the hard evidence of plant, animal, and mineral identifications, we are fully aware that many of our individual assertions are speculative, but we believe that they are plausible. We continually find associations that provide evidence for the hypothesis of an origin in sixteenth century New Spain. We both admit to academic deficiencies in linguistics and astronomy. Scholars in these fields are needed to complete the picture and we are frustrated at our inability to persuade them to join our quest. We hope that this book will encourage a community of scholars to complete the task of translating the *Codex*, which, in the final analysis, holds the key. We are convinced that this would be crucial to a fuller understanding of post-colonial Aztec history because this codex comes to us unfiltered by Spanish or Inquisitorial censors. Although we have found evidence to support our hypothesis, we remain open-minded scientists, and look forward to data to prove, disprove, or expand our understanding of the *Voynich Codex*.

West Lafayette, IN, USA
Dover, DE, USA

Jules Janick
Arthur O. Tucker

Acknowledgements

There are many who have helped us in our quest. Fernando Moreira and Elizabeth A. Flaherty contributed chapters. Special thanks are given to Anna L. Whipkey for her skills in formatting the many figures and tables. We recognize with appreciation many of our colleagues who reviewed various portions of this manuscript and provided useful comments, including Irwin L. Goldman, Kim Hummer, Robert Joly, Patricia Ryan, Thomas Ryba, Ralph Scorza, Theodore M. Sovinski, John H. Wiersema, and Steven C. Weller. We acknowledge the Interlibrary Loan services at Purdue University, Delaware State University, and the University of Delaware, with special thanks to Brenda Richards, who navigated Spanish documents, and to Susan Yost, Cynthia Hong-Wa, and Thomas Zanoni, who assisted with botanical identifications. The support of Rex Talbert and Mark Blumenthal is gratefully acknowledged for initially believing in us. Finally, appreciation is extended to the staff of Springer International Publishing, including Kenneth Teng, who approached us after reading an abstract, and also to Kari Jensen who skillfully carried out the line editing. We are grateful for the strong support provided by Amy Goldman and the Lillian Goldman Charitable Trust.

Contents

Part I
An Introduction to the Voynich Codex

Chapter 1
Origin and Provenance of the Voynich Codex

Arthur O. Tucker and Jules Janick

Origins

In 1912, an unnamed manuscript was purchased from a Jesuit college in Italy by Polish book dealer Wilfrid Voynich (1865–1930). As it is properly a codex, we call it the *Voynich Codex*. Wilfrid Voynich brought it to the USA in 1914 and attempted sell it – albeit unsuccessfully. At one time the codex was in the court of Holy Roman Emperor Rudolf II (1552–1612), who was a great collector of books and art for his Kunstkammer, or cabinet of wonders.

The origins of the *Voynich Codex* remain mysterious, and current dogma has been that it is a fifteenth century European manuscript as it was found in Italy and the vellum has been dated to ca. 1425. A strong indication that the codex must have been written after 1492 was based on the identification of a sunflower and capsicum peppers by Hugh O'Neill in 1944 (discussed in detail in Chap. 4). But this evidence was not accepted by most Voynich researchers, who were fixated on the hypothesis that the codex might be an Old World document. This assumption was accepted despite botanical evidence to the contrary and still persists (Clemens 2016). However, there were a few iconoclasts who suggested or promoted a New World origin of the *Voynich Codex*.

The Mexican Connection

The first mention of a New World origin for the *Voynich Codex* that we could discover was by Jacques B.M. Guy of the Telstra Research Laboratories in Clayton, Australia. In a Letter to the Editor of *Cryptologia* (1991b), entitled *Voynich Revisited,* the sunflower and capsicum identifications are provided (without a citation or mention of Hugh O'Neill) as evidence that the codex postdated the encounter of America by Christopher Columbus, the Italian explorer and navigator. Jacques

J. Janick, A. O. Tucker, *Unraveling the Voynich Codex*, Fascinating Life Sciences, https://doi.org/10.1007/978-3-319-77294-3_1

B.M. Guy states that a botanist colleague "*soon identified a passionfruit, followed by a few native Australian plants … not a single European plant, and all the rest unknown or fantastic*." He concludes that the *Voynich Manuscript* "*is authentic*" and then mentions that the plants in the 1552 Aztec Badanius [sic] Herbal also are unrecognizable, quoting Blunt and Raphael (1979). However, they are recognizable and were identified by Emmart (1940). Jacques B.M. Guy posits the query: "*Is then the Voynich manuscript a treatise written in Nahuatl by a Mexican (Aztec, Totonac or other convert) who had been taught some Latin*?" He then hedges his supposition and ends his letter as follows: "*but I would be very surprised, very surprised indeed if it turned out to be written in Nahuatl. Still it was a tempting thought, wasn't it?*"

The next mention of a Mexican connection was the online reports/blogs of James C. Comegys, a seventh- and eighth-grade science teacher at Martin Luther King, Jr. Middle School in Madera, California. James C. Comegys, who has a Master of Arts degree in linguistics, investigated the Mexican origin of the *Voynich Codex* in collaboration with his twin brother, John D. Comegys. These blogs are no longer on the internet and have not been read by us. However, there is evidence in the title of a 1999 oral presentation, *The congruence of Nahuatl grammar with Voynichese* by James C. Comegys and John D. Comegys, delivered at the Annual Central California Research Symposium at Fresno State University. In addition, James C. Comegys is listed as the author of a document entitled *Keys for the Voynich Scholar: Necessary Clues for the Decipherment and Reading of the World's Most Mysterious Manuscript Which is a Medical Text in Nahuatl Attributable to Francisco Hernández and his Aztec Ticiti* [sic] *Collaborators,* dated 31 May 2001. One copy of this document is filed in a remote Library of Congress warehouse in Maryland and is impossible to access it unless you are a US federal employee. This unusual document consisting of 104 typed pages reads like a personal journal and contains a number of abstracts and unpublished manuscripts without an index. The following statement is found on page iii: "*Unmistakable evidence within the Voynich Manuscript places its origin in Mexico; hieroglyphs, plants, architecture, etc. Moreover the evidence is consistent with the themes and methods of Francisco Hernández Expedition to New Spain in 1570. The best evidence that the Voynich is Mexican however is in grammar, syntax, lexicon and orthography of the document itself.*" We admit to difficulties in understanding the document's linguistic evidence, as explained in Chap. 11. The Voynichese symbols were deciphered and some plants were identified by names, but we remain skeptical about the following: peanut, maguey, Spanish lavender, *micaxihuatol,* cacao tree, cacao flower, and maize. The phytomorph on page 25 (folio 25v) is identified as poinsettia based on a sketch of a "*little dragon breathing fire, making obvious the speculation that the plant is bright red in color*." The plant, which is vegetative, remains unidentified by us, but the "little dragon" has been identified as the Mexican horned lizard (*Phynosoma taurus*), as explained in Chap. 6. James C. Comegys makes four references to Jacques B.M. Guy and mentions that they were in correspondence.

A third reference to the Mexican origin of the *Voynich Codex* was published in *HerbalGram*, a peer-reviewed journal, by Arthur O. Tucker and Rexford H. Talbert, submitted in December 2012 and published in 2013, the major evidence being that

37 plants, one mineral, and most animals were indigenous to the New World. This book greatly extends this hypothesis and provides additional evidence based on iconographic analysis. This hypothesis was supported by two subsequently published papers, one on botanical identification (Tucker and Janick 2016, see Chap. 4), and another on the presumed illustrator and authorship of the *Voynich Codex* (Janick and Tucker 2017, see Chap. 15).

Finally, an unpublished manuscript by John D. Comegys appeared in 2014 (copyright 2013) entitled *The Voynich Manuscript: Aztec Herbal from New Spain.* This manuscript concluded a Mesoamerican origin for the *Voynich Codex* based on its artwork and Voynichese script. John D. Comegys presents evidence that the codex is written in a form of humanist hand known as courtesan script, the national script of Spain in the sixteenth century, and that the artwork in the botanical drawings is Mesoamerican in style or heritage, including Aztec glyphs and root forms. The language is presumed to be a combination of Nahuatl and Spanish.

Literature

The origin and provenance of the *Voynich Codex* remain mysterious and tortuous. The literature is widespread and includes a huge web presence, several books, and a few scholarly articles. These are listed chronologically with comments in Table 1.1. A history of the codex can be found in Brumbaugh (1978), D'Imperio (1978a), Kennedy and Churchill (2006), and Kircher and Becker (2012), and there is a detailed chronology by Zandbergen (2016, 2017), with many dates based on his discoveries. We are convinced from botanical and zoological evidence (not analyses) that the *Voynich Codex* is a sixteenth century New World document from New Spain and the subject of its origins is in large part the focus of this book. Any manuscript that contains a sunflower and an armadillo cannot be a pre-Columbian Old World manuscript, and we are amazed that the many people involved in this work have not embraced this simple fact. However, the path of the document from Mexico to Europe needs to be explained. The symbols were partially decoded (Tucker and Talbert 2013), and various plants, animals, a mineral, and cities were deciphered. Various cognates with Mexican languages have been found. However, (alas) we have been unable to decipher most of the text, which we believe to be a mixed synthetic language or Nahuatl lingua franca rather than a cipher. Our hypotheses on the *Voynich Codex* are in large part based on the evidence of botanical, zoological, mineralogical, and geographic identification, supplemented by historical, iconographic, and biographical correlates.

Two works summarizing the *Voynich Codex* were found to be informative, fair, and balanced: *The Voynich Manuscript: An Elegant Enigma,* by D'Imperio (1978a/1979), and *The Voynich Manuscript: The Mysterious Code That has Defied Interpretation for Centuries,* by Kennedy and Churchill (2006). There have been three recent facsimile editions: *The Voynich Manuscript: A facsimile of the complete work*, published by the Palatino Press (2016); *The Voynich Manuscript*, edited by

Table 1.1 *Voynich Codex* hypotheses, 1921–2017

Date	Author of hypothesis (background)	Putative date of Voynich	Putative location of Voynich writing	Putative Voynich author	Putative Voynich language	Evidence to support hypothesis	Publication of hypothesis	Notes
1921 (1928)	William R. Newbold (professor of philosophy, University of Pennsylvania, PA, USA)	13th century	UK	Roger Bacon (1214–1294)	Medieval Latin	Claimed to find depiction of Andromeda Spiral Nebula.	Newbold (1928), Theroux (1994)	Announced hypothesis in 1921 lecture, before the College of Physicians and the American Philosophical Society; argument refuted by Manly (1931), Richard Salomon (1931)
1943	Joseph M. Feely (lawyer, Rochester, NY, USA)				Medieval Latin	None cited	Feely (1943)	"*His unmethodical method produced text in unacceptable medieval Latin, in unauthentic abbreviated forms*." (Tiltman 1967/2002)
1944	Hugh O'Neill (plant taxonomist and curator, LCU Herbarium, Catholic University of America, Washington, DC, USA)	Post-1492				Identified four plants, including sunflower and capsicum pepper from the New World.	O'Neill (1944)	Worked from incomplete black and white photostats, but said that six botanists agreed with his identification of the sunflower

1945	Leonell C. Strong (Yale School of Medicine, CT, USA)	ca. 1553	UK	Anthony Askham (astrologer, fl. 1553)	Medieval English	None cited	Strong (1945)	Worked from incomplete photocopies
1950, 1970	William Friedman (National Security Agency, Washington, DC, USA) [Unpublished, but cited by others]				Synthetic universal language	None cited	Tiltman (1967/2002), Zimansky (1970)	*"It was clear that the productions of these two men were much too systematic, and anything of the kind would have been almost instantly recognizable."* (Tiltman 1967/2002). Theory supposedly found in an envelope in the archives of the Philological Quarterly's editor (Zimansky 1970)
1962	William Friedman (National Security Agency, Washington, DC, USA)		*"Definitely European:"* UK, France, Italy, or Germany		Latin-based, may be English, French, Italian, or Teutonic	None cited	Friedman (1962), Zimansky (1970)	Newspaper article by Elizebeth Friedman, reported on her husband's findings, contrary to Tiltman and Zimansky; reported that a "Dutch" botanist named "Holm" did not agree with O'Neill's plant identifications, but did not provide any other information

(continued)

Table 1.1 (continued)

Date	Author of hypothesis (background)	Putative date of Voynich	Putative location of Voynich writing	Putative Voynich author	Putative Voynich language	Evidence to support hypothesis	Publication of hypothesis	Notes
1963	Hellmut Lehmann-Haupt (bibliographic – a consultant to H.P. Kraus, owner of MS 1962–1969)		Northern Italy, also possibly central or southern Italy or the Arab world		Arabic?	None cited	Letter to John Tiltman dated 1 November 1963; see D'Imperio (1978a, b)	
1967 (2002)	John H. Tiltman (National Security Agency, Washington, DC, USA)	1500–1641	Europe			None cited	Tiltman (1967/2002)	Presented 4 March 1967 to Baltimore Bibliophiles, later expanded and presented to the NSA in 1975 and 1976, but only released by the NSA in 2002: "*My analysis, I believe, shows that the text cannot be the result of substituting single symbols for letters in the natural order*"
1974	Robert S. Brumbaugh (professor of medieval philosophy, Yale University)			More than one writer (Currier A and Currier B)	Latin	Incorrect identification of plants	Brumbaugh (1974)	"*Given this start, I hope that someone whose botany is better than my own will work through the 10 pages of plant drawings …*"

1976 (2002), 2007	James R. Child (computer analyst, NSA, Washington, DC, USA)	ca. 1500	Northern Europe		Unknown North Germanic dialect	None cited	Child (1976/2002), Child (2007)	*"The distribution of vowel and consonant letters, some of which are surely Latin letters, makes a cipher improbable."* *"… the Voynich manuscript is in a natural language especially reminiscent of those of the Germanic family."*
1976 (1992)	Prescott Currier (linguist, UK)		UK?	Roger Bacon? or John Dee? (1527–1608)		None cited	Currier (1976)	Two authors, A and B
1976	William Ralph Bennett (professor of engineering and applied science and physics, Yale University)				Language similar to Hawaiian	None cited	Bennett (1976)	*"… it is worth mentioning that there actually are languages in some parts of the world that do have values of the entropy per character as low as those listed"*
1977	DENDAI [DICK] (Henry Ephron, American cryptanalyst)	16th century	Nola, Italy	Giordano Bruno	Latin	None cited	Dendai (1977)	
1978	Robert S. Brumbaugh (professor of philosophy, Yale University)		Khazar (Ukraine)		Forgery to fool Rudolf II, written in pseudo-Latin	None cited	Brumbaugh (1978)	

(continued)

Table 1.1 (continued)

Date	Author of hypothesis (background)	Putative date of Voynich	Putative location of Voynich writing	Putative Voynich author	Putative Voynich language	Evidence to support hypothesis	Publication of hypothesis	Notes
1978	John Stojko		Europe?		Slavic language of the Khazars (Ukrainian)	None cited	Stojko (1978)	
1978 (1979)	Mary E. D'Imperio (computer analyst, NSA, Washington, DC, USA)					None cited	D'Imperio (1978a/1979)	Primarily a review of the literature, but concentrates upon European sources almost exclusively and ignores possible sources in the New World
1978 (2009)	Mary E. D'Imperio (computer analyst, NSA, Washington, DC, USA)					None cited	D'Imperio (1978b/2009)	Found statistical evidence for more than one scribe
1979 (2009)	Mary E. D'Imperio (computer analyst, NSA, Washington, DC, USA)	20th century				None cited	D'Imperio (1979/2009)	"*It is hard to imagine any directly underlying natural language plain text whose characteristics can explain the phenomena adequately.*" & "*… an agglutinative language such as Turkish would provide an additional interesting test*"

1986	Michael Barlow (Quebec, Canada)	ca. 1163	Northern Europe	Wilfrid Voynich (1865–1930)		None cited	Barlow (1986)	Fraud
1987	Leo L. Levitov (doctor of medicine)			Cathar	Polyglot oral tongue	None cited	Levitov (1987)	Claimed that this is a liturgical manual of the Endura, or Cathar suicide rite
1991, 2004, 2006	Jacques B.M. Guy (Telstra Research Laboratories, Clayton, Australia)	16th century			Chinese, Nahuatl	None cited	Guy (1991a, b, c, 1997), Kennedy and Churchill (2004, 2006)	Proposed Marco Polo connection
1995	Eugene Newsom (, US Army Signal Corp Officer in World War II)	17th century	Italy	Tommaso Campanella (1568–1609)			Newsom (1995)	Self-published pamphlet by the author, who died in 2004
1996	Sergio Toresella (Italian historian of medicine)		Italy			None cited	Toresella (1996)	Alchemical herbal by author who had a psychiatric disturbance
1997	Jacques B.M. Guy (Telstra Research Laboratories, Clayton, Australia)				Two dialects of a true, natural language	None cited	Guy (1997)	Distribution of signs provide evidence for real language
1999	Antoine Casanova (computer science Ph.D. candidate, University of Paris)	1571–1577	Mexico		Two languages detected, possibly four at most	None cited	Casanova (1999)	"*We regard the manuscript as one single text or as a conglomerate of cryptograms endowed with six separate alphabets*"

(continued)

Table 1.1 (continued)

Date	Author of hypothesis (background)	Putative date of Voynich	Putative location of Voynich writing	Putative Voynich author	Putative Voynich language	Evidence to support hypothesis	Publication of hypothesis	Notes
1999, 2001	James C. Comegys (middle school teacher, California) and John D. Comegys (twin brother of J.C. Comegys)	1551–1586	Mexico	Francisco Hernández and Aztec collaborators	Classical Nahuatl	Incorrect plant identification by middle school students	Comegys and Comegys (1999), Comegys (2001)	Claimed that reading is from top to bottom, right to left
1999	Robert L. Williams CT, USA				Meaningless text in Old Greek	None cited	Williams (1999)	*"I believe the manuscript is written in a language invented by it's* [sic} *author, who used symbol/letter substitution cipher base on Greek"*
2000–2011	Philip Neal (researcher, natural language processing; technical author, information technology)				*"I have long believed that the Voynich manuscript is written in a cipher which used some kind of grid which restricted the occurrence of each character in certain positions with a Voynich 'word.'"*	None cited	Neal (2000–2011)	

2001, 2004, 2006, 2012	James Finn (independent researcher, North Carolina, USA)			Michel de Nostredame (Nostradamus) or his son Cesar	Hebrew	None cited	Finn (2001, 2004), Kennedy and Churchill (2004, 2006), Dragoni (2012)	Includes the end of the world
2001	Gabriel Landini (School of Dentistry, University of Birmingham, UK)					None cited	Landini (2001)	By "*spectral analysis*" … "*The findings shown here favor the natural language theory.*"
2002, 2004, 2006	Edith Sherwood (organic chemist, Texas, USA)	Early 15th century	East Asia			Incorrect plant identifications	Sherwood (2002), Kennedy and Churchill (2004, 2006)	Da Vinci connection, birthing manual
2002–2003	Akinori Ito				Natural language	None cited	Ito (2002a, b)	"*VMS is written in some kind of natural language*"
2003	Zbigniew Banasik (Poland)	Some centuries BC			Pre-Manchu	None cited	Banasik (2003)	
2003	Dana Gibson (Middle East historian, Canada)	Pre-medieval	Middle East, recopied by European scribes during medieval times		Semitic, possibly Nabatean	None cited	Gibson (2003)	
2004	Manish Rajkarnikar (software engineer, Minnesota, USA)				Human language	None cited	Rajkarnikar (2004)	"*Based on above observation, it can be concluded that VMS has property similar to that of human text and is a bit different from gibberish*"

(continued)

Table 1.1 (continued)

Date	Author of hypothesis (background)	Putative date of Voynich	Putative location of Voynich writing	Putative Voynich author	Putative Voynich language	Evidence to support hypothesis	Publication of hypothesis	Notes
2004, 2006	Gerry Kennedy and Rob Churchill (English writers)	16th century				Incorrect plant and animal identifications	Kennedy and Churchill (2006)	"*Voynich manuscript … a beautiful object an enigmatic, alluring and enduring mystery that is, in the final reckoning, perhaps better left unsolved?*"
	Beatrice Gwynn (Dublin, formerly at Bletchley Park)	1500–1600	Europe (area between Rome, Avignon, Munich, Prague)		Middle High German	None cited	Kennedy and Churchill (2004, 2006)	Da Vinci connection, hygiene manual
	Petr Kazil (IT security auditor, Rotterdam)	20th century		Unknown and unknowable		None cited	Kennedy and Churchill (2004, 2006)	Hoax
	Tim Mervyn	1450–1500	Northern Italy	Edward Kelley (1555–1597), John Dee, and/ Francis Pucci	Cipher	None cited	Kennedy and Churchill (2004, 2006)	Used notes from his uncle, Peter Long, a senior figure in the British Signal Intelligence
	Matthew Platts			Unknown	Romance dialect	None cited	Kennedy and Churchill (2004, 2006)	Found system of herbal medicine
	Jorge Stolfi (computer scientist, Brazil)	1450–1499	Italy, Spain, southern Germany		Proto-Manchu or Jurchen (Chinese)	None cited	Kennedy and Churchill (2004, 2006)	Based on a theory proposed by Zbigniew Banasik

2004, 2013	Gordon Rugg (computer scientist, School of Computing and Mathematics, Keele University, UK) and Laura Aylward	1503–1631	France	Edward Kelley with the help of John Dee	Gibberish, (probably based upon English or Latin)	None cited	Rugg (2004a, b), Aylward and Rugg (2004), Kennedy and Churchill (2004, 2006), Rugg (2013)	Assumed writer employed a Cardan Grille and thus a hoax, but did not cite evidence to support this hypothesis; supported by Schinner (2007). Disproven by Hermes (2012)
2004–2017, 2006, 2017	René Zandbergen (satellite navigation expert, European Space Agency)	15th century	England? Alpine region of northern Italy?	Fringe scientist?	Unknown, perhaps Arabic or Sanskrit derivation? Hoax?	None cited	Zandbergen (2004–2017, 2016, 2017), Kennedy and Churchill (2004, 2006)	
2005	Lawrence and Nancy Goldstone (historians & novelists, U.S.)	1470–1608	UK	Roger Bacon	Cipher	None cited	Goldstone and Goldstone (2005)	*"With all the failures and dead ends, it becomes tempting to wonder if Newbold might have been correct back in 1921. More than eighty years later, no one has really done any better"*
2005	Ursula Papke and Dirk Weydemann (Germany)				Presented concepts, not encrypted text	None cited	Papke and Weydemann (2005)	
2005, 2006, 2009	Claudio Marcelo Dos Santos (journalist, Argentina)	1450–1480	Milan	Edward Kelley and John Dee	Gibberish	None cited	Dos Santos (2005, 2006, 2009)	Quotes Gordon Rugg repeatedly

(continued)

Table 1.1 (continued)

Date	Author of hypothesis (background)	Putative date of Voynich	Putative location of Voynich writing	Putative Voynich author	Putative Voynich language	Evidence to support hypothesis	Publication of hypothesis	Notes
2006	Nicholas Pelling (computer programmer, UK)	?		Antonio Averlino (1400–1469), encrypted by Circo Simonetta (1410–1480)	Cipher; later annotations of months in Occitan	None cited	Pelling (2006)	Reviewed by Buonafalce (2007): "*Some of his conclusions are rather extravagant. Despite his efforts, he does not arrive at any convincing decipherment of the encoded text*"
2008	Erhard Landmann (linguist)				Latin	None cited	Landmann (2008)	Claims that the text deals with spacecraft, space origin of man, and Germanic mythology
2009	Richard Rogers (US Army/Navy computer programmer)					None cited	Pelling (2006)	
2009	Robert Teague (novelist, US)					None cited	Teague (2009)	Claims that the script is written vertically by the use of a "Magic Rectangle"

2009, 2011	Sravanna Reddy (Dept. of Computer Science, University of Chicago) and Kevin Knight (Information Sciences Institute, University of Southern California)					None cited	Knight (2009), Knight and Reddy (2011a, b)	*"Some features – the lack of repeated bigrams and the distributions of letters at line-edges – are linguistically aberrant, which others – the word length and frequency distributions, the apparent presence of morphology, and most notably, the presence of page-level topics – conform to natural language-like text"*
2009, 2012	Erich von Däniken (novelist)	1404–1438	Northern Italy			None cited	Von Däniken (2009), Dragoni (2012)	Connected with the Book of Enoch
2010	Aldo Gritti (novelist)	Pre-1912		Wilfrid Voynich		None cited; as a novel, this mixes fact and fiction	Gritti (2010)	Forgery with relationships to the death of Pierre Curie, the murder of Rosa Luxemburg, and the sinking of the Titanic
2010	Erich H. Peter Roitzsch (computer scientist, Germany)		Asia			None cited	Roitzsch (2010)	

(continued)

Table 1.1 (continued)

Date	Author of hypothesis (background)	Putative date of Voynich	Putative location of Voynich writing	Putative Voynich author	Putative Voynich language	Evidence to support hypothesis	Publication of hypothesis	Notes
2011	Grezorgz Jaśkiewicz (computer scientist, Warsaw University of Technology)	Medieval				None cited	Jaśkiewicz (2014)	"*The study shows the most similar languages according to these characteristics of a natural language*," but only analyzed languages in three areas in Europe and Asia
2011	Angelica Garel (student, Simon Fraser Univ., Canada)	13th century	UK	Roger Bacon	Untranslatable art piece	None cited	Garel (2011)	
2011	Viekko Latvala ("prophet of God," businessman, Finland)				Sonic waves and vocal syllables; mixture of Spanish and Italian	None cited	Watson (2011)	Translation based on direct line to God
2011	Lincoln Taiz (professor emeritus, University of California, Santa Cruz) and Saundra Lee Taiz	1570–1610				None cited	Taiz and Taiz (2011)	Claimed balneological section based on plant water relations
2012	Fabrice Kircher and Dominique Becker (novelists, France)	1404–1438	Northern Italy	Unknown traveler	Polyglot mixture of German, Swedish, Dutch Latin, English, Gaelic, and Nahuatl	None cited	Kircher and Becker (2012)	

2012, 2014	Thomas E. O'Neil (historian and novelist, USA	700–800	Italy		Italian	Architectural images similar to San Biagio Montepulciano, Castle Montepo, and Mura di Scansano	O'Neil (2012, 2014)	
2012	Jürgen Hermes (linguist, Germany)	1st century			*"The words morphologically have a very regular structure"*	None cited	Hermes (2012)	
2012	Tim Ackerson	700–800			Pre-Welsh or Scottish	None cited	Dragoni (2012)	
	Dan Burisch			Roger Bacon (ca. 1219/20 to ca. 1292	Cipher in Hebrew	None cited	Dragoni (2012)	*"describes sort of alien technology from the future to create or change DNA with the sound"*
	Karel Dudek			Georg von Handsch Limuz (1529–1595)		None cited	Dragoni (2012)	
	Wayne Herschel					None cited	Dragoni (2012)	Writings from Jesus to Judas
	George Hoschel				Ancient Latin	None cited	Dragoni (2012)	Recipe book
	Volkhard Huth	1480–1500	Germany				Dragoni (2012)	
	Jody Maat				Old Dutch	None cited	Dragoni (2012)	
	Adam D. Morris			Hieronymus Reusner (1558–16--?)		None cited	Dragoni (2012)	Connected with Reusner's Pandora

(continued)

Table 1.1 (continued)

Date	Author of hypothesis (background)	Putative date of Voynich	Putative location of Voynich writing	Putative Voynich author	Putative Voynich language	Evidence to support hypothesis	Publication of hypothesis	Notes
2012	Rolando Hernandez Rivero (novelist, Italy)				Old Spanish with Latin and English	None cited	Dragoni (2012)	
	Richard SantaColoma (jeweler and historian, USA)	16th century		Cornelius Drebbel (1572–1633) or Francis Bacon (1561–1626)	Cryptographic	None cited	Dragoni (2012), SantaColoma (2017)	
	Dirk Schröder					None cited	Dragoni (2012)	Work of kabbalah with numerological powers
	Mark Sullivan				Latin	None cited	Dragoni (2012)	
	Mandy Tonks			Wilfrid Voynich		None cited	Dragoni (2012)	Forgery
2013	Diego R. Amancio, Osvaldo N. Oliveira, Jr., Luciano da F. Costa, Eduardo G. Altmann and Diego Rybski	15th century	?		Mostly compatible with natural languages and incompatible with random texts	None cited	Amancio et al. (2013)	

2013 (2005)	Johnathan Dilas (der Matrixblogger, Germany)				Dreams and visions written down in an unreadable language	None cited other than reference to split-brain research	Dilas (2013, 2005)	*"… we do not have the ability to decipher the manuscript because a certain area of the brain is not active in our daily lives. So when the manuscript is at hand, we are just as unable to read as it is impossible in a dream."*
2013	Claudio Foti (novelist, Italy)	Early 15th century	Italy	Gian Francesco Poggio Bracciolini (1380–1459)		None cited	Schmeh (2013)	
2013	Walter Grosse (secretary and novelist)					Incorrect plant identifications	Grosse (2013)	A language of plants inspired by a pre-Babel language
2013	Marcelo A. Montemurro and Damián H. Zanette (computer scientists, University of Manchester, UK and Centro Atómico Bariloche e Instituuto Balseiro, San Carlos de Bariloche, Ro Negro, Argentina)	13th or more likely 15th century	Castel del Monte, Apulia, Italy			None cited	Montemurro and Zanette (2013)	Real language sequence, not gibberish; found semantic word networks and other statistical features

(continued)

Table 1.1 (continued)

Date	Author of hypothesis (background)	Putative date of Voynich	Putative location of Voynich writing	Putative Voynich author	Putative Voynich language	Evidence to support hypothesis	Publication of hypothesis	Notes
2013	Giuseppe Fallacara and Ubaldo Occhinegro (architects, Italy)	ca. 1565	Central New Spain	Copy of an earlier document written for Frederick II		Architectural similarities of images to Castel del Monte	Fallacara and Occhinegro (2013)	Misquote or omit evidence to the contrary, i.e., claim that McCrone found gum arabic (when they reported it was an unknown gum, not gum arabic), and they ignore the findings of O'Neill
2013, 2016	Arthur O. Tucker (emeritus herbarium director, Delaware State University), Rexford H. Talbert (information research scientist, NASA), and Jules Janick (Purdue University)	Mid-16th century	New Spain	Spanish educated Aztec	Primarily extinct dialect of Nahuatl from central Mexico with borrowed words from Spanish, Mixtec, and Taino	Plant, animal, and mineral identifications with distribution from Texas, west to California, south to Nicaragua	Tucker and Talbert (2013), Tucker and Janick (2016)	Calligraphy similar to *Codex Osuna*, nymphs similar to sibyls in Casa del Déan, "bird glyph" found in Mesoamerican codices and civil documents
2014 (©2013)	John D. Comegys (twin brother of James C. Comegys)	16th century	Mexico	Unknown native	Classical Nahuatl and Spanish	None cited	Comegys (2013)	Creation of pdf on 22 March 2014, but title page says ©2013; calligraphy similar to *Codice Otlazpan*, *Codex Osuna*, and *Codice de Santa Maria Asuncion*

2013	Ute Packheiser (novelist, Germany)			Paolo dal Pozzo Toscanelli (1397–1482)	Italian	None cited; as a novel, it mixes fact and fiction	Packheiser (2013)	
2013, 2014	Stephen Bax (professor of applied linguistics, University of Bedfordshire, UK)	13th to 14th centuries	?			Incorrect plant identifications	Bax (2013, 2014)	Sound value of 14 syllables and 10 words determined based on planets and the Pleiades
2014–2016	Jutta Kellner (Germany)			Several writers	Latin written in a "cryptological alphabet"	Incorrect plant identifications	Kellner (2014)	
2014, 2016	Torsten Timm (novelist, Germany)	Late 16th century	UK		Pseudo text	None cited	Timm (2014, 2016)	
2014	Peter Roush and Bryce Shi (students, School of Electrical and Electronic Engineering, University of Adelaide, Australia)					None cited	Roush and Shi (2014)	"*Based on characteristics such as word length distribution and WRI, the text appears similar to languages such as Hebrew and Latin.*" NB Only compared with English, Latin, Hungarian, Hebrew, Russian, Italian, and Chinese

(continued)

Table 1.1 (continued)

Date	Author of hypothesis (background)	Putative date of Voynich	Putative location of Voynich writing	Putative Voynich author	Putative Voynich language	Evidence to support hypothesis	Publication of hypothesis	Notes
2014	Stephan Vonfelt (student, Université de Toulouse Le Mirail, Laboratoire, Lettres, Langages et Arts, France)				Chinese?	None cited	Vonfelt (2014)	*"In the light of its characters, the Voynich manuscript brings out an ordered landscape … These elements undermine the hypothesis of a forger pressed by money."*
2015 (2006)	Richard Carthago (novelist, Germany)	16th century		John Dee (1527–1608) & Edward Kelley (1555–1597)	Symbolic language	Incorrect plant identifications	Carthago (2006, 2015)	Title page says 2006, 2015. Translates only first 13 pages
2015	Brian Cham (research assistant, media/software student and freelance designer)				Artificial language	None cited	Cham (2015)	*"Since the Voynich Manuscript's text does not seem to fit a natural language in these tests, nor is it random, then it must be artificial, in which case there is no reason for the CLS to fit."*
2015	Alain Touwaide (historian of medicine, formerly Smithsonian Institution)			Possibly Wilfred Voynich		None cited	Touwaide (2015)	Possibly a fraud

2016	Bradley Hauer & Grzegorz Kondrak (Department of Computing Science, University of Alberta, Edmonton, Canada)	15th century			Hebrew	None cited; assumes that a dictionary of the language exists	Hauer and Kondrak (2016)	*"The closest language according to the letter frequency method is Mazatec, a native American language from southern Mexico. Since the VMS was created before the voyage of Columbus, a New World language is an unlikely candidate."* & *"The top-ranking languages according to the decomposition pattern method are Hebrew, Malay (in Arabic script), Standard Arabic, and Amharic, in this order."*
2016	Raymond Clemens (librarian, Yale University)	Medieval; dates from carbon-14 dating of vellum	Europe			None cited	Clemens (2016)	Repeats past claims, no original research
2017	Nicholas Gibbs (English historian & TV writer)	1403–1438	Mediterranean country		Latin ligatures for words	Similarity to medieval treatises, including 12th century medical treatises from Salerno, Italy	Gibbs (2017a, b)	Assumption that "*each character in the Voynich manuscript represents an abbreviated word and not a letter*" would greatly limit text

(continued)

Table 1.1 (continued)

Date	Author of hypothesis (background)	Putative date of Voynich	Putative location of Voynich writing	Putative Voynich author	Putative Voynich language	Evidence to support hypothesis	Publication of hypothesis	Notes
2017	Jules Janick (Purdue University) and Arthur O. Tucker (emeritus herbarium director, Delaware State University)	Mid-16th century	New Spain	Illustrator (*tlacuilo)* was Juan Gerson and author was Gaspar de Torres	Mixed synthetic language with contemporary languages of New Spain	Ligated initials "JGT" and signature "Gasp. Torres" on folio 1v	Janick and Tucker (2017)	Juan Gerson was the *tlacuilo* (painter) at Tecamachalco and Casa del Deán. Gaspar de Torres was a medical doctor, estate lawyer, linguist of New Spain, master of students at Colegio de Santa Cruz, and governor of Cuba
2017	Stephen Skinner (Dee scholar), Rafał T. Prinke (historian of alchemy and esotericism) and René Zandbergen (space-flight dynamics researcher, European Space Agency)	First half of 15th century	Northern Italy	Jewish herbalist, astrologer, and/or physician		Similarity to Ghibelline castle architecture	Skinner et al. (2017)	"*… complete absence of any Christian imagery,*" "*most herbal illustrations cannot be identified with any certainty,*" Hugh O'Neill "identifications were far from certain …"

NSA National Security Agency, *VMS* Voynich Manuscript, *WRI* word recurrence interval, *CLS* curve–line system

Clemens (2016) and published by the Beinecke Rare Book and Manuscript Library at Yale University and *The Voynich Manuscript: The World's Most Mysterious and Esoteric Codex,* published by Watkins with a foreword by Stephen Skinner and an introduction by Rafal T. Prinke and René Zandbergen (Skinner and Zandbergen 2017).

One of the difficulties of *Voynich Codex* studies is that many of the manuscripts are web documents. Web blogs are not considered published in the scientific sense and often are transitory in nature. We attempted in this book to restrict our citations to books and articles in scholarly journals, but most of these on the *Voynich Codex* are involved with cryptology and linguistics, such as Montemurro and Zanette (2013). We are reluctant to comment on self-published books or non-published internet comments and chatter on the codex that can be found on the vibrant internet, but we have made a number of exceptions in the work, including: Pelling (2006), Sherwood and Sherwood (2008), Velinska (2013), Comegys (2013), Bax (2014), Zandbergen (2017), and SantaColoma (2017).

Voynich Resources

Mary E. D'Imperio (1978a). *The Voynich Manuscript: An Elegant Enigma* This 104-page pamphlet, based on photostats of the *Voynich Codex* with 45 hand-drawn illustrations, is an invaluable resource with extensive details on cryptology. It contains an index. It is very descriptive and provides historical comparisons. She concludes that the codex is a product of the sixteenth century (see page 51 of the pamphlet). Short shrift is given to plant and animal identification, although Elizebeth Friedman is quoted on Hugh O'Neill's identification of two American species including her comment that the Dutch botanist Holm identified 16 European plants, but the published source for this assertion was not found. There is scant information on folio 86v.

Gerry Kennedy and Rob Churchill (2004, 2006). *The Voynich Manuscript: The Mysterious Code That Has Defied Interpretation for Centuries* This volume, written by two professional writers, is a very fair discussion of the various hypotheses of the *Voynich Codex.* Hugh O'Neill's identification of the sunflower is mentioned, but the comment is made that the medieval wild version would contain a smaller head. However, the sunflower with a large head was well known in sixteenth century Mexico (see discussion in Chap. 16). They conclude that a holistic approach to the *Voynich Codex* is required.

René Zandbergen is an independent scholar and has a master's degree and Ph.D. in aeronautics and space engineering from Delft, the Netherlands. He is the world's expert on the history and origins of the *Voynich Codex* and has made many valuable discoveries in this area. His findings are summarized in a foreword to a *Voynich Codex* facsimile (Skinner et al. 2017) and in a huge website, http://www.voynich.nu, which is essential reading for *Voynich Codex* researchers. He claims that there is no doubt that the codex is from northern Italy in the fifteenth century and has been written by "*one brain*," but admits he has changed his opinion in the past. He makes no differentiation between author and illustrator. He cites Hugh O'Neill's sunflower identification, but finds contrary identifications by nonbotanists highly convincing. He makes no reference to Tucker and Talbert (2013) or any allusion to a New World origin. There are no remarks about the volcanoes in folio 86v. He makes no reference to animal identification and refers to the critter in folio 25v as a dragon. He suggests various approaches to solving the decipherment problem and is concerned whether the *Voynich Codex* represents a meaningful text.

Provenance

The conventional wisdom is that the *Voynich Codex* was once part of the famous curiosity cabinet (Kunstkammer) of Rudolf II. Son of Maximilian II, Rudolf II was at one time, king of Germany, Bohemia, Hungary, and Croatia, and Holy Roman emperor from 1576 to 1612. He is known as the greatest art patron and collector in the world, and there are famous portraits of him (Fig. 1.1). After his death, the codex apparently had various owners and ended up in a monastery library. It was purchased in 1912 by the Polish book dealer Wilfrid Voynich (Fig. 1.2a) from Jesuits at the Villa Mondragone in Frascati, Italy, who were discreetly selling some of their holdings. Wilfrid Voynich was convinced that the work was written by the

Fig. 1.1 Two portraits of Emperor Rudolf II: (**a**) by Hans von Aachen, and (**b**) as Vertumnus, god of seasonal change, by Giuseppe Arcimboldo

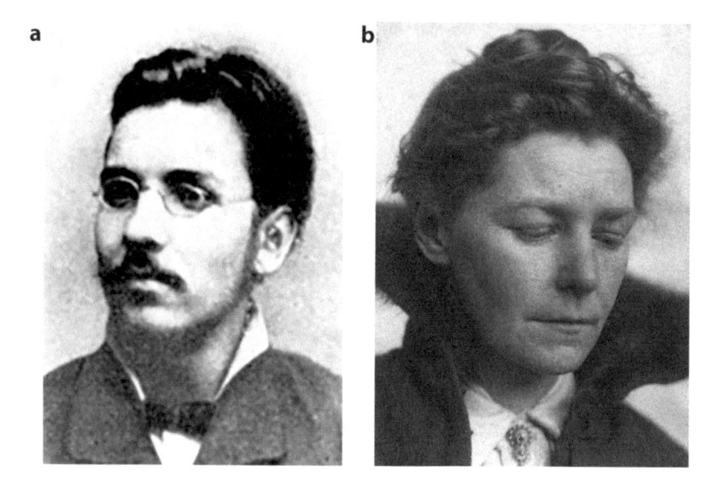

Fig. 1.2 Portraits of: (**a**) Wilfrid Voynich, 1895, and (**b**) his wife Ethel Boole

Englishman Roger Bacon (1214–1292), a Franciscan philosopher known as Dr. Mirabilis. At one time, Voynich valued the work at the then enormous sum of $160,000, but failed to find a buyer. Following the death in 1960 of Voynich's wife Ethel Lilian Boole (daughter of the famous mathematician George Boole; Fig. 1.2b), Voynich's secretary, the indefatigable Anne Nill (who was part owner), sold it to Hans Kraus, a book dealer, for $24,500. Kraus also failed to find a suitable buyer and donated the *Voynich Codex* to the Beinecke Rare Book and Manuscript Library at Yale University in 1969. Cataloged as Beinecke MS 408, the famous codex is now freely available on the web and not copyrighted.

The first historical record of the *Voynich Codex* was uncovered by René Zandbergen in a 1639 letter by Georg Baresch (1595–1662), an obscure alchemist from Prague, to Athanasius Kircher (1601–1680), a Jesuit priest and scholar (see Chap. 10). Apparently, Georg Baresch owned the codex, sent a copy of the script to Athanasius Kircher asking in vain for help, and commented that the codex represented "Egyptian science."

Evidence that the *Voynich Codex* was previously in the collection of Rudolf II is indirect. A 1665 letter from Johannes Marcus Marci (1595–1667), a Bohemian doctor, scientist, and rector of the University of Prague, to Athanasius Kircher claimed that Rudolf II purchased the codex for 600 gold ducats (about US $20,000 in the present day). A signature, Jacobi à Tepenecz, the ennobled name of Jacobus Horčicky (1575–1622), imperial chemist and personal doctor of Rudolf II, was discovered on the first page by Wilfrid Voynich, who applied chemicals to the page. It was later revealed by UV-induced visible fluorescence (Zyats et al. 2016: 31). The signature confirms that the codex was associated with the court of Rudolf II and existed between 1608, the date when Horčicky was ennobled by Rudolf II, and his death in 1622. There is little doubt that the *Voynich Codex* was once in the possession of Rudolf II. How he obtained it is still debated and mysterious.

Fig. 1.3 (**a**) John Dee and (**b**) Francisco Hernández

John Dee

From whom did Rudolf II obtain the *Voynich Codex*? Various hypotheses have been suggested and the evidence is circumstantial. One possible person is John Dee, 1527–1608 (Fig. 1.3a), first proposed by Wilfrid Voynich. John Dee was advisor to Queen Elizabeth I and signed his confidential letters to her with the cryptic "007" (which was the inspiration for British author Ian Fleming's James Bond). The supposed relationship of John Dee to the *Voynich Codex* is controversial, but remains part of the lore of the codex. Doubters included Prinke (n.d.) and Zandbergen (2016) who state that most Voynich scholars no longer support the John Dee connection. We are convinced that neither John Dee nor his occultist friend Edward Kelley (also spelled Kelly)was the author, but it is conceivable that either of them might have possessed the manuscript.

John Dee, mathematician, astronomer, and astrologer, devoted his life to the study of alchemy. He wrote on kabbalah, a form of Jewish mysticism, and the occult, and he was a learned man straddling the worlds of science, alchemy, and magic. He owned one of the largest libraries in England and was close to William Cecil (Lord Burghley), chief adviser to Queen Elizabeth I, and Francis Walsingham, principal secretary to the queen and head of the secret service. John Dee was invited to Poland and had a 1-h audience with Holy Roman Emperor Rudolf II in Prague Castle in 1584. As an agent of the British secret service, John Dee provides a possible link connecting the codex in Mexico and Central Europe via London, but the evidence is highly speculative.

There is a tenuous but intriguing connection between John Dee and the *Voynich Codex*. A 1672–1673 letter (Wilkins 1884) from Sir Thomas Browne (English polymath) to Elias Ashmole (celebrated English collector and namesake of the Ashmolean Museum in Oxford, UK) states that Arthur Dee (son of John Dee, born in 1579) lived with his father in Bohemia (present-day Czech Republic), both at Prague and in other parts of the country. The younger Dee had seen "*a booke lying by it containing nothing butt hieroglyphicks, which book his father bestowed much time upon; but I could not heare that he could make it out*" (Browne 1931). Arthur

Dee was between 5 and 10 years old during his father's sojourn in Bohemia from 1584 to 1589, and Prinke (n.d.) asserts that the text he observed was likely to be *Angelicus Opus,* a work in alchemy, which, according to John Dee's diary, contained nothing but hieroglyphics. Wilfrid Voynich assumed that John Dee sold the codex to Rudolf II based on a letter in Latin from Johannes Marcus Marci to Athanasius Kircher dated 1665 (which accompanied the *Voynich Codex*): "*Dr. Raphael, tutor in the Bohemian language to Ferdinand III, then King of Bohemia, told me the said book had belonged to the Emperor Rudolf and that he presented the bearer who brought him the book 600 ducats*" (Kennedy and Churchill 2006: 19–20). John Dee was fleeing the court of Rudolf II in 1586 and his diaries are remarkably scant, but his diary of 17 October 1586 does indicate that he had 630 ducats in his possession: "*after new disturbances and quarrels with Fr. Pucci, by reason of the money he desired of us, out of generosity and in the name of God, and as it were from the servants of God (rather than from us personally): Edward Kelley and I resolved, for the avoidance of the many scandals, which Pucci had invented and stirred up against us, by reason of his 800 florins, which he had earlier refused when we were ready to pay him, and when we showed him 630 ducats in the presence of God, so that he might accept what he deemed to be his* ..." (Fenton 2000).

If the *Voynich Codex* is a Mesoamerican manuscript, as we contend, its possible route from Mexico to Europe needs to be explained. If it arrived in Spain, it would have remained in the Spanish archives or may have been in private hands in Spain, as were many of these early manuscripts. The provenance of the *Codex Mendoza* of 1542, an Aztec pictorial history of rulers and conquests, is instructive. This codex, named after Antonio de Mendoza, viceroy of New Spain, who may have commissioned it, was shipped to Spain, but the fleet was attacked by French privateers (pirates), and the codex came into the possession of Andre Thevet, French Franciscan priest, cosmographer to King Henry II of France, and Brazilian explorer. It was later sold to the famous English geographer Richard Hakluyt for 20 French francs, passed to Samuel Purchase, then to his son, then to John Selden, and was eventually deposited into the Bodleian Library at the University of Oxford, where it remained in obscurity until 1831 (Berdan and Anawalt 1997). One hypothesis is that the *Voynich Codex* might have a similar story, and there is some confirmatory evidence, as some of the zodiac materials contain dates written by a later hand (see Chap. 8).

There are other potential routes to explain how a codex from New Spain could get to Europe. The famous navigator Sir Francis Drake (1540–1596), sea captain and privateer, had two forays to Mexico. In 1568, Drake, sailing with his second cousin Sir John Hawkins, went ashore in the port of San Juan de Ulúa (an island near present-day Veracruz, a Mexican state bordering the Gulf of Mexico), but the fleet was trapped. Drake and Hawkins escaped, but suffered a disastrous defeat by the Mexican fleet. Veracruz is a plausible home of the *Voynich Codex*, but it is unlikely that Sir Francis Drake could have obtained it in San Juan de Ulúa. However, on the secret circumnavigation voyage of 1577–1580, Drake's crew plundered Guatulco (or Huatulco), a Pacific Coast city on the Gulf of Tehuantepec in Oaxaca state, Mexico, ransacked the town, and robbed the home of Francisco Gomez Rengifo (a wealthy landowner) of 7000 pesos and "goods" (Nuttall 1914). The return voyage to England in 1580 (with the ship's hull filled with treasure) was extremely successful financially. The plunder would have been expected to be turned over to the queen's agents for distribution to the backers (of which the queen was one). Not only was this foray of Sir Francis Drake kept secret until after Elizabeth I's death, but the plunder was never inventoried, and we know from period accounts that Drake kept unknown goods for himself (Bawlf 2003; Fletcher 1854; Kelsey 1998, 2003; Sugden 2006; Thrower 1984; Walling 1907). There is another path to England. In 1585, Sir Francis Drake sacked the cities of Santo Domingo in Hispaniola, Cartagena (in present-day Colombia), and Saint Augustine (in present-day Florida), and on his return rescued colonists at Roanoke off the coast of what is now North Carolina. We believe that there is a possibility that the codex may have been in Santo Domingo and this will be explained in Chap. 15. These scenarios tie together Sir Francis Drake's forays into New Spain, John Dee, Rudolf II, and the *Voynich Codex*.

We speculate that John Dee, as secretary to Elizabeth I, might have received the *Voynich Codex* as a gift. If so, he could very well have sold it to Rudolf II or his agents (note that the letter from Johannes Marcus Marci to Athanasius Kircher says "bearer," not John Dee). This may explain his possession of the 630 ducats stated in John Dee's diary of 17 October 1586. Roberts and Watson (1990) suggest that the folio numbers in the *Voynich Codex* might be in John Dee's hand, but a comparison with the numerals in Dee's book catalog is not convincing (Kennedy and Churchill 2006: 72). However, it should be noted that John Dee was very possessive of his books and he repeatedly bought, not sold, them. Sir Francis Drake was extremely religious and despised John Dee for his "necromancy," but Drake did meet with Edward Kelley, the devious (and possibly mentally unstable) friend of John Dee who accompanied Dee to Prague and may have been the seller (French 1984; Halliwell 1842; Parry 2011; Skinner 2011).

Was the Voynich Codex in John Dee's Library?

Roberts and Watson (1990) asserted that the *Voynich Codex* was a non-cataloged book in the library of John Dee, listing it as DM 93 (Beinecke MS 408). However, they admitted that the external evidence that identified non-cataloged books (such as Dee's name, marks, or symbols) was inconclusive for the codex. Furthermore, they claimed that John Dee foliated (numbered the folios) throughout the *Voynich Codex*, which was confirmed by a comparison with diary entries found in the Bodleian Library, MSS Ashmole 1790 folio 9v (pages 172–173). An entry in the Beinecke Rare Book and Manuscript Library catalog noted by Prinke (n.d., footnote 15) offers confirmation: "*we thank A.G. Watson for confirming this identification through a comparison of the Arabic numerals in the Beinecke manuscript with those of John Dee in Oxford, Bodleian Library Ashmole 1790.*" However, this claim was questioned by Prinke based on the numbers 8, 1, and 4, which were illustrated on page 71 of *The Voynich Manuscript* (Kennedy and Churchill 2006). We have checked Ashmole 1790, but can only say that the numbering of the folios of Ashmole 1790 and the *Voynich Codex* are so variable that no firm conclusions can be drawn at this time.

Francisco Hernández

There is a third, more direct route of the *Voynich Codex* from New Spain to Spain to Rudolf II. On 17 March 1564, Rudolf (who was aged 11) and his brother Ernest arrived in Barcelona to be raised in the court of his uncle, King Philip II of Spain, and remained there until 1572. During this time, they had contact with Philip II's physician, Francisco Hernández de Toledo, 1514–1587 (Fig. 1.3b). Hernández was a Spanish naturalist and in 1567 became court physician to Philip II. In 1570, Francisco Hernández, at the request of the king, embarked on a seminal scientific expedition to New Spain to study medicinal plants. He traveled for 7 years, with his son collecting and classifying specimens, interviewing indigenous people through interpreters, and conducting medical studies. Three indigenous painters (baptized Pedro Vázquez and Antón and Baltasar Elías) prepared illustrations. Parts of Francisco Hernández' extensive descriptions of his findings were published in a translated collection entitled, *Plantas y Animales de la Nueva Espana, y sus Virtudes por Francisco Hernández, y de Latin en Romance por Fr. Francisco Ximenez* (Mexico 1615).

Francisco Hernández was suggested as author of the *Voynich Manuscript* by Comegys (2001), who proposed that the codex might be Mesoamerican and written in Nahuatl. His analysis, carried out in part with his twin brother John D. Comegys and assisted by his middle school class, was archived in a difficult-to-access manuscript. The translation was flawed. James C. Comegys assumed that the text was written from right to left, as in Hebrew, and he assumed that Francisco Hernández was either Jewish or a *converso* (convert). However, the text was left justified and a long poem by Francisco Hernández revealed that he was or became a devout Catholic (Varey 2000). Francisco Hernández had a long correspondence with Philip II while in New Spain and could have sent the *Voynich Codex* to the king, but there was no record of this. Francisco Hernández could conceivably have brought the codex with him on his return to Spain in 1577. This would have been 5 years after Rudolf II had returned to Vienna, although Rudolf II's ties with Spain continued after his return to Prague (Marshall 2006). In his will, Francisco Hernández left all his manuscripts to Philip II (Varey 2000).

Conclusion

There is direct and indirect evidence that places the *Voynich Codex* in the possession of the court of Rudolf II early in the seventeenth century. There are various plausible explanations to reconcile the route of a sixteenth century New World codex from New Spain to Europe, including: New Spain to London via Sir Francis Drake; French pirates intercepting Spanish ships; or Mexico to Spain via Francisco Hernández, appointed general *protomedico* of the Indies by Philip II. In London, the codex could have been transferred to Rudolf II by John Dee or Edward Kelley. Furthermore, agents of Rudolf II were dispersed throughout Europe and collected for his *Kunstkammer* (Fučiková et al. 1997); thus, Rudolf II would not have had to directly interface with the seller. The *Voynich Codex* passed through various hands and ended up in a monastery library of the Jesuits in Frascati, Italy, where Wilfrid Voynich purchased it in 1912. Its final resting place was the Beinecke Rare Book and Manuscript Library at Yale University, and a facsimile edition is now available from that library (Clemens 2016). The origins of the *Voynich Codex* are unraveled in later chapters of this work.

Literature Cited

Amancio, D.R., E.G. Altmann, D. Rybski, O.N. Oliveira, and L. da F. Costa. 2013. Probing the statistical properties of unknown texts: Application to the Voynich manuscript. *PLoS One* 8 (7): e67310. https://doi.org/10.1371/journal.pone.0067310.

Aylward, L., and G. Rugg. 2004. *Emergent properties in text generation using tables and grilles.* Unpubl. pdf http://www.scm.keele.ac.uk/staff/g_rugg/voynich/emergent3a.pdf. 3 Feb 2017.

Banasik, Z. 2003. *Zbigniew Banasuk's Manchu theory.* http://www.ic.unicamp.br/~stolfi/voynich/04-05-20-manchu-theo/. 23 May 2017.

Barlow, M. 1986. The Voynich manuscript – By Voynich? *Cryptologia* 10: 210–216.

Bauer, C.P. 2017. *Unsolved! The history and mystery of the world's greatest ciphers from ancient Egypt to online societies.* Princeton: Princeton University Press.

Bawlf, S. 2003. *The secret voyage of Sir Francis Drake, 1577–1580.* London: Penguin Books.

Bax, S. 2013. *The Voynich manuscript—Informal observations on some linguistic patterns.* http://stephenbax.net/wp-content/uploads/2014/01/Voynich-a-provisional-partial-decoding-BAX.pdf. 3 Feb 2017.

———. 2014. *A proposed partial decoding of the Voynich script.* Unpubl. pdf http://stephenbax.net/wp-content/uploads/2014/05/The-Voynich-manuscript-observations-on-linguistic-patterns-v3.pdf. 3 Feb 2017.

Bennett, W.R. 1976. *Scientific and engineering problem-solving with the computer.* Englewood Cliffs: Prentice-Hall.

Berdan, F.F., and P.R. Anawalt. 1997. *The essential codex Mendoza.* Berkeley: University of California Press.

Blunt, W., and S. Raphael. 1979. *The illustrated herbal.* London: Thames and Hudson.

Browne, T. 1931. Works. In G. Keynes (ed), vol. 6, p. 325.

Brumbaugh, R.S. 1974. Botany and the Voynich "Roger Bacon" manuscript once more. *Speculum* 49: 546–548.

———. 1978. *The most mysterious manuscript.* Carbondale: Southern Illinois Press.

Burisch, D. 2012. Cited in Dragoni (2012).

Buonafalce, A. 2007. Review of the curse of the Voynich. The secret history of the World's most mysterious manuscript by Nicholas Pelling. *Cryptologia* 31 (4): 361–362.

Carthago, R. 2006, 2015. Das Voynich Manuscript—Übersetzubg der ersten 13 Seiten. GMA mbH, Germany.

Casanova, A. 1999. Méthodes d'analyse du langage crypté: Une contribution à l'étude du manuscript du Voynich. Docteur thesis. Université de Paris. http://voynich.free.fr/a_casanova_these_19mars1999.pdf. 18 Jan 2014.

Cham, B. 2015. *Introduction to the curve-line system.* Unpubl. website https://briancham1994.wordpress.com/2014/12/17/curve-line-system/. 3 Feb 2017.

Child, J.R. 1976/2002. The Voynich manuscript revisited. National Security Agency/Central Security Service, Fort George G. Meade, Maryland. DOCID:636899.

———. 2007. Again, the Voynich manuscript. Unpubl. pdf http://web.archive.org/web/20090616205410/http://voynichmanuscript.net/voynichpaper.pdf. 3 Feb 2017.

Clemens, R., ed. 2016. *The Voynich manuscript*. New Haven: Yale University Press.

Comegys, J.C. 2001. Keys for the Voynich scholar: Necessary clues for the decipherment and reading of the world's most mysterious manuscript which is a medical text in Nahuatl attributable to Francisco Hernández and his Aztec Ticiti collaborators. J.C. Comegys, Madera, CA. Library of Congress MLCM 2006/02284 (R) FT MEADE.

Comegys, J.D. 2013. The Voynich manuscript: Aztec herbal from New Spain. (5 June 2017).

Comegys, J.C., and J.D. Comegys. 1999. The congruence of Nahuatl grammar with Voynichese. 20th Century Annual Central California Research Symposium, Fresno State University oral presentation.

Currier, P. 1976. Papers on the Voynich manuscript. In M.E. D'Imperio, ed. New research on the Voynich, proceedings of a seminar 30 November 1976. Transcribed by Jacques Guy & Jim Reeds in 1992. Unpubl. pdf http://www.voynich.nu/extra/img/curr_main.pdf. 3 Mar 2016.

D'Imperio, M.E. 1978a/1979. The Voynich manuscript: An elegant enigma. National Security Agency/Central Security Service, Fort George G. Meade, Maryland. ADA070618. [released 29 June 1979] http://www.dtic.mil/cgi-bin/GetTRDoc?AD=ADA070618 and http://www.nsa.gov/about/_files/cryptologic_heritage/publications/misc/voynich_manuscript.pdf. 29 Dec 2012.

———. 1978b/2009. An application of cluster analysis and multidimensional scaling to the question of "hands" and "languages" in the Voynich Manuscript. National Security Agency/Central Security Service, Fort George G. Meade, Maryland. DOCID:636902. [National Security Agency Technical Journal 23(3):59–75, 1978; released 23 April 2009] http://www.nsa.gov/public_info/_files/tech_journals/Application_of_Cluster_Analysis.pdf. (18 January 2014).

———. 1979/2009. An application of PTAH to the Voynich manuscript (U). National Security Agency/Central Security Service, Fort George G. Meade, Maryland. DOCID:3562200. [National Security Agency Technical Journal 24(2):65–91, 1979; released 3 June 2009] http://www.nsa.gov/public_info/_files/tech_journals/application_of_ptah.pdf. 18 Jan 2014.

Dendai, D. 1977. A burning question in re the Voynich Ms (slightly revised). *Cryptogram* 43(2):25, 46–48, 42(3):49, 51–52, 72.

Dilas, J. 2013[2005]. Das Voynich Manuskript. Unpubl. pdf http://www.matrixseite.de/Texte/Jonathan_Dilas_Das_Voynich_Manuskript-Essay.pdf. 3 Feb 2017.

Dos Santos, M. 2005. *El manuscrito Voynich: El libro más enigmático de todos los tiempos*. Madrid: Aguilar.

———. 2006. El manuscrito Voynich/the Voynich Manuscript (Ensayo Historico). Punto De Lectura.

———. 2009. *L'enigma del Manoscritto Voynich: Il più grande mistero di tutti tempi*. Roma: Edizioni Mediterranee.

Dragoni, L. 2012. Il manoscritto Voynich: Un po' di chiarezza sul libro più misterioso del mondo. Unpubl. pdf http://www.cropfiles.it/altrimisteri/Voynich.pdf. 1 Sept 2014.

Emmart, E.W. 1940. The Badianus manuscript (Codex Barberini, Latin 141) Vatican Library. An Aztec Herbal of 1552. Johns Hopkins Press, Baltimore.

Fallacara, G., and U. Occhinegro. 2013. Manoscritto Voynich e Castel del Monte: nuova chiave interpretativa del documento per inediti percorsi di ricerca (Trans. The Voynich manuscript and Castel del Monte: new interpretative keys to reading the manuscript for an innovative research approach). Gangemi, Roma.

Feely, J.M. 1943. Roger Bacon's Cipher: The Right Key Found, Rochester. [privately printed].

Fenton, E. 2000. *The diaries of John Dee*. Charlbury: Day Books.

Finn, J.E. 2001. The Voynich Manuscript. Extraterrestrial contact during the Middle Ages? http://www.bibliotecapleyades.net/ciencia/esp_ciencia_manuscrito02.htm. 23 May 2017.

———. 2004. *Pandora's hope: Humanity's call to adventure*. Baltimore: PublishAmerica.

Fletcher, F. 1854. *The world encompassed of Sir Francis Drake*. London: Hakluyt Society.

French, P.J. 1984. *John Dee: The world of an Elizabethan magus*. London: Routledge & Kegan Paul.

Friedman, E.S. 1962. The most mysterious MS. Still an enigma. Washington Post 5 August:E1, E5.

Fučiková, E., J.M. Bradburne, B. Bukovinská, J. Hausenblasová, J.L. Konečný, I. Muchka, and M. Šroněk, eds. 1997. *Rudolf II and Prague: The imperial court and residential city as the cultural and spiritual heart of Central Europe*. Prague: Prague Castle Administration.
Garel, A. 2011. The Voynich Manuscript through an intersemiotic approach. Unpubl. web page http://journals.sfu.ca/wl404/index.php/book/article/viewFile/29/31. 3 Feb 2017.
Gibbs, N. 2017a. Voynich manuscript: The solution. *The Times Literary Suppplement* September 5 https://www.the-tls.co.uk/articles/public/voynich-manuscript-solution/. 11 Sept 2017.
———. 2017b. Herbal remedies. *The Times Literary Supplement* September 8 5971:15–16.
Gibson, D. 2003. The Voynich question: A possible Middle East connection. Unpubl. web page http://nabatea.net/vconnect.html. 3 Feb 2017.
Goldstone, L., and N. Goldstone. 2005. *The friar and the cipher*. New York: Doubleday.
Gritti, A. 2010. *I custodi della pergamena proibita*. Milan: Rizzoli.
Grosse, W. 2013. *Le code du manuscript Voynich enfin décrypté*. Grenoble: Le Mercure Dauphinois.
Guy, J.B.M. 1991a. On Levitov's decipherment of the Voynich Manuscript. Cited in Kennedy and Churchill (2006).
———. 1991b. Voynich revisited. *Cryptologia* 15 (2): 161–166.
———. 1991c. Statistical properties of two folios of the Voynich manuscript. *Cryptologia* 15: 207–218.
———. 1997. The distribution of signs c and o in the Voynich manuscript: Evidence for a real language? *Cryptologia* 21: 51–54.
Gwynn, B. 2006. Cited in Kennedy and Churchill (2006).
Halliwell, J.O., ed. 1842. *The private diary of Dr. John Dee*. London: Camden Society.
Hauer, B., and G. Kondrak. 2016. Decoding anagrammed texts written in an unknown language and script. *Transactions of the Association for Computational Linguistics* 4: 75–86.
Hermes, J. 2012. Textprozessierung—Design und Applikation. Ph.D. Thesis, Universität zu Köln. http://kups.ub.uni-koeln.de/4561/. 6 Mar 2016.
Hoschel. 2012. Cited in Dragoni (2012).
Huth, V. 2012. Cited in Dragoni (2012).
Ito, A. 2002a. Effects of line context and prefix 40 upon contextual deviation of Voynich "words." Unpubl. pdf http://www.geocities.co.jp/Technopolis/7220/voy/paper021217.pdf. 3 Feb 2017.
———. 2002b. Observation of left and right entropy in Voynich MS. Unpubl. pdf http://www.geocities.co.jp/Technopolis/7220/voy/paper021209.pdf. 2 Feb 2017.
———. 2003. Relationship between red/green words and their characters. Unpubl. pdf http://www.geocities.co.jp/Technopolis/7220/voy/paper030313.pdf. 3 Feb 2017.
Janick, J., and A.O. Tucker. 2017. Were Juan Gerson the illustrator and Gaspar de Torres the author of the Voynich codex? *Notulae Botanicae Horti Agrobotanici Cluj-Napoca* 45 (2): 343–352. https://doi.org/10.15835/nbha45210693.
Jaśkiewicz, G. 2014. Analysis of the letter frequency distribution in the Voynich manuscript. Unpubl. pdf http://csp2011.mimuw.edu.pl/proceedings/PDF/CSP2011250.pdf. 31. August 2014. (Paper presented at the International Workshop CS&P'2011, Bialystok University of Technology.
Kazil, P. 2006. Cited in Kennedy and Churchill (2006).
Kellner, J. 2014. Das Voynich-Manuskript ist gelöst! Unpubl. Website http://voynich-manuskript.de/ms408/home/. 4 Feb 2017.
Kelsey, H. 1998. *Sir Francis Drake, the queen's pirate*. New Haven: Yale University Press.
———. 2003. *Sir John Hawkins, queen Elizabeth's slave trader*. New Haven: Yale University Press.
Kennedy, G., and R. Churchill. 2004. *The Voynich manuscript*. London: Orion.
———. 2006. *The Voynich manuscript: the mysterious code that has defied interpretations for centuries*. Rochester, Vermont: Inner Traditions. (First published in 2004 by Orion press, London).
Kircher, F., and D. Becker. 2012. *Le manuscrit Voynich décodé*. Agnières: SARL JMG editions.
Knight, K. 2009. The Voynich Manuscript. MIT. Unpubl. pdf http://www.isi.edu/natural-language/people/voynich.pdf. 1 May 2014.
Knight, K. and S. Reddy. 2011a. What we know about the Voynich manuscript. Unpubl. pdf http://www.isi.edu/natural-language/people/voynich-11.pdf. 30 Aug 2014.

———. 2011b. What we know about the Voynich manuscript. Pages 78–86 in workshop on language technology for cultural heritage, social sciences, and humanities, LaTeCH, Proceedings of the workshop. 24 June, 2011, Portland.

Landini, G. 2001. Evidence of linguistic structure in the Voynich manuscript using spectral analysis. *Cryptologia* 25: 275–295.

Landmann, E. 2008. The Voynich-Manuscript. Unpubl. pdf http://elifonaot.q32.de/cms/lib/exe/fetch.php?media = de:pub:2007:20071231e:20071231e_dasvoynichmanuskript.pdf. 4 Feb 2017.

Levitov, L.E. 1987. *Solution of the Voynich manuscript: A liturgical manual for the enduring rite of the Cathari heresy, the cult of Isis*. Laguna Park: Aegean Park Press.

Maat J. 2002. Cited in Dragoni (2012).

Manly, J.M. 1931. Roger bacon and the Voynich Ms. *Speculum* 6: 345–391.

Marshall, P. 2006. *The magic circle of Rudolf II: Alchemy and astrology in renaissance Prague*. New York: Walker & Co.

McCrone Associates. 2009. Materials analysis of the Voynich manuscript analysis. Project MA475613. beinecke.library.yale.edu/sites/default/files/voynich_analysis.pdf. Accessed 5 June 2017.

Mervyn, T. 2006. Cited in Kennedy and Churchill (2006).

Montemurro, M.A., and D.H. Zanette. 2013. Keywords and co-occurrence patterns in the Voynich manuscript: An information-theoretic analysis. *PlosOne* 8 (6): e66344. https://doi.org/10.1371/journal.pone.0066344.

Neal, P. 2000–2011. The Voynich Manuscript. Unpubl. web page http://www.voynich.net/neal/. 17 Jan 2013.

Newbold, W.R. 1928. *The cipher of Roger Bacon*. Philadelphia: Univ. of Pennsylvania.

Newson, E. 1995. *A split in the mystery curtain*. Bigelow: Eugene Newsom.

Nuttall, Z. 1914. *New light on Drake: A collection of documents relating to his voyage of circum-navigation 1577–1580*, 350–359. London: Deposition of Francisco Gomez Fengifo, Factor of the Port of Guatulco. The Hakluyt Society.

O'Neil, T. 2012. Voynich manuscript: The code unchopped. Createspace Independent Publishing https://www.createspace.com/.

———. 2014. *Voynich manuscript. The code unchopped*. Volume II. Independent Publishing. https://www.createspace.com/.

O'Neill, H. 1944. Botanical observations on the Voynich MS. *Speculum* 19: 126.

Packheiser, U. 2013. *Das Voynich manuscript*. FP Web & Publishing. http://www.utep.de.

Palatino Press. 2016. *The voynich manuscript: A facsimile of the complete work*.

Papke, U., and D. Weydemann. 2005. Geheimnisvollstes Manuskript der Welt entschlüsselt. http://www.ms408.de/downloads/Pressemitteilung.pdf. 25 May 2017.

Parry, G. 2011. *The arch-conjuror of England, John Dee*. New Haven: Yale University Press.

Pelling, N. 2006. *The curse of the Voynich: The secret history of the World's most mysterious manuscript*. Surbiton, Surrey: Compelling Press.

Platts, M. 2006. Cited in Kennedy and Churchill (2006).

Prinke R.T. n.d. Did John Dee *really* sell the Voynich MS to Rudolf II? juy http://main2.amu.edu.pl/~rafalp/WWW/HERM/VMS/dee.htm.7 July 2017.

Rajkarnikar, M. 2004. Analyzing the Voynich manuscript. Unpubl. pdf http://www.d.umn.edu/~tpederse/Courses/CS5761-SPR04/Projects/rajk0007.pdf. 3 Feb 2017.

Rivero, R. H. 2012. Cited in Dragoni (2012).

Roberts, R.J., and A.G. Watson, eds. 1990. *John Dee's library catalogue*. London: Biographical Society.

Roitzsch, P. 2010. *Das Voynich-Manuskript: Ein ungelöstes Rätsel der Vergangenheit*. Münster: Verlagshausaus Monsenstein und Vannerdat OHG.

Roush, P., and B. Shi. 2014. Projects: 2014S-144 Cracking the Voynich Manuscript code. Unpubl. web page https://www.eleceng.adelaide.edu.au/students/wiki/projects/index.php/Projects:2014S1-44_Cracking_the_Voynich_Manuscript_Code. 6 Feb 2017.

Rugg, G. 2004a. An elegant hoax? A possible solution to the Voynich manuscript. *Cryptologia* 28: 31–46.

———. 2004b. The mystery of the Voynich manuscript: New analysis of a famously cryptic medieval document suggests that it contains nothing but gibberish. *Scientific American* 291 (1): 104–109.

———. 2013. *Blind spot: Why we fail to see the solution right in front of us*. New York: HarperOne.

Saloman, Richard. 1931. 386. Manly, John Mathews, Roger Bacon and the Voynich MS. In; Speculum 6, S.345–391. 5 Taf. Schrift and Sparache Kulturwissenschaftliche Bibliographie zum Nachleben der Antike 1:96.

SantaColoma, R. 2017. The 1910 Voynich theory. http://www.santa-coloma.net/voynich_drebbel/voynich.html. 7 Apr 2017.

Schinner, A. 2007. The Voynich manuscript: Evidence of the hoax hypothesis. *Cryptologia* 3: 95–107.

Schmeh, K. 2013. A milestone in Voynich manuscript research: Voynich 100 conference in Monte Porio Catone, Italy. *Cryptologia* 37: 193–203.

Sherwood, E. 2002. Leonardo and the Voynich Manuscript. www.edithsherwood.com/voynich_author_da_vinci/. 5 June 2017.

Sherwood, E., and E. Sherwood. 2008. The Voynich botanical plants. http://edithsherwood.com/voynich_botanical_plants/index.php. 30 June 2013.

Skinner, S. 2011. Dr. John Dee's spiritual diary (1583–1608). Singapore: Golden Hoard.

Skinner, S., R.T. Prinke, and R. Zandbergen. 2017. *The Voynich manuscript: The world's most mysterious and esoteric codex*. London: Watkins.

Stojko, John. 1978. *Letter to God's eye: The Voynich Manuscript for the first time deciphered and translated into English*. Vantage Press, New York.

Stolfi, J. 2006. Cited in Kennedy and Churchill. (2006).

Strong, L.C. 1945. Anthony Askham, the author of the Voynich manuscript. *Science* 191: 608–609.

Sugden, J. 2006. *Sir Francis Drake*. London: Pimlico.

Sullivan, M., 2012. Cited in Dragoni (2012).

Taiz, L., and S.L. Taiz. 2011. The biological section of the Voynich manuscript: A textbook of medieval plant physiology. *Chronica Horticulturae* 51 (2): 19–23.

Teague, R. 2009. Cracks in the ice: Toward a translation method for the Voynich Manuscript. Unpubl. pdf http://www.as.up.krakow.pl/jvs/library/4-5-2008-01-12/Cracks%20in%20the%20Ice.pdf. 3 Feb 2017.

Theroux, M. 1994. Deciphering "the most mysterious manuscript in the world." The final word? *Borderlands* 50 (2): 36–43.

Thrower, N.J.W., ed. 1984. *Sir Francis Drake and the famous voyage, 1577–1580*. Berkeley: University California Press.

Tiltman, J. H. 1967. The Voynich Manuscript: "The most mysterious manuscript in the world." National Security Agency/Central Security Service, Fort George G. Meade, Maryland. DOCID:631091. [released 23 April 2002] www.nsa.gov/public_info/_files/tech_journals/voynich_manuscript_mysterious.pdf. 5 June 2017.

Timm, T. 2014. How the Voynich Manuscript was created. Unpubl. pdf http://arxiv.org/ftp/arxiv/papers/1407/1407.6639.pdf. 3 Mar 2016.

———. 2016. Co-Occurrence Patterns in the Voynich Manuscript. Unpubl. pdf [http://arxiv.org/ftp/arxiv/papers/1601/1601.07435.pdf. 3 March 2016.

Toresella, S. 1996. Gli erbari degli alchimisti. Pages 31–70 in L. Saginati, ed. Arte farmaceutica e piante medicinali. Erbari, vasi, strumenti e testi dalle raccolte liguri. Pacini Editore, Pisa.

Touwaide, A. 2015. Il manoscritto più misterioso – L'erbario Voynich. In *Villa Mondragone "seconda Roma"*, ed. M. Formica, 141–159. Roma: Palombi Editori.

Tucker, A.O., and J. Janick. 2016. Identification of phytomorphs in the Voynich codex. *Horticultural Reviews* 44: 1–64.

Tucker, A.O., and R.H. Talbert. 2013. A preliminary analysis of the botany, zoology, and mineralogy of the Voynich manuscript. *HerbalGram* 100: 70–85.

Varey, S., ed. 2000. *The Mexican treasury: The writings of Dr. Francisco Hernández*. Stanford: Stanford University Press.

Velinska, E. 2013. The Voynich Manuscript: Plant id list. ellievelinska.blogspot.com/2013/07/the-voynich-manuscript-plant-id-list.html. 5 June 2017.
Von Däniken, E. 2009. History is wrong. Trans. N. Quaintmere. Franklin Lakes: New Page Books.
Vonfelt, S. 2014. The strange resonances of the Voynich manuscript. Unpubl. pdf http://graphos-copie.free.fr/publications/Voynich_en.pdf. 23 Jan 2017.
Walling, R.A.J. 1907. *A sea-dog of Devon…A life of Sir John Hawkins*. London: Cassell and Co..
Watson, L. 2011. 'Prophet of God' claims mysterious manuscript's code has been cracked. Daily Mail. http://www.dailymail.co.uk/news/article-2069481/Prophet-God-claims-mysterious-manuscripts-code-cracked.html. 16 May 2017.
Williams, R.L. 1999. A note on the Voynich manuscript. *Cryptologia* 23: 305–309.
Wilkins, S., ed. 1884. The works of Sir Thomas Browne. Vol. 3. London: George Bell and Sons.
Woolley, B. 2000. *The Queen's conjurer: The life and magic of John Dee*. London: Flamingo.
Zandbergen, R. 2004–2017. The Voynich Manuscript. www.voynich.nu/. 6 June 2017.
———. 2016. Earliest owners. In *The Voynich manuscript*, ed. R. Chandler, 1–9. New Haven: Yale University Press.
———. 2017. The history of the Voynich MS. www.voynich.nu/history.html. 5 June 2017.
Zimansky, C.A. 1970. William F. Friedman and the Voynich manuscript. *Philological Quarterly* 49: 433–442.
Zyats, P., E. Mysak, J. Stenger, Lemay M. -R, A. Bezur, and D.D. Driscoll. 2016. Physical findings. In *The Voynich manuscript*, ed. R. Chandler, 23–37. New Haven: Yale University Press.

Chapter 2
The Voynich Codex

Jules Janick and Arthur O. Tucker

Codex Foliation

The *Voynich Codex* is composed of folded sheets of vellum called folios. An individual folio is composed of two pages. The front side is called *recto* (r) and the reverse side is *verso* (v). Each folio is numbered in the upper corner on the recto side from to 1 to 116 with Arabic numbers in a different ink and penmanship from the text (obviously a late addition after a rebinding). Folios are referred to by their individual numbers, such as 1r, 1v, 2r, 2v, and so on up to 116v. Some folios consist of two to four pages pasted together and folded, and those are called foldouts.

The folios are bound in sets (quires) of four folios (eight pages) each and sewn together (Pelling 2006). Numbers have been penned in the lower right corner of the last page of each quire to ensure they would be put together in the correct sequence to create a bound volume. As the quire numbers use Latin abbreviations, they were probably done in the scriptorium of some monastery where the codex was stored in the seventeenth and eighteenth centuries. The folio size is 23.5 × 16.2 cm, but there are numerous foldouts with two to four pasted pages. Fourteen folios are missing (12, 59–64, 74, 91, 92, 97, 98, 109, and 110). At present, the codex contains 102 folios. Many of those folios are foldouts with pages pasted together. Adjusting for the missing folios and the foldout sections, the work is equivalent to 262 pages. Gerry Kennedy and Rob Churchill (2006: 10), citing Wilfrid Voynich, also estimated an equivalence of not less than 262 pages. Improper binding has altered the original sequence and some of the sections are not contiguous. The contents of the *Voynich Codex* are summarized in Table 2.1.

J. Janick, A. O. Tucker, *Unraveling the Voynich Codex*, Fascinating Life Sciences, https://doi.org/10.1007/978-3-319-77294-3_2

Table 2.1 *Voynich Codex* contents

Folio/s[a]	No. pages	Contents and notes
1v	1	Text + paragraph markers
1v–57r	108	Herbal (folios, 12 missing)
57v	1	Magic circle with key-like sequences of symbols
58r, 58v	2	Text only (+ a few stars)
65r, 65v	2	Herbal
66r	1	Text with key-like sequences
66v	1	Herbal
67r–69v	14	Cosmological
70r–73v	12	Zodiac circles
75r, 75v	2	Balneological
76r	1	Text with key-like sequences
76v–84v	17	Balneological
85r	1	Text only
85v	3	Cosmological (two foldout pages), one page, text only
86r	2	Text only (one page) + spewing cones, figures, birds (one page)
86v	6	Rosette, kabbalah map
87r, 87v	2	Herbal
88r–89v	8	Pharmaceutical (9 total)
89v–90v	4	Herbal (folios 91, 92 missing)
93r–96v	10	Herbal (folios 97, 98 missing)
99r–102v	12	Pharmaceutical (12 total)
103r–115	23	Recipe (text + stars) (folios 109, 110 missing)
116v	1	Text in a different hand + doodle
Total	234	
Summary	*No. pages*[a]	*Percentage*
Herbal	127	54.3
Pharmaceutical	20	8.5
Balneological	19	8.1
Zodiac	12	5.1
Cosmological	16	6.8
Rosette/kabbalah/map	6	2.6
Recipe	23	9.8
Others[b]	11	4.7
Total pages[c]	234	100.0

The number of pages includes pages in foldouts

[a]Includes foldouts, but does not include 14 missing folios (28 pages: folios 12, 59–64, 74, 91, 92, 97, 98, 109, 110)

[b]Others: first (folio 1r) and last page (folio 116v) = two pages; folio 57v with key-like sequences in figure; two pages of text with key-like sequences (folios 66r, 76r); five other pages with text (folios 58r, 58v, 85r, 85v(3), 86r(1)); one page with spewing cones (folio 86r)

[c]Adjusting for 14 missing folios (28 pages) and foldout sections, the overall work is equivalent to 234 + 28 = 262 pages

Facsimiles

The *Voynich Codex* is available online from the Beinecke Rare Book and Manuscript Library at Yale University (beinecke.library.yale.edu/collections/highlights/voynich-manuscript). In addition, there have been a number of printed facsimiles. Three are worthy of mention. A facsimile of the complete work from the Palatino Press (2015) lists folio numbers on the bottom of each page and we have used this version to assign folio numbers to all images in the present work. However, since then, two more precise facsimile editions have appeared. One is a remarkable version from the Beinecke Library edited by Raymond Clemens (2016) and accompanied by essays. It duplicates the foldout pages in their actual size and format. However, the foldout pages are not labeled, which makes them difficult to cite. This was corrected in a facsimile entitled *The Voynich Manuscript: The World's Most Mysterious and Esoteric Codex*, with a foreword by Skinner and an introduction by Prinke and Zandbergen (2017). Although the foldout pages are not presented in their original sizes, each foldout page is labeled. However, the folio numbers of the foldouts are different from the Palatino facsimile in some cases, as described in Table 2.2.

Table 2.2 Folio designations of foldout pages in Voynich Codex facsimiles

Palatino Press (2015)[a]	*The Voynich manuscript*: *The world's most mysterious and Esoteric codex* (2017)[a]
67 recto (1, 2)	67r(1), 67r(2)
67 verso (1, 2)	67v(2), 67v(1)
68 recto (1, 2, 3)	68r(1), 68r(2), 68r(3)
69 verso (1, 2, 3)	69v(1), 70r(1), 70r(2)
70 recto	70v(2)
70 verso	70v(1)
71 verso (1, 2, 3, 4)	71v, 72r(1), 72r(2), 72r(3)
72 recto (1, 2)	72v(3), 72v(2)
72 verso	72v(1)
73 recto	73r
85 verso (1, 2, 3)	85r(2), 86v(4), 86v(6)
86 recto (1, 2)	86v(5), 86v(3)
86 verso	85v & 86r
88 verso (1, 2, 3)	88v, 89r(1), 89r(2 + piece)
89 recto (+piece)	89v(2 + piece)
89 verso (1, 2, 3)	89v(1), 90r(1), 90r(2)
90 recto	90v(2)
90 verso	90v(1)
94 verso (1, 2, 3)	94v, 95r(1), 95r(2)
100 verso (1, 2, 3)	100v, 101r(1), 101r(2)
101 recto	101v(2)
101 verso (1, 2, 3 + piece)	101v(1), 102r(1), 102r(2)

[a]Numbers in parentheses refer to page number designations in foldouts

Sections

The *Voynich Codex* has been divided by previous workers into sections known by convention as Herbal (or Botanical), Pharmaceutical (or Pharma), Balneological ("bathing" or Biological), Astrological, Cosmological, and Recipe.

Herbal (Folios 1v–11v, 13r–57r, 65r–65v, 66v, 87r–87v, 89v–90v, 93r–96v)

Most herbal folio numbers contain a single plant image (Fig. 2.1), but in four cases (folios 40v, 42r, 43v, and 87v), there are two images on the same page, including one (40v) with clearly two stages of the same plant. Pages 89v and 94v are foldouts and each comprises three pages pasted together. There are 131 plant images on 127 pages. The text surrounds the images indicating that it was applied after the images were drawn.

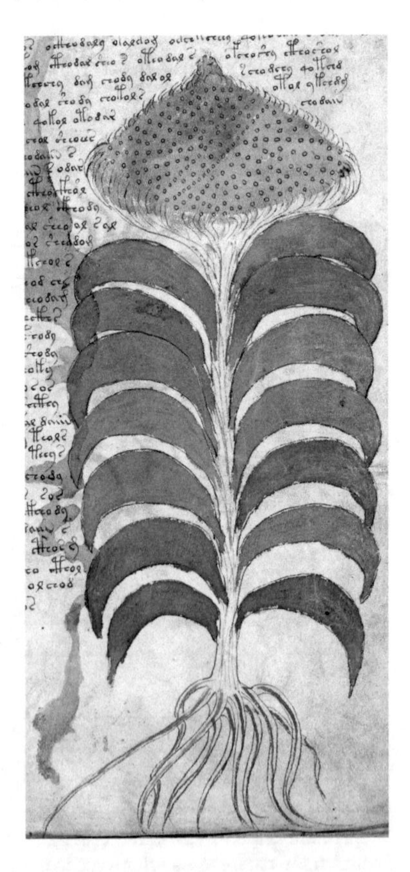

Fig. 2.1 An illustration of a sunflower in the Herbal section (folio 93r), first identified by Hugh O'Neill (1944)

Pharmaceutical (Folios 88r–89v, 99r–102v)

Twenty pages including foldouts contain multiple images of plant parts plus various elaborate containers, which are clearly modifications of apothecary jars (Fig. 2.2). There are 228 plant images, of which 26 are mainly roots, which are often not very distinctive. Unless they have unique features (such as nodules or attached leaves) or a name to provide a clue, they will forever remain what the plant taxonomists commonly call *indet.*, which is an abbreviation for the Latin *indeterminavit*, meaning that "*the specimen has not been, or cannot be, identified by the determinavit – the person who is verifying the specimen*" (Bridson and Forman 1992).

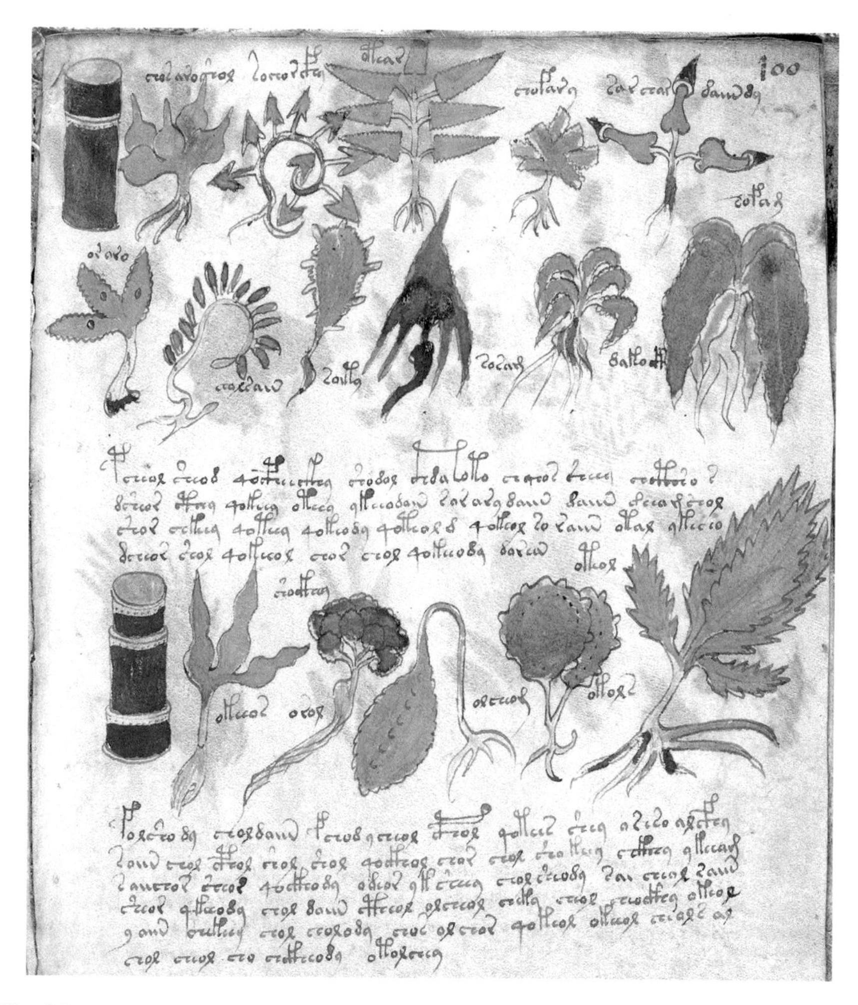

Fig. 2.2 A typical Pharmaceutical page (folio 100r) shows what appear to be apothecary jars (*majolica*) that resemble holy oil stocks, and 16 images of various plant parts with names in Voynichese symbols

Most significantly, many of the plants and jars are labeled in Voynichese script. Taken together the Herbal and Pharmaceutical sections consist of approximately 359 images of plants or plant parts. Of the 234 pages in the *Voynich Codex*, 144 (61.5%) are devoted to plants. Thus, the greatest part of the *Voynich Codex* is an herbal. In two cases, a plant in the Herbal section is also found in the Pharmaceutical section. In both sections, there are whimsical arrangements of roots with faces and animal-like appearances (see Chap. 14, Fig. 14.7).

There are 42 apothecary jars, many quite elaborate, of which 33 are labeled with Voynichese symbols. The jars are divided into two types based on the top openings: 25 with circular flat openings and two-, three-, or four-jointed sections in folios 88r, 99, 100, 101, and 102 (Fig. 2.3a, b), and 17 elaborate ones with narrow openings, which might be spouts, in folios 88 and 89 (Fig. 2.3e, f). We surmise that the jars with large circular openings were for solid material (dried leaves, berries, fruits, or roots),

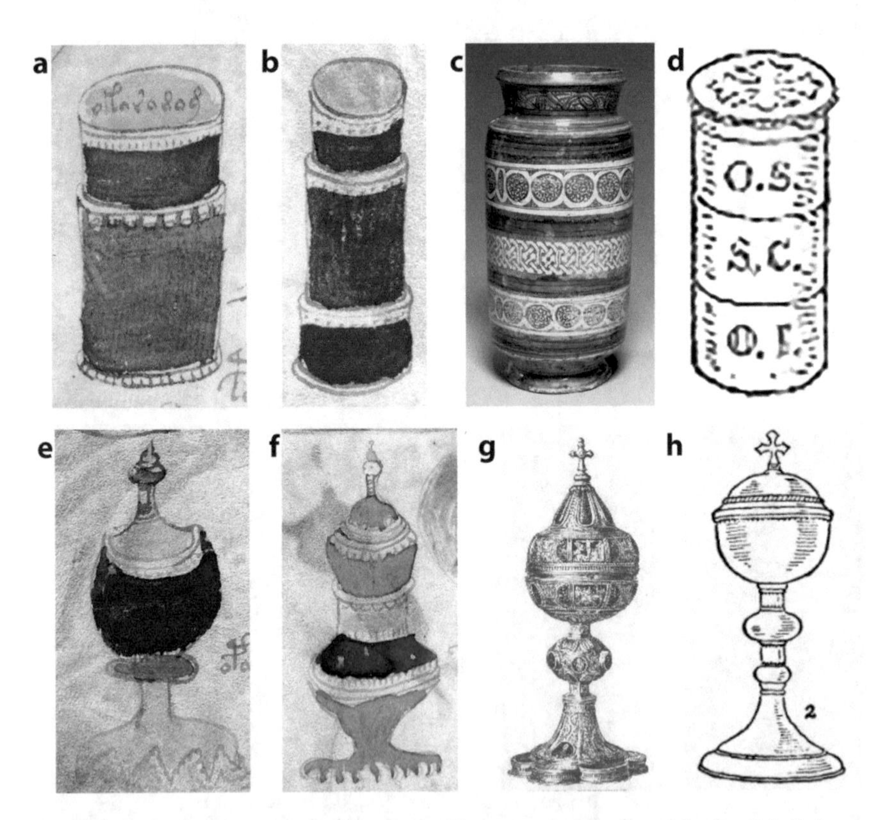

Fig. 2.3 Representative apothecary jars in the Pharmaceutical section of the *Voynich Codex* compared with medieval majolica, holy oil stocks, and ciboria: (**a**) flat-opening, two-jointed jar, folio 99r(1); (**b**) flat-opening three-jointed jar, folio 100r(2); (**c**) medieval majolica apothecary jar; (**d**) medieval holy oil container (O.S. *Oleum Sanctum*, S.C. *Sanctum Chrisma*, O.I. *Oleum Infirmorum*); (**e**) jar with a narrow lid, perhaps a spout, in two sections, folio 89r(3); (**f**) jar with a narrow lid, perhaps a spout, in four sections, folio 89r(1); (**g**) ornamented Portuguese ciborium; (**h**) classic medieval ciborium

whereas those with the small openings might have been for liquids, including oils. The flat opening jars appear to have been inspired by holy oil stock containers used in Roman Catholic ceremonies (Fig. 2.3d), whereas those with narrow openings (Fig. 2.3g, h) were inspired by ciboria reserved for the Eucharist (Tucker and Talbert 2013). Most medieval apothecary jars are ceramic and highly decorated. Those with curved sides are called *arborello* (Drey 1978).

Balneological (Folios 75r, 75v, 76v–84v)

There are 218 nude nymphs (all appearing to be the same shapely, blonde lady, often in groups) on 19 pages, cavorting or bathing in shallow pools with strange plumbing or vascular systems with green water (see Chap. 7, Fig. 7.4). This section is the most bizarre of the *Voynich Codex* (Fig. 2.4).

Cosmological (Folios 67r–69v, 86v)

There are 17 pages of "magic" circles, with several showing various combinations of sun, moon, and planets (Fig. 2.5). Folio 86v, known as the Rosette, is equivalent to six pages and has 11 circles (see Chap. 10). One non-astronomical magic circle, folio 57v (see Chap. 5, Fig. 5.3), includes a central circle with four female heads and four rings with text surrounding it, of which the first and the third ring from the center consist of the same series of symbols presumed to be an alphabet and considered to be a key-like sequence.

Fig. 2.4 Nymphs cavorting in a pool in the Balneological section (folio 78r)

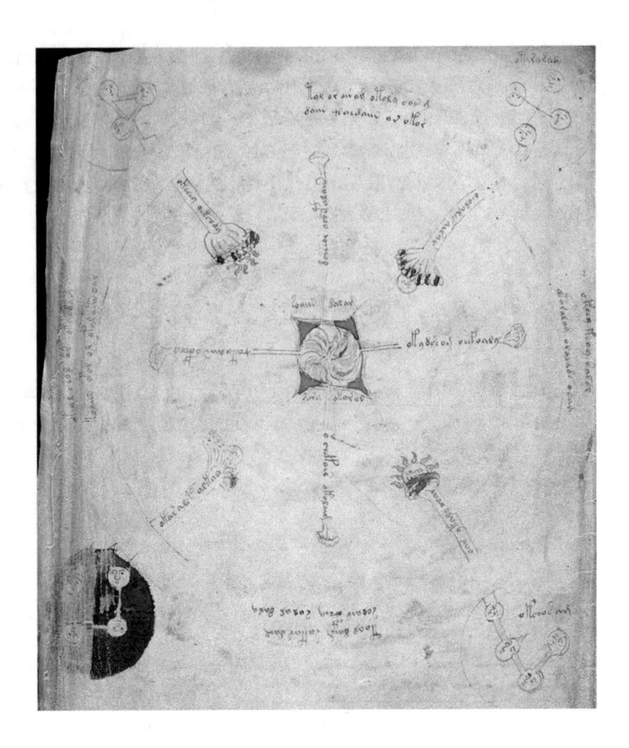

Fig. 2.5 A magic circle, folio 67v(1), showing symbols for the Earth, sun, moon, and a planet (presumably Venus). The four circles in the corners with faces may be T-O maps, clearly indicated by the lower left circle, which is in color (see Chap. 9)

Astrological (Folios 70r–73v)

This section includes 12 zodiac circles (70r–73v) surrounded by nude or clothed nymphs: 299 total (269 nude and 30 clothed). The nymphs are labeled with names (see Chap. 8) and surrounded by rings with words.

Recipe (Folios 103r–116r)

This section consists mainly of 23 pages of text that may be medicinal recipes or possibly poems or incantations. Each phrase, assumed to be a sentence, is highlighted by a six- to eight-pointed star in the left margin (Fig. 2.6).

Other Folios

There are a number of folios that do not fit any section. The first page, folio 1r, contains painted paragraph markers resembling "bird glyphs" often found in sixteenth century documents in New Spain (Fig. 2.7), more properly called *calderón* (Mackenzie 1997: 12). These include:

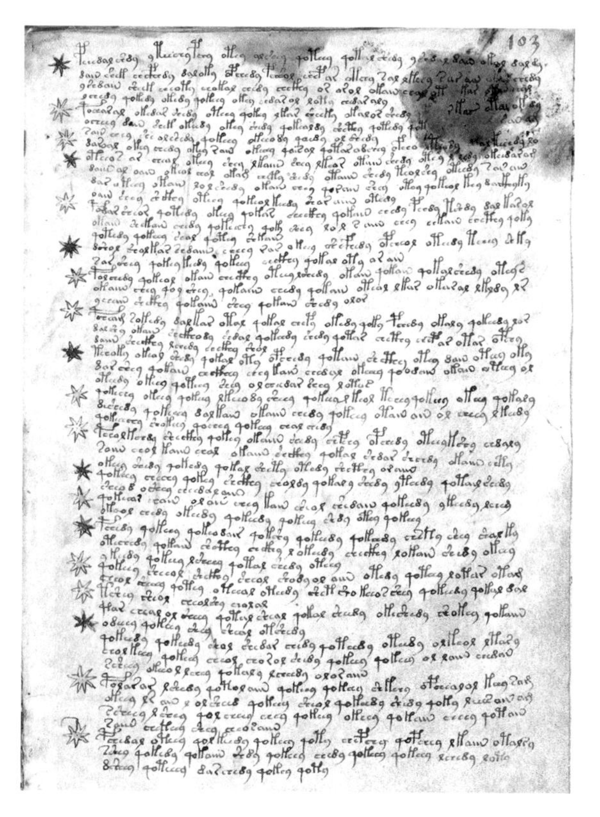

Fig. 2.6 The first page of the Recipe section (folio 103r)

Fig. 2.7 Use of paragraph markers (so-called "bird glyph" or *calderón*) in the *Voynich Codex* (folio 1r) and documents of New Spain in the sixteenth to seventeenth centuries

1. Trial records of Pedro Ruiz Calderón while under prosecution by the Inquisition for practicing black magic written in Mexico City, 1540 (Janus 2012).
2. *Codex Mendoza* of 1541, folio 65r (Berdan and Anwalt 1997).
3. *Codex Osuna* of 1563–1566, folio 34r (Chávez Orozco 1947)
4. Bernardino de Sahagún's *Ejercicio en Lengua Mexicana* of central Mexico, 1574 (Newberry Library 2007a)
5. *Codex Aubin* of 1576, folio 55r (Anonymous 1576)
6. *Marquesado del Valle Codex* of 1550–1600, folio 29r (Anonymous 1550–1600)
7. *Registry of Matrimony* from Tlatelolco, Mexico City, 1632–1633 (Newberry Library 2007b)

Along with this mark, scribes also used stars to delineate paragraphs, common to sixteenth–seventeenth century Spanish documents.

Four pages contain only text: folios 58r, 58v, the right side of 85v, and the left side of 86r. Four pages contain a list of alphabet symbols, perhaps a key: folio 57v (see Chap. 5, Fig. 5.3), 66r (also contains a naked, swollen, apparently ill nymph in the lower left corner), 76r, and 86r (left page). The right half of folio 86r has multiple bizarre elements, including spewing, phallic-like images, two birds, two hidden nude figures, and a T-O map (see Chap. 9, Figs. 9.4 and 9.5, and Chap. 10, Fig. 10.3) combined with much text (Fig. 2.8). The last incomplete folio, 116v, apparently written by a different hand, is sometimes referred to as the "*michiton oladabas*" page, based on a supposed translation of the first two words in its text. The folio also contains a nude and a sheep doodle.

The various disparate sections of the *Voynich Codex* and the indecipherable language continue to fascinate. We think it is probable that the illustrator of the images and author of the text are separate people as the text surrounds the illustrations in many cases. All the artwork is clearly by the same person, demonstrated by the consistency of drawing and especially the whimsical nature of many of the doodles and sketches interspersed throughout the manuscript. We conclude that the artist and author are polymaths with wide ranging interests.

Pigments, Pigment Binder, and Composition Analyses

The *Voynich Codex* has a very limited palette. The text is brownish-black and many of the illustrations are washed with four colors: blue, red, green, and brown. McCrone Associates Inc. of Westmont, IL, USA, analyzed these pigments (including the white paint in the face of folio 70, which we cannot locate) in an internal report to the Beinecke Rare Book and Manuscript Library in 2009. Their conclusions are summarized here with our comments in the hope that it will shed light on the origins of the codex. This analysis was not published in a refereed scientific journal, the materials and methods were not revealed, and the online results were not backed by data. Thus, we do not know if the conclusions were correct. Regardless of the limitations of this internal report, it is the only one done so far and thus relevant to examine the uses of these pigments in the Old and New Worlds to help provide evidence as to the origins of the *Voynich Codex*.

Fig. 2.8 Folio 86r(2) containing spewing "phallic-like" images with two hidden figures, two birds, a T-O map in the middle, and a quire number (14^9 = 14th) in the lower right corner

Pigment analysis was carried out by Dr. Alfred Vendl at the Institutsvorstand, Institut für Kunst und Technologie, Universität für angewandte Kunst in Vienna, Austria in consultation with ProOmnia representatives. Samples were collected using a fine-pointed tungsten needle and a micro scalpel, and techniques involved a combination polarized light microscopy, infrared spectroscopy, micro X-ray diffraction, energy-dispersive spectrometry, and scanning electron microscopy. Their conclusions are summarized below with our comments.

Inks

Most of the ink samples for text and drawings were chemically similar and contained iron, sulfur, calcium potassium, and carbon with trace amounts of copper, and occasionally zinc, suggesting an iron gall ink. Iron gall ink is traditionally made by the reaction of iron sulfate ($FeSO_4$) with the bulbous outgrowths (galls) that result from the insect infestation of oak or other plants and contain gallotannins. When heated, the gallotannins convert to a brown solid. It should be noted that although the McCrone analysis refers to "black inks," most of the text color is not black, but brownish. The text in the *Codex Cruz-Badianus* (Emmert 1940), mostly brown with titles in red, was also determined to be iron gall ink (Zetina et al. 2008). We conclude that the ink analysis has not determined if the *Voynich Codex* was written in Europe or Mesoamerica.

Green Pigment

The green paint was tentatively identified as a copper and copper–chlorine resinate. The presence of chlorine suggested that the crystalline material might be atacamite plus a gum binder. Atacamite was initially named after the Atacama region of Chile, where it is most abundant, but is only found in any appreciable amounts in Australia, Bolivia, Mexico, Peru, Namibia, Russia, and the United States. Paratacamite (which may be confused with atacamite) also occurs as a breakdown product of bronze when chlorides are available ("bronze disease"). Atacamite was only rarely used in European paintings because of the difficulty of obtaining enough as a pigment (Naumova and Pisareva 1994; Richardin et al. 2011). Clearly, the use of atacamite does not rule out a New World origin for the *Voynich Codex*, and, in fact, its presence in Mexico is supporting evidence for its origin there.

Blue Pigment

The blue paint was unambiguously identified as ground azurite and minor amounts of cuprite, a copper oxide. Azurite ($Cu_3(CO_3)_2(OH)_2$) is a secondary copper mineral found in the oxidized zones of copper-bearing ore deposits, chiefly by the action of carbonated water on copper minerals or by copper sulfate solutions reacting with limestone. Azurite is often pseudomorphed to malachite ($Cu_2(CO_3)(OH)_2$), and the two are often found together, along with cuprite (Cu_2O). Azurite is found in many locations including Mexico (Mottana et al. 1978).

Red–Brown Pigment

The red–brown paint was identified as red ochre and the crystal phases consisted of hematite, iron sulfide, and possibly minor amounts of lead sulfide and palmierite in a gum medium. Red ochre (with hematite) is a common pigment worldwide and is found in Mesoamerica (Ortega et al. 2001; Luján and Chiari 2012; Kerpel 2014). The Aztecs called hematite *tlahuitl*.

Pigment Binders

The pigment binder was characterized by McCrone Associates as an unknown gum, but gum arabic was excluded. The spectra included several sharp peaks in the region of 1100–1000 cm^{-1}, which was unexpected, suggesting the possibility of other unidentified constituents. Pigments are bound together with a variety of media that also help adherence to the substrate. Thus, many pigment binders have also been used as glues. A possible binder might be "superglues" from orchid pseudobulbs, called *tzacutli* (Nadal et al. 2007).

Composition

The in-house report from McCrone Associates (2009) describes the *Voynich Codex* as being bound in vellum with vellum pages, whereas the Beinecke Library web page for the *Voynich Codex* identifies it as parchment. Parchment is made from the hides of sheep or calves. Vellum, a superior and more expensive form of parchment, is made from the skin of very young or stillborn calves.

We conclude that the pigment and gum analyses do not offer any evidence for the precise location of the origins of the *Voynich Codex*. The McCrone Associates' analysis is not incompatible with a New World origin.

One of the startling finds of the McCrone report, based on ultraviolet examination of folio 1r, was confirmation of a signature, Jacobi à Tepenecz, reported by Wilfrid Voynich after he performed a chemical wash. This name refers to Jacobus Horčicky, a wealthy pharmacist, imperial chemist, and physician to Rudolf II. Horčicky was granted minor nobility in 1608, presumably for curing the emperor of a grave illness, allowing him to amend "de Tepenecz" to his name. Thus, this signature dates the *Voynich Codex* to the possession of a courtier of the court of Rudolf II between 1608 and his death in 1622. The signature was later confirmed by UV-induced visible fluorescence (Zyats et al. 2016: 31).

Dating of the Vellum

In 2011, the University of Arizona issued a press release on the age of the *Voynich Codex* vellum using carbon dating (Stolte 2011). Dr. Gregory Hodgins of the University's Department of Physics performed the analysis in 2009 using an accelerator mass spectrometer to measure the ratio of ^{14}C (carbon-14) to ^{12}C (carbon-12) in four snippets of the manuscript. The samples dated from 1404 to 1438. Subsequently, Hodgins made several oral presentations and announcements with additional data, but the information has not been published (Zandbergen 2016).

Estimates on the statistical variability of the McCrone data have been provided by Sherwood (2010) and Zandbergen (2016). Sherwood stated: *"I have come to my personal conclusion that the animal(s) whose skins were used to make the parchment for the Voynich Manuscript were probably killed sometime during the first half of the fifteenth century."*

This date of the vellum's age, at first sight, would seem to undermine our hypothesis that the *Voynich Codex* is a sixteenth century manuscript written in Mexico. However, the age of the vellum does not determine the age of the manuscript, as the *Voynich Codex* could well be a palimpsest (Greek, *palímpsestos*, "scratched again," or "scraped again") in which the text has been either scraped or washed off so that the page can be reused for another document. Organic acids, such as citric, ascorbic, and oxalic acids, were sometimes used as mild bleaches and could also erase the text (Hirshfeld 2009).

Although no thorough multi-spectral analysis had been performed on the *Voynich Codex*, McCrone Associates (2009) performed a UV examination of folio 1r. Their remarks are as follows: "*Many parts of this page exhibit a blotchiness consistent with chemical staining which, it was said, had been applied in order to enhance the writing. It may also have been the long term result of washing away some of the original writing*." This is explained in a 1921 paper by John Manley who reported that Wilfrid Voynich, finding traces of writing at the bottom of the first page, applied a chemical commonly used by paleographers to develop faded writing, and found Jacobi à Tepenecz, the named assumed by Jacobus Horčicky in 1608 when he was ennobled. Clearly, a complete multi-spectral analysis of the *Voynich Codex* is needed, to determine if it is a palimpsest, as we surmise.

The removal of previous writing on parchment or vellum for later use is well documented in the case of the *Archimedes Codex* (Netz and Noel 2007; Hirshfeld 2009). Two teams determined that a priceless mathematical work by Archimedes (287–212 BCE), written in Constantinople, was based on erasure and overwriting of a thirteenth-century monk's prayer book, reflecting the value of parchment or vellum at that time. More than 105 out of 3,300 volumes (or 3.2%) in the library of the St Catherine's Monastery on the Sinai Peninsula have been shown to be palimpsests (Powell 2016; Soskice 2010). Some palimpsests themselves have been overwritten; thus, a Syriac Christian text was erased and overwritten in Arabic, and then subsequently erased and rewritten in Hebrew (Soskice 2010). In New Spain, in times of scarcity, use was made of already printed paper, which was sold as a waste product (Reyes 1948, cited in Ramirez 2013: 39 Vol. 1). Reused parchment was available at the scriptorium of El Colegio de Santa Cruz de Tlatelolco in New Spain (Gravier 2011). Researchers from the Bodleian Library at the University of Oxford have demonstrated that the *Codex Selden*, also known as the *Codex Añute* (1560) and written on a deer hide, was a palimpsest written over a pre-colonial manuscript (Snijders et al. 2016).

Conclusion

The mysterious *Voynich Codex* is an incomplete text of 116 folios and 262 potential pages illustrated in color with the text in symbolic form surrounding the images. The content is comprehensive in scope with sections on herbal medicine, ritual bathing, astrology, and astronomy. The age of the vellum has been dated to ca. 1425, but the date of the text is unknown. Parchments were overwritten to create palimpsests and old parchments were reused in New Spain. A complete multi-spectral analysis of the *Voynich Codex* is needed, but we have had no success in accomplishing this. Pigment and gum analyses are not incompatible with a New World origin. In addition, two paragraph markers on the first page ("bird glyphs" or *calderón*) and apothecary jars can be matched to common items in New Spain. We find nothing in the *Voynich Codex* that would negate its origin in New Spain in the sixteenth century.

Literature Cited

Anonymous. 1550–1600. Marquesado del Valle Codex. http://www.wdl.org/en/item/9681/. 7 July 2013.

———. 1576. Codex Aubin/Códice Aubin 1576/Códice de 1576/Historia de la nación mexicana/ Histoire mexicaine. Mexico. http://www.britishmuseum.org/research/search_the_collection_database/search_object_details.aspx?objectid=3008812&partid=1&searchText=Aubin+Codex¤tPage=1. 28 Jan 2013.

Berdan, F.F., and P.R. Anawalt. 1997. *The essential Codex Mendoza*. Berkeley: University of California Press.

Bridson, D., and L. Forman. 1992. *The herbarium handbook*. London: Kew.

Chávez Orozco, L. 1947. *Codice osuna*. Mexico: Ediciones del Instituto Indigenista Interamericano.

Clayton, M., L. Guerrini, and A. de Ávila. 2009. *Flora: The Aztec herbal*. London: Royal Collections Enterprises.

Clemens, R., ed. 2016. *The Voynich manuscript*. New Haven: Yale University Press.

Dale, J.D.H. 1875. *The sacristan's manual*. 3rd ed. London: T. Booker.

Drey, R. 1978. *Apothecary jars*. London: Faber and Faber.

Emmart, E.W. 1940. *The Badianus manuscript*. Baltimore: Johns Hopkins Press.

Gravier, M.G. 2011. Sahagún's codex and book design in the indigenous context. In *Colors between two worlds*, ed. G. Wolf and J. Connors, 156–119. Milan: The Florentine Codex of Barnardino de Sahagún. Officina Libraria.

Griffenhagen, G., and M. Bogard. 1999. *History of drug containers and their labels*. Madison: American Institute of the History of Pharmacy.

Hind, A. 1963. *A history of engraving & etching from the 15th century to the year 1914*. New York: Dover Publications.

Hirshfeld, A. 2009. *Eureka man: The life and legacy of Archimedes*. New York: Walker & Co.

Huson, P. 2004. *Mystical origins of the tarot, from ancient roots to modern usage*. Rochester: Destiny Books.

Janus, O. 2012. Album: A priest who performed black magic. Live Science. http://www.livescience.com/22964-calderon-priest-black-magic.html. 12 Mar 2017.

Kennedy, G., and R. Churchill. 2006. *The Voynich manuscript*. Rochester: Inner Traditions First published in 2004 by Orion press, London.

Kerpel, D.M. 2014. *The colors of the new world: Artists, materials, and the creation of the Florentine codex*. Los Angeles: Getty Research Institute.

Luján, L.L., and G. Chiari. 2012. Color in monumental Mexica sculpture. *Res: Anthropology and Aesthetics* 61 (62): 330–342.

Mackenzie, D. 1997. *A manual of manuscript transcription for the dictionary of the old Spanish language*. 5th ed. rev. R. Harris-Northall. Madison: Hispanic Seminary of Medieval Studies.

Manly, J.J. 1921. The most mysterious manuscript in the world. Did Roger Bacon write and has the key been found. *Harper's Monthly Magazine* 143 (July): 186–197.

McCrone Associates. 2009. Materials analysis of the Voynich manuscript analysis. Project MA475613. beinecke.library.yale.edu/sites/default/files/voynich_analysis.pdf. Accessed 5 June 2017.

Mottana, A., R. Crespi, and G. Liborio. 1978. *Simon & Schuster's guide to rocks and minerals*. New York: Simon and Schuster.

Nadal, L. F., F. S. Olguín, and L. Navarijo. 2007. Un excepcional mosaico de plumaria Azteca: El Tapacáliz del Museo Nacional de Antropología. *Estudio de Cultura Náhuatl (México: UNAM Instituto de Investigaciones Históricas)* 38:85–100.

Naumova, M.M., and S.A. Pisareva. 1994. A note on the use of blue and green copper compounds in paintings. *Studies in Conservation* 39: 277–283.

Netz, R., and W. Noel. 2007. *The Archimedes codex*. London: Westfield and Nicolson.

Newberry Library. 2007a. The Aztecs and the making of Colonial Mexico: Christianizing the Nzahua. Bernardino de Sahagún, Ejercicio en Lengua Mexicana, Central Mexico, 1574. http://publications.newberry.org/aztecs/s3i5.html. 12 Mar 2017.

———. 2007b. The Aztecs and the making of Colonial Mexico: Christianizing the Nzahua. Registry of Matrimony from Tlatelolco, Mexico City: 1632–1633. http://publications.newberry.org/aztecs/s4i12.html. 12 Mar 2017.

O'Brien, W.A. 1933. *Sacristy and sanctuary*. New York: Benziger Brothers.

O'Neill, H. 1944. Botanical observations on the Voynich MS. *Speculum* 19: 126.

Ortega, M., J.A. Ascencio, C.M. San-Germán, M.E. Fernández, L. López, and M. José-Yacamán. 2001. Analysis of prehispanic pigments from "Templo Mayor" of Mexico City. *Journal of Journal of Materials Science* 36:751–756.

Palatino Press. 2015. *The Voynich manuscript: A facsimile of the complete work*. www.palatinopress.com

Pelling, N. 2006. *The curse of the Voynich: The secret history of the World's most mysterious manuscript*. Surbiton, Surrey: Compelling Press.

Powell, E.A. 2016. Recovering hidden texts. *Archaeology* 69 (2): 38–43.

Richardin, P., V. Mazel, P. Walter, O. Laprévote, and A. Brunelle. 2011. Identification of different copper green pigments in renaissance paintings by cluster-TOF-SIMS imaging analysis. *Journal of the American Society of Mass Spectrometry* 22: 1729–1736.

Romero Ramirez, M.E. 2013. Limp, laced-case binding in parchment on sixteenth-century Mexican printed books. Ph.D. thesis, Camberwell College of Arts, University of the Arts London. Vols. 2.

Sherwood, E. 2010. Analysis of radiocarbon dating statistics in reference to the Voynich manuscript. http://www.edithsherwood.com/radiocarbon_dating_statistics/radiocarbon_dating_statistics.pdf. 30 Dec 2012.

Skinner, S., R.T. Prinke, and R. Zandbergen. 2017. *The Voynich manuscript: The world's most mysterious and esoteric codex*. London: Watkins.

Snijders, L., T. Zaman, and D. Howell. 2016. Using hyperspectral imaging to reveal a hidden precolonial Mesoamerican codex. *Journal of Archaeological Science: Reports* 9: 143–149.

Soskice, J. 2010. *The sisters of Sinai: How two lady adventurers discovered the hidden gospels*. New York: Vintage Books.

Stolte, D. 2011. UA experts determine age of book 'nobody can read.' http://uanews.org/node/37825. 25 July 2012.

Tucker, A.O., and R.H. Talbert. 2013. A preliminary analysis of the botany, zoology, and mineralogy of the Voynich manuscript. *HerbalGram* 100: 70–85.

Wintle, S. 2017. Master of the Banderoles. http://www.wopc.co.uk/netherlands/master-of-the-banderoles. 2 Mar 2017.

Zandbergen, R. 2016. The radio-carbon dating of the Voynich MS. www.voynich.nu/extra/carbon.html. 5 June 2017.

Zetina, S., J.L. Ruvalcaba, T. Falcón, E. Hernández, C. González, E. Arroyo, and M. López Estéticas. 2008. Painting syncretism: A non-destructive analysis of the Badiano Codex. Pages 1–10 in 9th international conference on NDT of Art, Jerusalem, Israel, 25–30 May 2008.

Zyats, P., E. Mysak, J. Stenger, M.-F. Lemay, A. Bezur, and D.D. Driscoll. 2016. Physical findings. In *The Voynich manuscript*, ed. R. Clemens, 23–37. New Haven: Yale University Press.

Chapter 3
An Historical Context for the Voynich Codex: Aztec Mexico and Catholic Spain

Fernando A. Moreira

Spanish Encounter the New World

The Spanish encounter of the West Indies by Christopher Columbus in 1492 followed by the Cape route to India by the Portuguese explorer Vasco da Gama in 1498 were pivotal events of world history and are convenient dates to mark the beginning of the modern era. After the return of Columbus to Spain in 1493, the humanist Italian tutor to the Spanish court, Peter Martyr d'Anghiera, soon recognized that a New World had been discovered. Although before 1519, the Spanish were confined to Cuba, Hispaniola, and Panama, the invasion of what is now central Mexico by the conquistador Hernán Cortés brought about an explosive battle between the medieval culture of Catholic Spain and the Aztecs of central Mexico. The Aztecs created enormous temples, possessed sophisticated art, studied astronomy, practiced a complex religion, and utilized a logophonetic writing system; yet, they lacked beasts of burden and even the wheel. The culture was destroyed as a result of war, death, and cultural genocide via the clash with Catholic Spain. Following Cortés' conquest of central Mexico (1519–1521), a new culture developed from the formation of *Nueva España* (New Spain), a New World colonial territory north of Panama.

The culture of New Spain in the sixteenth century was brought about by the encounter between two violent civilizations: the medieval Catholic Spanish culture replete with firearms, horses, and religious fervor associated with strangling and burning of heretics, and the equally sophisticated Aztec culture of central Mexico with its vast temple cities, monumental art and literature, and its complex religion associated with ritual killing and cannibalism. The indigenous civilization was overwhelmed by a combination of superior military technology and its susceptibility to new human diseases introduced by the conquistadors. The *Voynich Codex* reflects the fusion of these two civilizations.

This chapter briefly reviews the history of central Mexico before and after the arrival of the Mexica from Aztlán, their legendary home, to the formation of the Aztec Triple Alliance in 1428, up to its eventual demise and transformation by

J. Janick, A. O. Tucker, *Unraveling the Voynich Codex*, Fascinating Life Sciences, https://doi.org/10.1007/978-3-319-77294-3_3

Cortés' forces. The story of New Spain is complex, with viewpoints presented by the conquering forces and those who sympathized with the conquered. Objectivity, although difficult, is the prime goal.

History of the Aztecs

The Aztecs (an endonym after their legendary home, Aztlán) lived in central Mexico from the thirteenth to sixteenth centuries. They referred to themselves as Mexica and their name is enshrined in the country's present name. Oddly enough, the term Aztec was never mentioned by Cortés or Bernardino de Sahagún (1499–1590), the Franciscan friar who became famous for documenting Aztec life. Aztec was used by an eighteenth century Jesuit and later popularized by German naturalist Alexander van Humboldt in the nineteenth century. The Aztecs were also known as Tenochca, the people of Tenochtitlan, their sumptuous capital city on Lake Texcoco, now the site of Mexico City where the lakes have long been drained. The Mexica spoke Nahuatl, and thus can be referred to as Nahuas. Nahuatl and its variants are part of the Uto-Aztecan language family group that extends from Oregon to Panama. Various ethnic groups spoke it and the language still is spoken by more than a million people. Many Nahuatl-derived words from Mexico are now common in English, such as avocado, chili, chocolate, coyote, and tomato. The Aztecs were latecomers to central Mexico, but as the dominant rulers at the time of the conquest, they have been the focus of intense study.

Sahagún (ca. 1499–1590) was born Bernardino de Rivera in Sahagún, Spain. He arrived in New Spain in 1529 and participated in the evangelization process of the indigenous population. Although he was first a missionary, he became fluent in Nahuatl and spent about 50 years recording Aztec ethnohistory, compiling the 2,400-page *Historia General de las Cosas de Nueva España* (*General History of the Things of New Spain*, Sahagún 1951–1982) completed in 1569, still the most influential work on the ethnic history of the Aztecs. The manuscript ended up in Florence and is thus known as the *Florentine Codex* (Wolf and Conners 2011). The work was carried out with the help of Nahua scribes trained at El Colegio de Santa Cruz de Tlatelolco (also called Colegio de Santa Cruz, in present-day Mexico City). Along with indigenous artists, they were responsible for the 2468 illustrations.

Pre-conquest Mexican Society

The history of civilizations in central Mexico is linguistically and ethnically complex. For example, from 150 to 750 CE one of the largest cities of the time grew to some 20 km^2. It was characterized by enormous temple pyramids that remain a popular tourist attraction in Mexico. For reasons yet understood the city fell out of history.

Fig. 3.1 Founding the city of Tenochtitlan. (Source: Diego Durán's *Historia de las Indias de Nueva España e Islas de Tierra-Firme* 14v, ca. 1579)

To the Aztecs, the ruins were considered a sacred place, referred to as *Teotihuacan*, meaning the "place of the gods."

From 1100 to 1400 CE, annual rainfall declined in Mesoamerica, shifting maize agriculture southward. For good or bad, the Nahua, guided by a priest, fled their ancestral lands of Aztlán and migrated south. The Mexica were one of seven tribes to leave and make it to the central basin of Mexico. Around the late 1240s, they arrived at Chapultepec in the already populated Valley of Mexico, present-day Mexico City and its suburbs (Smith 1984: 174). Two powerful city-states, the Colhuacan and the Azcapotzalco, ruled this region. When the Mexica were expelled from Chapultepec, refuge was offered by the Colhuacan ruler and the Mexica were eventually assimilated into that culture. In the fourteenth century, the Mexica eventually settled on a reedy island in Lake Texcoco, directed by their ruler Tenoch, who was supposedly guided by the god Huitzilopochtli. It was there that they built the glorious capital city, Tenochtitlan. Legend claims that they encountered an eagle devouring a snake (or bird) perched on a cactus, which grew from a rock in the middle of the lake, a reference to Aztlán, their northern legendary home (Fig. 3.1). Although pre-conquest Aztec art did not include a snake in the aforementioned scene, it did include an *atl tlachinolli*, an Aztec symbol of war.

By the early fifteenth century, Tenochtitlan was a tributary to Azcapotzalco, where the Mexica served as soldiers in their wars (Carmack et al. 2006: 89). In 1428, the Aztec Empire was created as a result of the formation of the Triple Alliance, composed of the city-states Tenochtitlan, Texcoco (home of the Acolhua), and Tlacopan (a Tepanec kingdom, which toppled Azcapotzalco). By the early sixteenth century, the expansionistic empire had encompassed much of Mesoamerica as far south as Soconusco near present-day Guatemala. Aztec merchants (*pochteca*) were present in areas that the Aztec army had not yet conquered. By 1491, Tenochtitlan dominated

Fig. 3.2 Map of Tenochtitlan. (Source: Friedrich Peypus' *Praeclara de Nova maris Oceani Hyspania Narratio*, ca. 1524)

the Triple Alliance with a population of about 200,000 people and covered some 13 kilometers squared on a strict grid layout. Figure 3.2 shows a woodcut of Tenochtitlan from 1524 that was published alongside Cortés' second letter to Holy Roman Emperor Charles V (also known as Carlos I of Spain). About 50 city-states existed in the central basin and some 450 were subject to the empire. The communication and trade systems were sufficiently developed by the time Moctezuma II (also spelled Motecuhzoma II or Montezuma II) was in power, and the arrival of the Europeans was no revelation. An early document revealed that in 1512 Moctezuma II had sent emissaries to Guatemala to warn the K'iche' kingdom of the impending battle with the Spanish, spotted on the Gulf Coast (Carmack 2001: 171).

The Triple Alliance was an informal or hegemonic empire, ruled indirectly. It was composed of multilinguistic and multiethnic peoples, with Nahuatl as the lingua franca. The empire also was not physically conterminous, as it had detached tributary provinces, such as Soconusco. The Aztecs did not interfere with the internal workings of their subjugated polities, often reinstating the same rulers they defeated. The empire was a tribute system and by no means a single system of government. The Aztec Empire was organized into various ethnic political units (city-states) called *altepetl*.

Politics and Economy

The Aztecan economic system comprised two branches: commercial and political. Food production was vital in feeding the city's population along with 1 million others who lived in the central basin. The chief foods cultivated were corn, beans, squash, amaranth, chia, chili, and maguey. They had diverse systems of agricultural production, including: *chinampa* (lake gardens) on lake shores, *tlacolol* (swidden agriculture, or rotating crops) along foothills, terrace farming on hills, and complex irrigation along river beds (Carmack et al. 2006: 91). Chinampas would produce some seven times a year and provided 50% of the dietary needs of Tenochtitlan. The rest were provided via tributes from the provinces. Protein was supplemented mainly by turkeys, domesticated dogs, and game meat (such as tapir, deer, rabbit, and peccary, a New World pig). Secondary food sources were locust, fish, crustaceans, water birds, grubs, lizards, spirulina algae, and water fly eggs (Harris 1998: 230). The noble class ate the flesh of human sacrificial victims.

A sophisticated system of craft manufacturing existed, involving feather working, textiles, metallurgy, lapidaries, and paintings. The production of these goods was mainly controlled by the elites. Artisans lived communally in special residences. Craftsmen would also be responsible for producing the many public works and ornamental facades. The nobles in the central basin controlled agricultural surplus, with less control in the provinces. The *pochteca* (long-distance traders) served as importers, wholesalers, and retailers, with enclaves as far south as Panama. Transportation of goods and travel in Mexico was by foot or canoes, the latter which crowded Lake Texcoco (Berdan 2014: 113). Peasants would carry their own merchandise to sell at the markets, whereas the elite could have caravans of porters.

The market was the economic lifeline of Aztec society. It served as the local and the international venue for commonplace and exotic objects, and was administered by government trade officials, who determined fair pricing for goods and collected tributes and taxes. The most prominent regional market was at Tenochtitlan's sister city, Tlatelolco. It featured local produce and served as the place to acquire goods from distant regions. Markets at Xochimilco and Texcoco and throughout the basin of Mexico were also important. Some markets appeared to specialize, although not exclusively in specific commodities. The Texcoco market was known for its cloth, Coyoacán for its wood products, and Acolman for its dogs. Land in general terms was not a marketable commodity and was usually granted or transferred, sometimes entirely with the tenant occupants (Carmack et al. 2006: 95).

Aztec Culture

Linguistics

Language diversity in Mexico is unparalleled, from language families that dominate the Mexican highlands and beyond, to those small and unrelated. Language relationships are usually described metaphorically by the use of a phylogenetic tree and

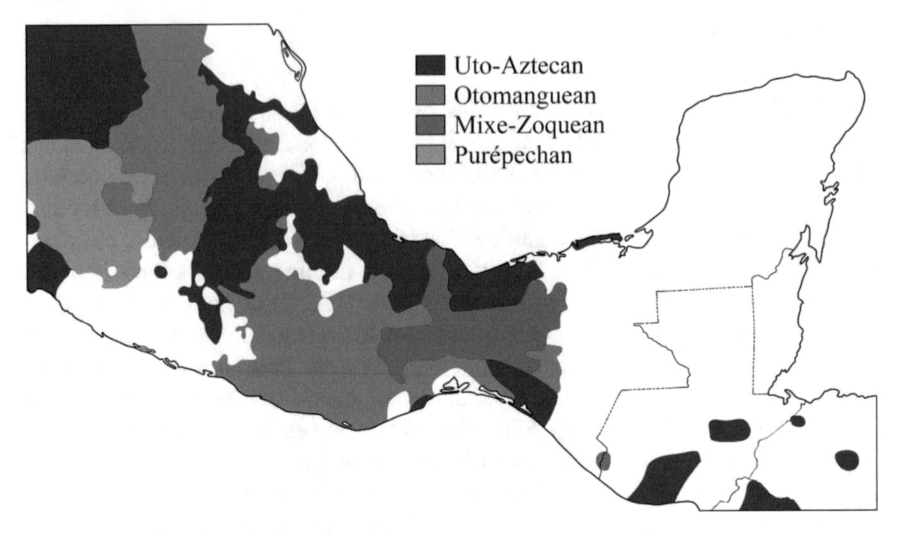

Fig. 3.3 Language family distribution map of Mexico

are classified using comparative linguistics. Language families known as monophyletic units are groups of languages that are related by descent from a common ancestor. These can be further divided into smaller phylogenetic elements denoted as "branches" of a family. The language families found in central Mexico are the following: Uto-Aztecan, Otomanguean, Mixe-Zoquean, and Purépechan (Fig. 3.3 shows the language distribution areas). It should be noted that identifying a language group does not entail the identification of a cultural group. Instead, native groups themselves use language, geographic region, and membership of particular communities for self-classification.

Ancient Uto-Aztecan is a geographically broad language family that extends from Oregon to Panama. This family contains 58 languages; yet, colonial sources advocate a larger size. Linguistic studies suggest that its origins might stem from southeast California, Arizona, and northwest Mexico. The family is divided into two generally recognized groups, Northern and Southern Uto-Aztecan. The latter group is of most interest and is further broken down to: Pimic, Taracahitic, and Corachol-Aztecan (Campbell 1997: 134). Corachol-Aztecan is the family that contains Nahuatl (Classical Nahuatl), along with Pipil (Nawat), Pochutec (extinct), Cora, and Huichol. Pipil is well known for its lateral affricate "tl" /t͡ɬ/. Nahuatl was spoken primarily in central Mexico, notably by members of the Aztec Empire. It also has some 30 surviving dialects, according to the Mexican Institute of Indigenous Languages.

Absolutive noun suffixes, with the exception of Cora/Huichol, are typical in Uto-Aztecan grammar. These suffixes have no apparent semantic value and appear in nouns as heads of lexical entries and, depending on the linguistic environment, can be dropped. A noteworthy absolutive noun suffix present in Classical Nahuatl takes the form "-tl" after vowels and "-tli" after consonants, *a:-tl* (water) and *no:ch-tli* (cactus fruit) respectively. Classical Nahuatl is an extinct language from central

Mexico and was considered the lingua franca in the sixteenth century. This particular form survived in colonial texts and is likely a sociolect, a language variety associated with a specific social group, in this case, the aristocracy. Pajapan Nahuatl, also known as Isthmus Nahua, has various phonological differences from Classical Nahuatl, such as the simplification of the lateral fricative to a dental stop, illustrated in the word for person, *tla:catl* versus *ta:gat,* for example.

Religion

State and religion were conflated in Aztec society and the *tlatoani* (king) stood at the top of both. Immediately below him were the equally important offices of the *Quetzalcoatl Totec Tlamacazqui* and the *Quetzalcoatl Tlaloc Tlamacazqui*, which headed the two sanctuaries at the top of the Templo Mayor, one of the main Aztec temples in Tenochtitlan. The former was dedicated to Huitzilopochtli and the latter to Tlaloc. The activities of their offices alternated between the dry and wet seasons. Both priestly offices were elected by the *tlatoani*, the *tlatoque* (Aztec city-state rulers), and the great judges, based on the qualities of a person's purity and righteousness. These were the only pair of priests that could marry and have a family and were usually of noble lineage. Below them was the *Mexicatl Teohuatzin*, who oversaw all rituals, in addition to the *calmecac*, the school for male nobility. He had two assistants, the *Huitznahua Teohuatzin* and the *Tecpan Teohuatzin*. The first was dedicated to ritual, whereas the second was concerned with education. They both administered various *teopantlalli*, temple lands, and oversaw the selection of *ixiptla*, deity impersonators (Aguilar-Moreno 2007: 153–154).

The *tlenamacac* were the fire priests. They carried out all the human sacrifices and were the only ones who could cut the victims and remove their hearts. They also performed the *Toxiuhmolpilia*, the so-called New Fire ceremony performed at the end of every 52-year cycle, based on the ascent and descent of the star cluster known as the Pleiades, that ensured the sun's rebirth and the motions of the heavens for the next cycle. Priestesses also existed in the religious institution. They were called *cihuatlamacazque* and participated in fertility festivals, but could not participate in those that involved sacrifices. Various other offices existed from teacher priests (*tlamatini*), to those in charge of carrying effigies of the gods during war campaigns (*tlamacaztequihuaque*), to the celibate jaguar priests (*oceloquacuilli*), who were responsible for sorcery.

In 1428, Itzcoatl and Tlacaelel, the first *tlatoanis* (kings) of the empire, rewrote history and burned all the books, changing the religious system to support the political institution and legitimizing its authority. Huitzilopochtli was the principle deity, the patron of war and sacrifice. The Aztecs had the sole responsibility for feeding the sun with the blood of the sacrificed to guarantee its preservation. This justified the aggressive expansion of the empire, taking in sacrificial victims to provide nourishment to the gods. Creation was a cyclic process with periods of time that ended catastrophically called "suns." It is important to note that various creation stories and gods overlapped and often contradicted one another, providing symbolic expressions for the various groups in Aztec society.

Astronomy

Aztec astronomy, like that of most civilizations, was pseudo-religious and its study was conducted by priests and the nobility, notably during sunset and at midnight. Yet, unlike other cosmological world views, the Aztecs believed that the heavens were extremely disordered, and that doom could occur at any time. The astronomers looked to the sky because it was intrinsically linked to the agricultural cycle and was pivotal for creating a calendar system. They kept track of the lunar month, the solar year, solstices, equinoxes, and even the orbiting cycle of Venus (Totonametl), and Mercury (Piltzintecuhtli). Movements related to the sun were the most significant. According to the Aztec world view, cyclical phenomena from the heavens were linked to earthly events, from the conquests of their enemies to the accessions of *tlatoanis* (kings). The most severe cosmic events were eclipses, which typically meant that a universal or local disaster was imminent. As far as planets go, Venus, the most important, had two key aspects, the "morning star" and "evening star." The former was good fortune, whereas the latter could cause illness or death. A coincidental numerical relationship existed between the sun and Venus. With a synodic period of 584 days, it took exactly five of these periods to equal eight solar years, 5 + 8 is 13, which alluded to the 13-day periods of the *tonalpohualli* (260-day calendar).

Aztec constellations were often associated with deities. Yohualitqui Mamalhuaztli (Orion's Belt), Citlaltlachtli (Gemini), Colotl Ixayac (likely Scorpio), and Tianquiztli (the Pleiades) approximately marked the cardinal direction (Aguilar-Moreno 2007: 308). The most prominent constellation group was Tianquiztli, the "marketplace." Sahagún notes that during the New Fire ceremony, the priests would climb up the "hill of stars" and wait for the Pleiades to pass the zenith, marking the fifth cardinal point. Once the star cluster did so, the priests were convinced that the movement of the heavens had not ceased, and that the world would continue for another 52 years. They practiced horizon astronomy, which essentially meant that their observations focused on the east–west horizon, the oscillating path of the sun that went over and under *tlalticpac* (the earthly layer). Not surprisingly, the fifth sun was known as 4 Olin Tonatiuh by the Aztecs, which literally meant "sun movement," a semantic clue of what drove Aztec astronomy.

Botanical Gardens

About 26,285 vascular flora species exist in Mexico alone, and some are endemic (Frodin 2001: 263). Although Mexicans were associated with gruesome rituals and warfare, they were pioneering horticulturists and lovers of aesthetics. In 1457, a massive garden named Texcotzingo was constructed on a hill by the ruler Nezahualcoyotl, the *tlatoani* of Texcoco (who was also responsible for the garden retreat at Chapultepec). It was more than just a botanical garden cultivating various flora from the empire, but represented a map of their political domain and was a hydrological engineering accomplishment. It contained paths, bas-relief sculptures, and shrines.

Aqueducts transferred water from Mount Tlaloc into pools and irrigation systems, located at Texcotzingo and beyond. Not everything was purely functional, as waterfalls, fountains, and baths also adorned the park. Certain sectors of Texcotzingo, owing to their southwestern placement and the mist produced from pools, created an suitable micro-environment for growing tropical plants at high altitudes.

Status rivalry was not limited to conquest, but also to beauty. Hence, in response to Nezahualcoyotl's creation, Moctezuma I reconstructed the ancient Huaxtepec, a semitropical botanical garden. It was a large, orderly "six-mile circumference" garden, which was hydraulically fed by a crossing river, constructed at a much lower altitude. The garden was filled with ornamental, aromatic, and medicinal plants, with coastal imports such as vanilla, cacao, and magnolia trees. According to the conquistador Díaz del Castillo, it was the best garden he and Cortés had seen. It served as a place of visual and olfactory pleasure. Recreational gardens and nurseries were commonplace in the Aztec world, some served purely for pleasure, whereas nurseries supplied seeds and tree saplings for parks, in addition to flowers and greens to decorate temples, which could be sold at central markets. The value placed on beautiful plants was so extreme that commoners were banned from planting some of them by decree (Evans 2004: 477).

Bathing

Hygiene was essential among the Aztecs. The cities had drainage systems and access to fresh water and sanitation. Human excreta and urine were also collected and disposed of. The *temazcalli*, literally "bathhouse," was used for steam cleaning and it had other applications, such as medicinal and ritual cleansings. The Aztecs used the root of a plant they called *xiuhamolli* (likely *Ipomoea murucoides*) or simply *amolli* (likely *Sapindus saponaria*), which was, according to Sahagún, a cleanser used to remove dirt. The Jesuit Clavijero also reported that the fruit of the *copalxocotl* tree (probably *Cytocarpa* sp.) was used as soap. The Aztecs bathed often and sometimes many times a day and made use of natural bodies of water, such as lakes, rivers, and ponds. Nezahualcoyotl, for example, used rock-carved baths in his pleasure garden at Texcotzingo. The conquistadors took special note of the level of cleanliness of Moctezuma I and the Aztecs in general. The Iberians of the time had an inherent fear of water, which they attributed to disease and nakedness, considered to be unchristian (Constable 2014: 257).

Medical Knowledge

To be ill in the Aztec world meant to have diminished your *tonalli*, that is your "life force," that unique, ineffable quality that determined your destiny. Deities, magic spells, and natural causes could impair the *tonalli* and thus Aztec medicine involved a variety of curing techniques from religious rituals to herbal remedies (Aguilar-Moreno 2007: 359). Therefore, it is not surprising that two types of doctors existed in the Aztec world: the *tepati*, who used herbals, and the *ticitl*, the sorcerers (Schendel et al. 2014). The medical knowledge of the Aztecs impressed the Spanish crown so much that they sent emissaries to New Spain to study the herbs, stones, trees, and roots used by Aztec physicians (Schendel et al. 2014). Such works as the *Codex Cruz-Badianus*, written in 1552, had 185 pictures of plants alone, whereas Sahagún's work had 149 herbs, and the *Historia de las plantas de Nueva España* by Francisco Hernández included some 1000 plants, most of which were medicinal (Gates 2000: 119).

Art

The Aztecs used art primarily to disseminate religious, military, and political ideology. They lacked the concept of art created "in and of itself" (Aguilar-Moreno 2007: 178). Craftsmen who worked with gold, silver, and pearls were called *tolteca*, after the ancient people recognized as their antecedents, and they occupied a privileged position among Aztec society. Stone monumental art was used to exemplify mytho-historical events and concepts of life and death, and the style focused on the "*abstraction of the whole, and realism in detail*" (Aguilar-Moreno 2007: 190). Typical stone structures included *cuauhxicalli*, which held the hearts and blood of victims and were decorated in jaguar/eagle iconography. Others were celebratory stone structures dedicated to conquests and the expansion of the great temple.

Terracotta sculptures that emphasized nature deities and commoners also existed in Aztec culture, but were kept to a minimum. Stone took preference in elite life. According to the Aztec world view, wood was conceptually inferior to stone; yet, its composition made it a perfect material for objects, such as god effigies, spear-throwers, masks, and notably the Aztec *teponaztli,* a horizontal drum. These drums were so lavishly decorated that only the codices rivaled them. As mentioned previously, feather work was produced by artisans, who created shields, fans, and headdresses. Although only a small number of mural paintings have survived, they did adorn temples with themes that ranged from processions of warriors to rain deities. Colors included blue, red, ocher, white, and black (Van Tuerenhout 2005: 219).

Writing and Poetry

The Aztec writing system was similar to Mixtec writing. Classified as logophonetic, it was a system that used logographic symbols, and symbols that represented actual sounds. To the Aztecs, this system of writing served the following functions: calendrical, accounting, and naming of people, places, and events. Nahuatl words, because of the language's polysynthetic nature (root/many affixes), would be

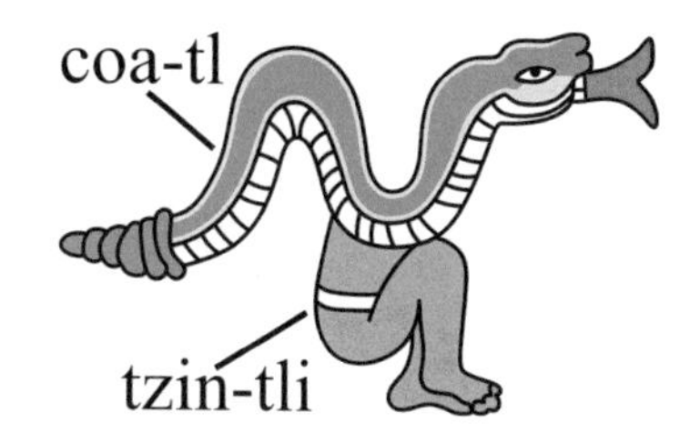

Fig. 3.4 The compound toponym *Coatzinco*. (Source: Redrawn by author from the *Codex Mendoza*, folio 42r, ca. 1542)

written in groups of logograms, often in conflated form. Due to the complexity of representing some words in logograms, the Aztecs used symbols for their phonetic values, which is called rebus writing. For example, the toponym *Coatzinco,* which means "place of little snake," was created from the logograms for snake, *coa-tl,* and the buttocks, *tzin-tli* (Fig. 3.4), the latter being used as its root form, which is a homophone of the diminutive "-tzin."

The prestige the spoken word had is made clear by the word *tlahtoa*, which means "to speak." The words for ruler, ambassador, poet, and scribe were derived from this term (Van Tuerenhout 2005: 237). The *calmecac* (school for Aztec elites) taught the Aztec nobility how to read, but commoners seldom had this privilege; therefore, most of the empire was illiterate. Being *tlacuiloque* (indigenous scribes and painters) was hereditary. The skills, which were passed from father to son, were highly prized by the Aztecs. They wrote primarily on *amate*, paper made from the bark of various species of *Ficus*, but they used other media as well.

The Aztecs made distinctions between prose and poetry, calling prose *tlahtolli* and poetry *xochitl cuicatl* (meaning "flower song"). Prose, written in annals and chronicles, included mytho-historical accounts from an *altepetl* (city-states) perspective, such as the *Codex Chimalpopoca* and the *Florentine Codex*. There were four categories of *xochitl cuicatl*: sad songs, songs of spring, war songs, and plain songs (Karttunen and Lockhart 1980: 21). The poetic themes ranged from natural phenomena, to gods, to love, and to ancient traditions, and they were always sung (Aguilar-Moreno 2007: 282). Music complemented poetic sessions, with drums as the foremost instrument. Religious poetry was hard to comprehend if one was not a priest as metaphors and allusions were often used.

Post-conquest Mexican Society: New Spain

The Conquest

Hernán Cortés (1485–1547), born in Medellin, Spain, was the son of minor nobility. He was not born to greatness, but achieved it. At 14, he studied Latin with his uncle in Salamanca for 2 years. In 1504, he sailed for Santo Domingo, on the island of Espanola (present-day Hispaniola), seeking adventure and wealth. Cortés was a man of action and soon received land rights (*encomienda*). He was twice appointed magistrate of Santiago. In 1511, he became a member of an expedition to conquer Cuba and was appointed secretary to the governor, Diego Velazquez. News of Juan de Grijalva's foray into the mainland with evidence for wealth and gold encouraged Cortés to accept the offer of captain general of a new expedition and he moved quickly before the governor could change his mind.

On 10 February 1519, Cortés disregarded orders from Cuba Governor Velasquez and sailed off to what would become New Spain, hiring reinforcements along the way. Equipped with nine caravels, about 500 soldiers, and 16 horses, he landed in Cozumel, an island off the Caribbean coast of Mexico. Cortés, having heard rumors of Castilians from the people of Campeche, had sent two indigenous messengers to the mainland. They encountered Gerónimo de Aguilar in a canoe, along with the two messengers. Aguilar, shipwrecked in 1511, had been a slave for 8 years. He had learned Yucatec Mayan and became an interpreter for Cortés (Díaz del Castillo 2012: 34).

Having defeated the Chontal Maya, Cortés founded Santa María de la Victoria at Potonchán, the capital of Tabasco. Adhering to tradition, the defeated Chontal offered gifts of gold, jade, and animal skins, along with 20 young women (among them was Malinalli or La Malinche). Malinalli took the Christian name Marina, and through her knowledge of Nahuatl and Chontal Maya also served as an interpreter for Cortés, later becoming the mother of his child. At first, interpretation depended on Aguilar's Mayan to Spanish, but his role ended as Marina learned Spanish.

Cortés arrived at the Totonac town of Cempoala and was greeted by dignitaries, who were then persuaded to rebel against their Aztec overlords. By July 1519, men loyal to Diego Velasquez planned to seize a ship and head back to Cuba. In defiance, Cortés retaliated by sinking all the ships, permanently stranding his men. Cortés with his army marched in search of the great empire "where the sun sets." Emissaries of Moctezuma II were sent to persuade the Spaniards to abandon their march to Tenochtitlan. During his march, Cortés made alliances with the Nahua of Tlaxcala (called Tlaxcalans). The unconquered Tlaxcalans were surrounded by the Aztec Empire; thus, they were fixed in a never-ending war. Initially, the Tlaxcalans fought the invaders, under the banner of Xicotencatl II Axayacatl. But eventually he was persuaded to be their ally and traveled with Cortés to Cholula, a large ceremonial center considered a sacred place for the Aztecs.

Compared with the Aztecs, the Spanish had technologically advanced weaponry, such as cannons, armor, horses, and war dogs. For reasons unknown, Cortés ordered an attack and, with the help of Tlaxcalan warriors, massacred 3000 people and set Cholula ablaze. After many attempts to dissuade the Spaniards from reaching Tenochtitlan, Moctezuma II gave in to pressure and requested that they come. On 8 November 1519, Cortés and his forces entered the capital. The Lienzo de Tlaxcala, a painted canvas created in central Mexico around 1552, clearly shows both Cortés and Marina in Tenochtitlan (Fig. 3.5). According to Sahagún, the rulers of Texcoco and Tlacopan accompanied Moctezuma II and they welcomed Cortés on the great causeway, but by mid-November Moctezuma II had been taken prisoner. When Cortés heard of a large force sent by Velasquez to arrest him, he set off to surprise them and left Pedro de Alvarado in charge of holding Tenochtitlan.

In May 1520, Cortés returned with some 3000 Tlaxcalan and Spanish soldiers and found that Alvarado had slaughtered the nobility who had attended the Toxcatl

Fig. 3.5 November 1519, Cortés and Marina in Tenochtitlan. (Source: *Lienzo de Tlaxcala*, ca. 1552, Mesolore)

celebrations. The population rose up against the Spanish and their allies. Moctezuma II was made to speak to calm the population, but was subsequently slain in the process. On 30 June of the same year, the Spaniards and their allies escaped for Tlacopan through the causeway that linked the cities together, pursued by Aztec warriors on foot and in canoes. Many Spaniards drowned or were killed. This was known as the *Noche triste*, the night of sorrows, for when Cortés realized the carnage, he wept for the loss of his men. By the time the Spaniards and allies reached Tlaxcala, some 860 Spanish soldiers and 1000 Tlaxcalan warriors were lost.

The Aztecs sent emissaries to Tlaxcala to negotiate the handover of the Spanish forces, but the Tlaxcalans refused to surrender them to Cuitláhuac, the newly appointed *tlatoani* (king) of Tenochtitlan. Looking for allies, Cortés solidified an alliance with the Tlaxcalans, making heavy concessions, agreeing to hand over Cholula to the Tlaxcalans, and even exempting them from future tribute, but after the fall of Tenochtitlan, this treaty was annulled. Having defeated various Aztecan tributary states armies, Cortés gained status, and therefore more native allies. Once the Spanish forces had regrouped, the Tlaxcalans contributed no less than 10,000 troops to help Cortés bring Texcoco into his fold. As they entered the city, they found it deserted and the Tlaxcalans were permitted to sack it.

Cortés installed the "pro-Spanish" Tecocoltzin as ruler of Texcoco and the city served as the military base for the Spanish invasion (Brokaw and Lee 2016: 138, 218). Tecocoltzin sent messengers throughout the empire and urged its members to become vassals of the Spanish king, supporting the Spanish cause (Brian et al. 2015: 2). Cuauhtémoc succeeded Cuitláhuac (after his death from smallpox) and became the final *tlatoani* (king) of the empire. He resisted the Spanish forces for 8 months, but by 13 August 1521, he was caught fleeing across Lake Texcoco. Instead of executing him, Cortés treated Cuauhtémoc with reverence. But then Cortés also tortured him and burned his feet, hoping that the king would reveal the location of gold treasure (which Cortés eventually acquired, albeit a disappointing amount) (Pemberton 2001). By 1525, Cuauhtémoc was taken south on Cortés' expedition to Honduras. Fearing a conspiracy, Cortés had Cuauhtémoc and the rulers of Texcoco and Tlacopan executed by hanging.

Encomienda and Slavery

The *encomienda* system was formally introduced into New Spain in 1503. It was a dependency relation practice where Spanish kings would grant an area to a Spaniard (the *encomendero* or grant recipient), commonly a conquistador, with the task of protecting and evangelizing a group of natives living in that area. In exchange, the *encomendero* would receive products and services; essentially, he had the right to collect tribute. The *encomienda* was a grant. There was no title attached to the land and its natives could not be legally controlled or sold. The Spanish *encomenderos* formed the first administration systems before there was an established government. Owing to the crown's fear of the formation of a hereditary aristocracy, the inheritability of the *encomienda* was restricted to a privileged few, such as the descendants of Cortés. The former *encomendero*, Bartolomé de las Casas, seeing the results of the *encomienda* system, was appalled by it and staunchly fought it, as it essentially resulted in the systematic slavery of the native population. By 20 November 1542, the New Laws were signed by Holy Roman Emperor Charles (Carlos) V, abolishing the *encomienda* system, and shifting responsibility for the indigenous population under the crown as tribute-paying subjects. The law was so unpopular that even the viceroy of New Spain resisted and then ignored it. By the 1550s up to the mid-seventeenth century, the *encomienda* was replaced by the *repartimiento*, an obligated paid work term.

The Church

Apart from pacifying the New World, the Reyes Católicos set out to effectively and systematically evangelize the natives of their new realm. Cortés requested that Franciscan and Dominican friars be brought to New Spain, and by June 1524, a

Fig. 3.6 The Franciscan twelve apostles in a mural at the monastery of San Miguel Arcángel in Huejotzingo. (Source: Enhanced photo from Wikimedia Commons)

group of 12 Franciscan missionaries arrived in Mexico City. They were to become known as the Twelve Apostles of Mexico and were led by Martín de Valencia. Figure 3.6 shows all 12 missionaries facing a cross in a monastery mural from Huejotzingo. Allegedly, as they entered the city, and in front of Aztec nobility, Cortés threw himself at their feet. Although unusual behavior, it was meant to show the nobility how to conduct themselves in front of friars (Trexler 1997: 146). The apostles' primary focus was in the valleys of Puebla and Mexico, mainly Tlaxcala, Huejotzingo, and Texcoco. Fray Toribio de Benavente, known simply by his Tlaxcalan appellation Motolinía (Nahuatl for "poor" or "afflicted"), was the most prominent of the 12 disciples (Trigger et al. 2000: 219).

Motolinía was a notable ethnographer. He gathered valuable yet rather unsystematic data on the conquest-period Nahua people of central Mexico. Like Bartolomé de las Casas, Motolinía fought for indigenous rights; yet, unlike Casas, and as expressed in a letter to Holy Roman Emperor and King of Spain Carlos V, he did not attack the *encomienda* system and staunchly defended the conquest as being necessary for evangelization (Zorita 1994: 295). The Franciscans had a long-term strategy for evangelization: the early conversion of children from noble lineages, as converting adults proved to be difficult. Although before the Twelve Apostles' arrival, the conversion problem was elucidated in 1523 when Cortés ordered the execution of a high-ranking Cholulan noble along with six others. They had refused to convert to Christianity; thus, they met their deaths by savage dog attacks (Fig. 3.7). According to Motolinía, the Franciscans embellished Christian festivals with entertainment and dance, using participation in rituals as a privilege to attract nobles to the new faith. Once the nobles were lured, the commoners would follow suit (Schroeder and Poole 2007: 39). Unlike their Dominican counterparts, they would baptize as many of the local population as possible and would achieve this by learning their local languages.

Diego Durán, a notable Dominican friar, noted that once the locals received the sacrament of marriage from the church, they also would seek out a native elder and be married under local customs (Schroeder and Poole 2007: 33). Therefore, the church adopted some of the local traditions, hybridizing the sacrament and retaining

Fig. 3.7 A Cholulan priest/noble gets killed for rejecting conversion. His death was ordered by Cortés in 1523. (Source: Manuscrito del aperreamiento, Doc. No. 374, ca. 1560)

more locals. By the late 1530s, Motolinía boasted, though likely an exaggeration, that Franciscan friars performed wedding rites for up to 500 couples a day. Marriage was meant to regulate sexual relations. The clergy in New Spain was minimal and not enough of them were present on the ground to enforce Christian doctrines, let alone was anyone fluent in the local languages. To counter this, Christianized local officials called *licenciados* enforced the marriage doctrine in their respective communities. Motolinía claimed that there was one in every community.

The Holy Inquisition

The development of the Mexican Inquisition began with the so-called Monastic Inquisitions, trials conducted by leaders of monastic religious orders, such as the Franciscans and Dominicans. Fray Martin de Valencia, of the aforementioned Twelve Apostles, first exercised the office of the inquisitor (Chuchiak 2012: 10). The Inquisition transitioned when the bishop of Mexico, Juan de Zumárraga, holding the office of "Protector of the Indians," inherited the jurisdiction; yet, never established the tribunal himself. In 1536, he launched an Episcopal Inquisition on all heresy claims (156 cases in his time in office, not just on natives, but mestizos and Europeans as well). Zumárraga indicted Don Carlos Ometochtzin for idolatry and concubinage, and the Texcoco lord was burned at the stake in 1539. The bishop was reprimanded by the *Santo Oficio* (Holy Office) for his severe handling of the Texcoco lord trial and eventually he relinquished his title of apostolic inquisitor.

The Spanish Inquisition or, more formally, the *Tribunal del Santo Oficio de la Inquisición,* was a natural response from the monarchs of Aragon and Castile to the Papacy's political control via the Papal Inquisition, which controlled the religious affairs in Aragon (Pérez 2005: 33). Established in 1478, and lasting well over 350 years, its purpose was to ensure orthodoxy from converts. Arguably, the Inquisition was created to curb the influence of the opposition and increase the political and economic wealth of the monarchs. The Tribunal of the Inquisition in New Spain was officially established by decree in 1569 by King Philip II as an extension of the Spanish Inquisition, and in 1570 it was amended to exclude the indigenous population, as they were classified as "savages of limited responsibility in matters of faith" (Bunson and Bunson 2014: 207).

The *Provisorato de Indios y Chinos* was a provisional Episcopal court presided over by a chief ecclesiastical judge, called a *provisor*. He was responsible for the office whose sole objective was to monitor and make inquiries about religious crimes committed by the local indigenous population and had full jurisdiction over these transgressions (Chuchiak 2012: 352). This was essentially a separate Indian Inquisition. Before the formal tribunal, the indigenous people had suffered the greatest number of cases against them; yet, after 1571, and although cases were still accumulating, no formal legal trials occurred against them (Chuchiak 2012: 11). Archbishop Pedro Moya de Contreras was the first official inquisitor general of New Spain. In 1571, he exercised an *auto-da-fé*, a Portuguese term for "act of faith," and, depending on the charge, the accused were publicly executed.

Crypto-Jews, mulattos, and blacks suffered extensively during this period. An infamous case was that of Luis de Carbajal y Cueva, a former Jew and Christian convert from Spain, who became a wine trader and eventual soldier in New Spain. He became wealthy from the aboriginal slave trade and managed to bring his family to Nuevo Leon. Rumors began to circulate that he and his family had relapsed to the Jewish faith. Although Luis de Carbajal y Cueva died before his sentence could be executed, the remainder of his family suffered greatly and after a large *auto-da-fé*, most were burned at the stake.

Establishment of Schools and Colegios

Before its collapse, Tenochtitlan had two academic systems: the *calmecac* and the *telpochcalli*. Young Aztec male nobility were educated in the *calmecac* and institutions were located in each district of the city. The nobles received meticulous religious and military training, so that they could become priests or high-ranking warriors. The commoners, although not exclusively, attended the *telpochcalli*, where they received basic training.

European primary schools had been present in New Spain as early as 1523 (in Texcoco and later in Tenochtitlan) and provided literacy and music instruction. Figure 3.8 shows a pupil being trained to sing. By 1536, the Colegio de Santa Cruz was established at the site of an earlier *calmecac* (school for Aztec elites) to train boys from the Aztec nobility for the clergy. The *Colegio* was strongly opposed by the Dominicans, who questioned the orthodoxy and ability of the natives to learn Christian doctrine, concluding that they should not be baptized and therefore could

Fig. 3.8 Student being trained to sing. (Source: Sahagún's *Florentine Codex*, Book 10, folio 19r, ca. 1577)

not enter the priesthood. The Franciscan friar Sahagún was a staunch proponent of higher learning for the indigenous population. He defended their competence and eventually taught at and became dean of the Colegio de Santa Cruz. Franciscans, aided by indigenous assistants, trained natives to be literate in Nahuatl, Spanish, and Latin, and provided instruction in music, logic, philosophy, and indigenous medicine. Unfortunately, by 1555, the First Council of Mexico decreed as part of the *limpieza de sangre* that indigenous people, along with mestizos, Moors, and mulattos, were banned from serving as priests (Kirk and Rivett 2014: 293). Although this ban was reversed by Pope Gregory XII, it was largely ignored by New Spain, and the ban was upheld by the Third Provincial Council of Mexico in 1585.

By the sixteenth century, New Spain had at least a half dozen academic institutions. In addition to the already mentioned Colegio de Santa Cruz, the Nahua elite had La Escuela de San José de los Naturales, founded in 1529 by Fray Pedro de Gante (Gómez Canedo 1982). In 1532, Zumárraga founded a school for mestiza (female) orphans called El Colegio de Nuestra Señora de la Caridad (Bazarte 1989), and El Colegio de San Juan de Letrán in 1548, to educate illegitimate children, mainly mestizos (males) (Méndez Arceo 1990). La Real y Pontificia Universidad de México was founded by royal decree in 1551, which was the first university north of Lima, Peru (Gómez Canedo 1982; Méndez Arceo 1990).

Fray Pedro de Gante (1480–1572) was born in Ghent, Belgium, as Peeter Van Der Moers (also spelled Moere, Moor, or Muer). Evidence points to kinship ties with Carlos V, as Gante was allegedly an illegitimate son of Holy Roman Emperor Frederick III, or that of his son Maximilian I. As a Franciscan, Gante traveled to New Spain leaving on May 1st, 1523, and arriving 9 months before the Twelve Apostles (Torre Villar 1974:2). He travelled with Fray Juan de Tecto from the

Sorbonne and Fray Juan de Ayora from Scotland. Gante started on his mission to evangelize the local population and taught arithmetic, music, and Christian doctrine, among other things, at La Escuela de San José de los Naturales, where he also became fluent in Nahuatl (Mills et al. 2002). He avoided ordination and remained a lay brother.

Disease

Estimates place the population in Mexico before European contact at 25 million, and it was reduced to a mere 3 million. The main cause was disease. The demographic collapse of sixteenth-century Mexico is one of the highest in recorded history. Smallpox struck between 1519 and 1520, initially killing 5–8 million people. According to Sahagún, the plague came during the month of *Tepeilhuitl* (the 13th month of the Aztec calendar) and pustules spread throughout plague-infected people's faces and bodies. Victims neither could walk "*nor stretch themselves out on their backs*." Sahagún noted that the plague could also cause blindness. The *Florentine Codex* shows a rather disturbing scene of natives covered in maculopapular rashes (skin breakouts with small, red lumps; Fig. 3.9). The disease was so effective that there was no one to take care of the sick and some simply starved to death. An additional 7 million died in the epidemics of 1545 and 1576. Both epidemics were caused by *cocoliztli* (Nahuatl for "pest"), which appears to have been hemorrhagic fevers transmitted by rodents. Some of the *cocoliztli* symptoms

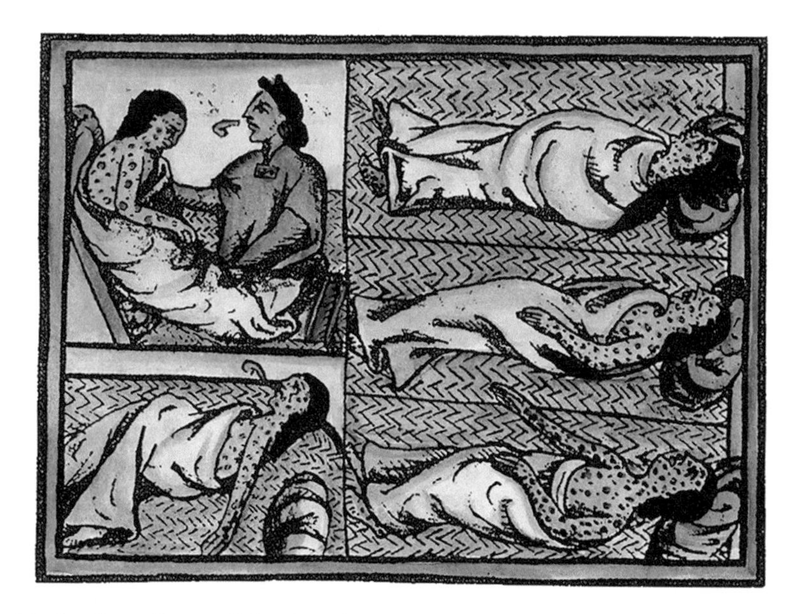

Fig. 3.9 Smallpox. (Source: Sahagún's *Florentine Codex*, Book 12, folio 54, ca. 1585)

included high fever, black tongue, dysentery, large nodules behind the ears, and profuse bleeding from the nose, eyes, and mouth. Victims typically died within 4–5 days (Acuna-Soto et al. 2002).

Evidence suggests that during the mid-sixteenth century, precipitation levels might have been extremely low. (In fact, they were the lowest in the past 500 years). These droughts inflamed the impact of infectious disease on the human population. The 1545 and 1576 epidemics of *cocoliztli* were caused by indigenous hemorrhagic fevers transmitted by rodent hosts and aggravated by extreme drought conditions (Acuna-Soto et al. 2002). A bacterial pathogen associated with *cocoliztli* has been now identified as *Salmonella enterica* (Vágene et al. 2018).

Jewish and Moorish Influences

The year 1492 is best known for Christopher Columbus' expedition (under the Spanish banner) and unprecedented encounter with the New World, which led to its eventual conquest. Yet, this was only part of the Spanish crown's realizations that year. In January, and after 10 years of fighting with Moorish forces, Abu Abdallah Muhammad XII (also called Boabdil) surrendered the city of Granada to the Castilian forces of Ferdinand II of Aragon and Isabella I of Castile. The Reconquista had effectively concluded, ending almost 770 years of Islamic presence in the Iberian Peninsula. By the end of March of the same year, the Spanish crown had signed the Alhambra Decree ordering the expulsion of all Jews present in the kingdoms of Castile and Aragon. By 1500, converted Moors known as Moriscos had also been ordered to leave Spain.

Conversos, Jews who had converted to Catholicism, were permitted to stay. The rest either converted, left for the Maghreb, the Ottoman Empire, or Portugal, or were summarily executed. Those who fled to Portugal suffered a delayed persecution, for in 1497, the king of Portugal, Manuel I, decreed that all Jews had to convert to Christianity, leading to expulsions and massacres. Spanish Jews, known as Sephardim, who did not leave the Iberian Peninsula and hid their religion, were known as crypto-Jews, but more generally as the pejorative *Marrano* (literally "pig/swine" in Spanish), with likely etymology from the Arabic *muḥarram,* meaning "forbidden thing."

Anti-Semitism in the Iberian Peninsula was nothing new. The 1478 Inquisition was preceded by the 1391 massacre of Jews in Seville, which moved farther northward to Toledo and Barcelona, among other places. By 1449, Pedro Sarmiento of Toledo created the racial exclusion laws called *Sentencia-Estatuto*, banning *conversos* from holding private/public office, church benefices, and the right to testify in court unless a *converso* could prove four generations of Christian blood (Gorsky 2015: 175). *Limpieza de Sangre*, rather, the cleansing of the blood, was institutionalized racism that forbade people of Jewish or Muslim ancestry to hold office or participate in public affairs, regardless of religion. This concept made its way to

New Spain and developed as racial purity between Spaniards and the indigenous population. By 1501, Queen Isabella had ordered the governor of Hispaniola, Nicolás de Ovando, to block Jews, Moors, and New Christians from entering the New World (Baron 1970: 136), although some *conversos* did manage to seek passage, either by bribes or through loopholes. At times, *probanza* (certification) was not enforced, as one could travel as another person's servant. Yet, before the Portuguese rebellion of 1640, Mexico City alone had 15 synagogues, while there where three in Puebla and others in Guadalajara, Vera Cruz, Zacatecas, and Campeche (Hernández 2014: 17).

Before the expulsion, the kabbalah had in part been developed in Spain. In its simplest terms, it is an esoteric Judaic school of thought. The *Zohar*, the *Book of Splendor*, was written in Aramaic and is generally thought of as the book of kabbalah. It circulated around 1280 and was likely written by Rabbi Moses de León in the state of the Crown of Castile. As Christians began to study the kabbalah during the Renaissance, they felt that it proved Christianity, noting compatible analogies. For example, the Holy Trinity as interpreted in the *Zohar*: "*The Ancient of Days has three heads. He reveals himself in three archetypes, all three forming but one. He is thus symbolized by the number Three* ..." The *editio princeps* of the Syriac New Testament printed in Vienna in 1555 contains a woodcut of the crucifix, the Sephirotic Tree, and the Apostle John, who was present during the crucifixion of Christ (Wilkinson 2007: 182). The discovery of the New World was a sign that the world was coming to an end. The map on folio 86v strongly resembles the Sephirotic Tree.

Sephardic Jews may have influenced some aspects of the Virgin herself, via the Shekhinah, the divine presence (Hernández 2014: 17). In the 13th century and as a Jewish response to the growing significance of the Virgin Mary, the kabbalah overtly feminized Shekhinah entirely as the archetype of the divine woman. It was identified with the moon and considered the "wife of Moses" (Lancaster 2006: 111, 157). The Virgin of Guadalupe's antecedent was the Virgen del Coro, housed in the Guadalupe Monastery in Extremadura, Spain. The latter also stands on a crescent moon, wears a blue cloak with 47 stars, and emanating from her person is a mandorla, a *vesica piscis*-shaped aureole. The main iconographical difference is the expected presence of baby Jesus, which is lacking in the Spanish version. This would be less offensive to *conversos* who remained loyal to Judaic beliefs. Hernández argues that the mandorla is akin to the fire of the burning bush encountered by Moses, which is the divine presence of God. Mary, for her part, is the intermediary of God and mankind (Hernández 2014).

Despite the Spanish rejection of Jews and Judaism in addition to Moors and Islam, their influence, although indirect, was enormous. The Hebrew Bible, with its emphasis on Jerusalem, was a powerful symbol, especially to the Franciscans as demonstrated by Motolinía's establishment of Puebla de los Angeles as the Celestial City of Jerusalem. Moorish culture was influential as a result of its contributions to science, mathematics, astronomy, and architecture.

Conclusion

The clash of the Spanish and Aztec cultures was a tragic episode in human history. Both were extremely violent. The Spanish burned, strangled, and tortured. The Aztecs performed human sacrifices and religious cannibalism. Yet, both cultures possessed magnificent art and literature, and some exemplary qualities. The destruction of the Aztec culture was due not only to the advanced military technology of the Spanish, but it was aided by the systematic Christianization of the locals and the introduction of plagues and diseases into a population without resistance. The result was almost total cultural genocide. Nevertheless, the influence of the Aztec culture survived in the fused culture of New Spain. The *Voynich Codex* can be seen as a reflection of this sixteenth century transformation, whose reverberations are still felt.

Literature Cited

Acuna-Soto, R., D.W. Stahle, M.K. Cleaveland, and M.D. Therrell. 2002. Megadrought and megadeath in 16th century Mexico. *Emerging Infectious Diseases* 8: 360–362.

Aguilar-Moreno, M. 2007. *Handbook to life in the Aztec world*. New York: Oxford University Press.

Baron, S.W. 1970. *Social and religious history of the Jews: Late middle ages and era of European expansion, 1200–1650: Catholic restoration and wars of religion*. New York: Columbia University Press.

Bazarte, A. 1989. El Colegio de Niñas de Nuestra Señora de la Caridad. In *Imágenes de lo cotidiano: Anuario conmemorativo del V Centenario del Descubrimiento de América*, vol. 1. México: Universidad Autónoma Metropolitana.

Berdan, F.F. 2014. *Aztec archaeology and ethnohistory*. Cambridge, UK: Cambridge University Press.

Brian, A., B. Benton, and P. García Loaeza. 2015. *The native conquistador: Alva Ixtlilxochitl's account of the conquest of new Spain*. University Park: Pennsylvania State Press.

Brokaw, G., and J. Lee. 2016. *Fernando de Alva Ixtlilxochitl and his legacy*. Tucson: The University of Arizona Press.

Bunson, M., and M. Bunson. 2014. *Our Sunday visitor's encyclopedia of Catholic history*. Huntington: Our Sunday Visitor Publishing.

Campbell, L. 1997. *American Indian languages: The historical linguistics of native America*. Oxford: Oxford University Press.

Carmack, R. 2001. *Evolución del reino K'iche'*. Guatemala: Fundación Cholsamaj.

Carmack, R., J.L. Gasco, and G.H. Gossen. 2006. *The legacy of Mesoamerica: History and culture of a native American civilization*. 2nd ed. Upper Saddle River: Prentice Hall.

Chuchiak, J.F. 2012. *The inquisition in new Spain, 1536–1820: A documentary history*. Baltimore: The Johns Hopkins University Press.

Constable, O.R. 2014. Cleanliness and convicencia: Jewish bathing culture in Medieval Spain. In *Jews, Christians and Muslims in medieval and early modern times: A festschrift in honor of Mark R. Cohen*, 257–269. Leiden: Brill.

De la Torre Villar, E. 1974. Fray Pedro de Gante, maestro y civilizador de América. *Estudios de Historia Novohispana* 5: 9–77.

De Sahagún, B. 1951–1982. *Florentine Codex. General history of the things of New Spain*. [1540–1585] 12 vol. Trans. A.J.O. and Anderson C.E. Dibble. Salt Lake City: University Utah Press.

Díaz del Castillo, B. 2012. *The true history of the conquest of new Spain*. Indianapolis: Hackett Publishing Company Inc..

Evans, S.T. 2004. Aztec palaces. In *Palaces of the ancient New World*, 7–58. Washington, DC: Dumbarton Oaks Research Library and Collection.

Frodin, D.G. 2001. *Guide to standard floras of the world: An annotated, geographically arranged systematic bibliography of the principal floras, enumerations, checklists and chorological atlases of different areas*. Cambridge, UK: Cambridge University Press.

Gates, W. 2000. *An Aztec herbal*. New York: Dover Publications. (First publ. Maya Society. 1939. Baltimore, MD).

Gómez Canedo, L. 1982. *La educación de los marginados durante la Época Colonial: Escuelas y colegios para Indios y Mestizos en la Nueva España*. México: Editorial Porrúa.

Gorsky, J. 2015. *Exiles in Sepharad: The Jewish millennium in Spain*. Philadelphia: University of Nebraska Press.

Harris, M. 1998. *Good to eat: Riddles of food and culture*. Long Grove: Waveland Press, Inc..

Hernández, M. 2014. *The virgin of Guadalupe and the conversos: Uncovering hidden influences from Spain to Mexico*. New Brunswick: Rutgers University Press.

Karttunen, F., and J. Lockhart. 1980. La estructura de la poesía nahuatl vista por sus variantes. *Estudios de Cultura Nahuatl* 14: 15–64. México: Instituto de Investigaciones Históricas, Universidad Nacional Autónoma de México.

Kirk, S., and S. Rivett. 2014. *Religious transformations in the early modern Americas*. Philadelphia: University of Pennsylvania Press.

Lancaster, B.L. 2006. *The essence of kabbalah*. London: Arcturus Publishing Limited.

Méndez Arceo, S. 1990. *La Real y Pontificia Universidad de México: antecedentes, tramitación y despacho de las reales cédulas de erección*. México: Universidad Nacional Autonoma de Mexico.

Mills, K., W.B. Taylor, and S.L. Graham. 2002. *Colonial Latin America: A documentary history*. Washington, DC: Rowman & Littlefield Publishing Group.

Pemberton, J. 2001. *Conquistadors: Searching for El dorado: The terrifying Spanish conquest of the Aztec and Inca empires*. Melbourne: Canary Press.

Pérez, J. 2005. *The Spanish inquisition: A history*. New Haven: Yale University Press.

Schendel, G., J. Amézquita, and M.E. Bustamante. 2014. *Medicine in Mexico: From Aztec herbs to betatrons*. Austin: University of Texas Press.

Schroeder, S., and S. Poole. 2007. *Religion in new Spain*. Albuquerque: University of New Mexico Press.

Smith, M. E. 1984. The Aztlan migrations of the Nahuatl chronicles: Myth or History? Ethnohistory 31(3):153–186.

Trexler, R.C. 1997. *The journey of the magi: Meanings in history of a Christian story*. Princeton: Princeton University Press.

Trigger, B.G., R.E. Adams, W.E. Washburn, and M.J. MacLeod. 2000. *The Cambridge history of the native peoples of the Americas*. Cambridge, UK: Cambridge University Press.

Vágene, A.J., A. Herbig, M.G. Campana, N.M. Roblers Garcia, C. Warinner, S. Sabin, M.A. Spyrou, A. Andrades Vatueña, D. Huson, N. Tuross, K.I. Bos, and J. Krause. 2018. *Salmonella enterica* genomes from victims of a major sixteenth-century epidemic in Mexico. *Nature Ecology & Evolution*. https://doi.org/10.1038/s41559-017-0446-6.

Van Tuerenhout, D.R. 2005. *The Aztecs: New perspectives*. Santa Barbara: ABC-CLIO.

Wilkinson, W.J. 2007. *Orientalism, Aramaic, and kabbalah in the Catholic reformation: The first printing of the Syriac new testament*. Leiden: Koninklikjke Brill NV.

Wolf, G., and J. Connors, eds. 2011. *Colors between two worlds. The Florentine Codex of Barnardino de Sahagún*. Milan: Officina Libraria.

Zorita, A. 1994. *Life and labor in ancient Mexico: The brief and summary relation of the lords of New Spain*. Trans. B. Keen. Norman: University of Oklahoma Press.

Part II
Evidence for Mesoamerican Origins

Chapter 4
Phytomorph and Geomorph Identification

Arthur O. Tucker and Jules Janick

Plant Images in the Voynich Codex

The *Voynich Codex* contains an estimated 359 plant images or phytomorphs: 131 in the Herbal section, plus 228 in the Pharmaceutical section. The phytomorphs in the Herbal section are often quite bizarre and whimsical. They appear to be drawn by the same hand using a pen for outlines and then rather crudely tinting the forms with a few basic mineral pigments in green, brown, blue, or red. The roots are quite stylized and strange, often in the shape of geometric forms or animals. The leaf shapes are clearly exaggerated. The stems often seem to be inserted onto other stems or roots and have been erroneously referred to as "grafted." However, the floral parts are frequently quite detailed and helpful for identification. The plants in the Pharmaceutical section are much reduced, often confined to a shoot with a single leaf or only roots. Furthermore, these images are regularly associated with names in the Voynichese symbolic script. A careful analysis of the images leads us to conclude that the artist was particularly concerned with certain features significant to identification.

This chapter is based on three published works:

1. A paper by Rev. Dr. Hugh O'Neill (1944) that identified two New World phytomorphs, a sunflower and chili peppers, in the *Voynich Codex*.
2. A paper by Arthur O. Tucker and Rexford H. Talbert (2013) that identified 37 phytomorphs in the *Voynich Codex* as indigenous to the New World.
3. A paper by Tucker and Janick (2016) that extended the list to 58 phytomorphs.

This chapter extends the list to 60 phytomorphs. Although many of the *Voynich Codex* illustrations, at first sight, could be considered bizarre or whimsical (see Chap. 14, Fig. 14.7), most contain morphological structures that permit botanical identification. Many enthusiasts have attempted to analyze the plants of the *Voynich Codex*, but few are knowledgeable plant taxonomists or botanists, despite their large

J. Janick, A. O. Tucker, *Unraveling the Voynich Codex*, Fascinating Life Sciences, https://doi.org/10.1007/978-3-319-77294-3_4

web presence. Most of the plant identification has been predicated on the conclusion that the *Voynich Codex* is a fifteenth century European manuscript (Friedman 1962), such as web reports by non-botanists Ellie Velinska (2013) and Edith Sherwood and Erica Sherwood (2008), who identified the plants of the *Voynich Codex* as Mediterranean. This is discussed more fully in Chap. 16 (in-text, and in Table 16.4 and Fig. 16.7).

The first exception to the conclusion that the Voynich plants were European was a short, remarkable 1944 paper in *Speculum* (a refereed journal of the Medieval Academy of America) by the distinguished plant taxonomist Hugh O'Neill (1894–1969), former director of the LCU Herbarium at the Catholic University of America (CUA) in Washington, D.C. From black-and-white photostats provided by Theodore C. Petersen (1883–1966) at CUA, O'Neill identified two Mesoamerican plants in the *Voynich Codex*, a sunflower and capsicum pepper. He was qualified to make this identification because he was familiar with the flora of Mexico and allied regions. Hugh O'Neill collected 8000 herbarium specimens in British Honduras (present-day Belize), Guatemala, and Nicaragua in 1936, and subsequently wrote a paper on the Cyperaceae of the Yucatan Peninsula (O'Neill 1940). Besides acquiring numerous specimens from Ynes Mexia and other Mexican collectors for the LCU Herbarium, he also directed the dissertation of Brother B. Ayres in 1946 on *Cyperus* in Mexico (Tucker et al. 1989). O'Neill was so well regarded by his colleagues in plant taxonomy that five species were named after him: *Calyptranthes oneillii* Lundell, *Carex oneillii* Lepage, *Eugenia oneillii* Lundell, *Persicaria oneillii* Brenckle, and *Syngonanthus oneillii* Moldenke.

The Voynich community never accepted Hugh O'Neill's identifications, presumably because they lacked botanical expertise, but principally because they invalidated their strong beliefs in their own hypothesis of a fourteenth or fifteenth century European origin. Despite O'Neill's documented background in plant taxonomy, his expertise was called into question by cryptologist Elizebeth Friedman, who wrote in 1962: "*Although a well-known American botanist, Dr. Hugh O'Neill, believes that he has identified two American plants in the illustration no other scholar has corroborated this, all agreeing that none of the plants depicted is indigenous in America. Sixteen plants, however, have been independently identified as European by the great Dutch botanist Holm.*" Mysteriously, there was only one mid-twentieth century plant taxonomist named Holm: Herman Theodor Holm (1854–1932), but he was Danish-American and was only in the CUA faculty from July 1932 until he died in December of that year. This Holm spent almost his entire career researching the plants of the Arctic and the Rocky Mountains and had no documented expertise in Mesoamerican plants (Stafleu and Cowan 1979).

O'Neill's discovery should have had powerful repercussions for *Voynich Codex* studies, but it did not make an impression until Tucker and Talbert (2013) identified a New World origin for 37 plants, seven animals, and one mineral in the *Voynich Codex* and concluded that it originated in sixteenth century Mexico. In this chapter, the number of plants identified is expanded to 60 phytomorphs (57 species).

Plant Identification

In the analysis mentioned below, botanical images (phytomorphs) of the Herbal and Pharmaceutical sections of the *Voynich Codex* are combined by family and species in alphabetical sequence of the angiosperms, incorporating the folio number in the codex. Multiple phytomorphs occur on each folio of the Pharmaceutical section. On each page, the plants are numbered from left to right, starting from top left. Some folios, such as folio 101v, are a trifold; thus, the section of the folio number is indicated in parenthesis, e.g., folio 101v(3). In the Pharmaceutical section, where there are many images, the plant is specified by a number (#) from left to right, and top to bottom (see Chap. 2, Fig. 2.2, where there are 16 plants; plant #8 is the third plant in the second row).

The nomenclature below follows a concordance of the cited revisions, and/or the Germplasm Resources Information Network (GRIN) (USDA, ARS 2015), and/or the collaboration of the Royal Botanic Gardens, Kew, and the Missouri Botanical Garden (Plant List 2013). The text closely follows Tucker and Janick (2016).

Fern: Ophioglossaceae

Folio 100v#5. *Ophioglossum palmatum* (Fig. 4.1) Hugh O'Neill (1944) identified the eusporangiate fern *Botrychium lunaria* (L.) Sw., moonwort, in the *Voynich Codex* on folio 100v. However, O'Neill did not designate the phytomorph number, only the folio number, but only folio 100v#5 would match with a member of the Ophioglossaceae. The fronds with both a fertile and a sterile portion, the fertile portion bladelike and palmately lobed into multiple segments (Fig. 4.1a), suggest instead a specimen of the eusporangiate fern *Ophioglossum palmatum* L. The photograph

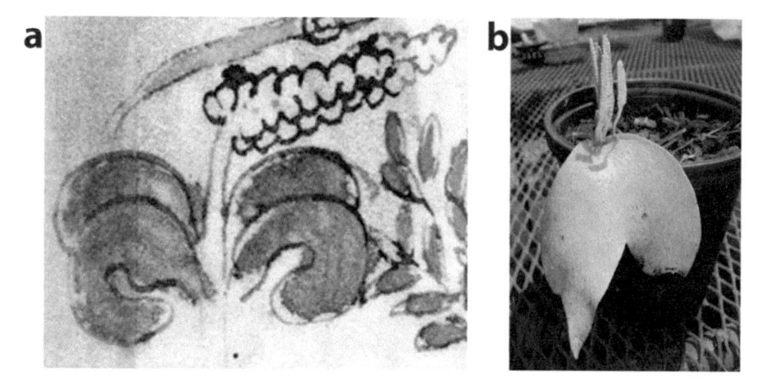

Fig. 4.1 *Ophioglossum palmatum*: (**a**) folio 100v#5; (**b**) *O. palmatum*. (Source: Robbin Moran, New York Botanical Garden)

of this species (Fig. 4.1b) confirms the identification. This species is epiphytic in dense, wet forests at low to middle elevations from Florida to Brazil (Mickel and Smith 2004).

Gymnosperm: Taxodiaceae

Folio 100r#15. *Taxodium* Sp., cf. *Taxodium mucronatum* (*T. huegelii, T. mexicanum*)? (Fig. 4.2) This phytomorph (Fig. 4.2a) is very crude, but appears to be either the cones or whole plant outlines of the Mexican cypress, *Taxodium mucronatum* Ten. (*T. huegelii* hort. ex P. Lawson & C. Lawson; *T. mexicanum* Carrière). The cones and forked tree trunk of *T. mucronatum* are shown (Fig. 4.2b, c). This species is often multi-trunked in older specimens, such as the Tule tree, or El Árbol del Tule (ca. 2000 years old), on the grounds of a church in Santa María del Tule in the Mexican state of Oaxaca. The Nahuatl name for *T. mucronatum* is *ahoehoetl/aueuetl/ahuehuetl, ahuehuecuahuitl,* or *ahuehuete/ahuehuetl* (Hernández et al. 1651; Hernández 1942; Dressler 1953; Díaz 1976; Farfán and Elferink 2010).

Angiosperms: Asparagaceae/Agavaceae

Folio 100r#4. *Agave* sp., cf. *A. atrovirens* (Fig. 4.3) Phytomorph #4 on folio 100r (Fig. 4.3a) appears to be a pressed specimen of an *Agave* sp. with leaves bearing a toothed edge. It is quite possibly *Agave atrovirens* Karw. ex Salm-Dyck (Fig. 4.3b),

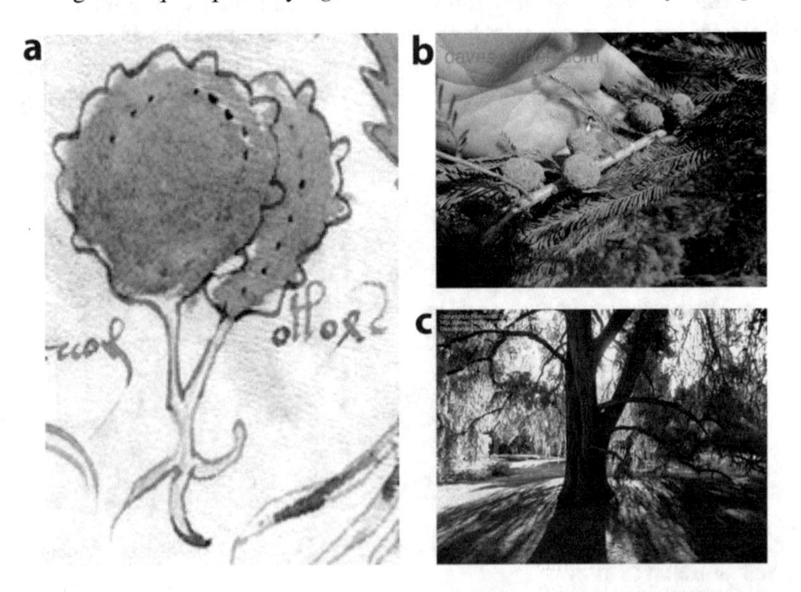

Fig. 4.2 *Taxodium* sp.: (**a**) folio 100r#15; (**b**, **c**) grouped strobili (cones) and forked tree trunk of *T. mucronatum* respectively. Source: Geoff Stein

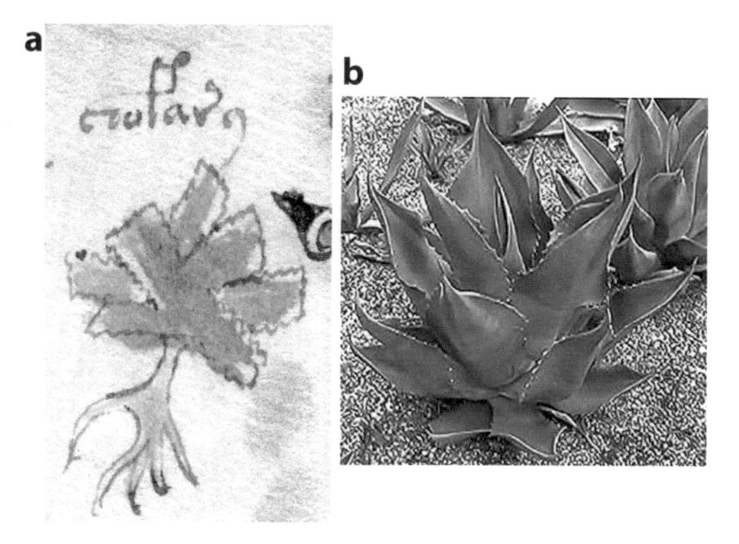

Fig. 4.3 *Agave* sp.: (**a**) folio 100r#4; (**b**) *Agave atrovirens*. (Source: Jeno Kapitany)

which was a source for beverages, such as pulque, mescal, and tequila in sixteenth century colonial New Spain (Hough 1908; Dressler 1953). In Hernández et al. (1651) and Hernández (1942), this was called *metl* (Dressler 1953) or *metl coztli/ mecoztli*. In the *Florentine Codex* (Sahagún 1963), this was known as *macoztic metl*.

Apiaceae

Folio 16v. *Eryngium* sp., cf. *E. heterophyllum* (Fig. 4.4) Probably the most phantasmagoric phytomorph in the *Voynich Codex* is the *Eryngium* sp., portrayed in folio 16v (Fig. 4.4a). The inflorescence is colored blue, the leaves red, and the rhizome ochre, but the features verge on a stylized appearance rather than the botanical accuracy of the *Viola bicolor* of folio 9v. This lack of technical attention makes identification beyond genus difficult if not impossible. However, a conjecture might be *Eryngium heterophyllum* Engelm (Mathias and Constance 1941). This species, native to Mexico and Arizona, New Mexico, and Texas in the USA, has similar blue inflorescence and involucral bracts (Fig. 4.4b), and stout roots, and it also develops rosy coloring on the stems and basal leaves (Fig. 4.4c). However, *E. heterophyllum* has pinnately compound leaves, not peltate, but the leaves subtend the inflorescence and cover the stem, suggesting that the phytomorph might have been drawn from a dried, fragmented specimen. This lack of concern about the shape of the leaves also plagues identifications in the *Codex Cruz-Badianus* (Clayton et al. 2009). Today, Wright's eryngo, or Mexican eryngo (*Eryngium heterophyllum*), is used to treat gallstones in Mexico and has been found experimentally to have a hypocholesteremic effect (Navarrete et al. 1990).

Fig. 4.4 *Eryngium* sp.: (**a**) folio 16v; (**b**) inflorescence of *E. heterophyllum*. (Source: Blooms of Bressingham®); (**c**) leaves of *E. comosum* R. Delaroche illustrating anthocyanin accumulation. (Source: Pedro Tenorio-Lezama)

Folio 41v, *Lomatium* sp., cf. *L. dissectum* (Fig. 4.5) Folio 41v (Fig. 4.5a) shows a plant that bears a dense fruiting umbel with schizocarps that are flat, which is characteristic of the Apiaceae. These schizocarps have a dark center and white wings. The growth habit is a rosette bearing coppery tuberous roots. The leaves are deeply dissected and palmate. The best match would be biscuitroot, a *Lomatium* sp., most probably *dissectum* (Nutt.) Mathias & Constance (Fig. 4.5b–d). This is native from British Columbia to Mexico. The fruits are flat schizocarps with papery wings born in a full umbel. The growth habit is a rosette and it bears tuberous roots that are coppery when cleaned. The leaves vary from coarsely to deeply dissected, but are best described as pinnatifid, not palmate. This confusion of palmate versus pinnate is common in both the *Voynich Codex* and the *Codex Cruz-Badianus*.

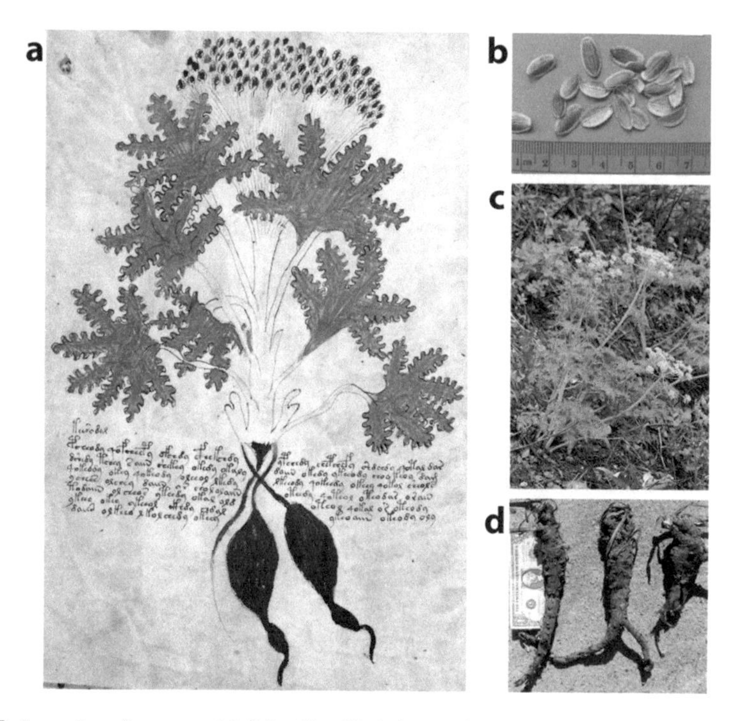

Fig. 4.5 *Lomatium dissectum*: (**a**) folio 41v; (**b**) fruits; (**c**) foliage and flowers; and (**d**) tuberous roots

Apocynaceae

Folio 100r#14. *Gonolobus chloranthus* (Fig. 4.6) Phytomorph #14 on folio 100r (Fig. 4.6a) appears to be the ridged fruit of an asclepiad, possibly the Mexican species *Gonolobus chloranthus* Schltdl (Fig. 4.6b). The *tlalayotli* in the *Florentine Codex* (Sahagún 1963: pl. 488a), with a similar illustration of the fruit (but with smooth ribs), is nominally accepted as the related species *G. erianthus* Decne. or *calabaza silvestre*. The roots of *G. niger* (Cav.) Schult. are used today in Mexico to treat gonorrhea (González Stuart 2004).

Araceae

Folio 100r#2. *Philodendron mexicanum* (Fig. 4.7) Phytomorph #2 on folio 100r (Fig. 4.7a) appears to be a vining aroid with hastate leaves, ripped from a tree, most probably *Philodendron mexicanum* Engl. (Fig. 4.7b, c) known as *huacalazochitl* in

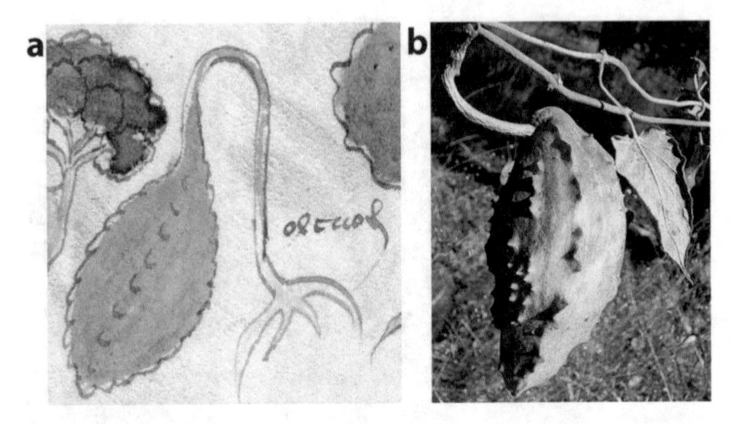

Fig. 4.6 *Gonolobus chloranthus*: (**a**) folio 100r#14; (**b**) *G. chloranthus* fruit. (Source: Guadalupe Cornejo-Tenorio and Guillermo Ibarra-Manríquez, The Field Museum)

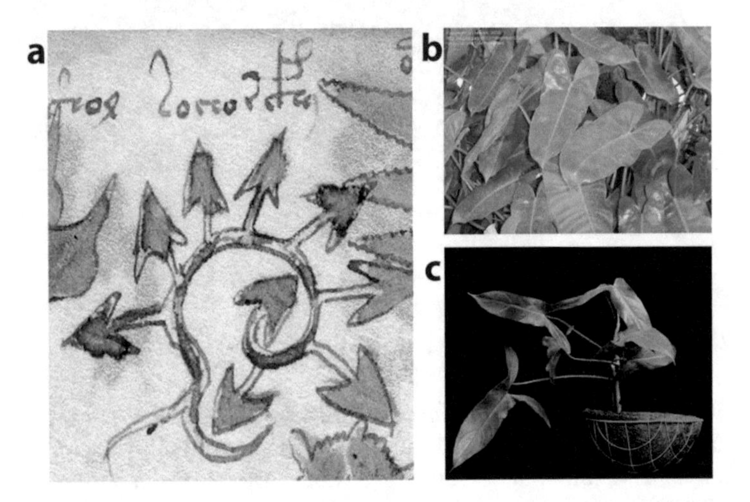

Fig. 4.7 *Philodendron mexicanum*: (**a**) folio 100r#2; (**b**) leaves of *P. mexicanum*. (Source: Dave's Garden user tathisri); (**c**) leaves of *P. mexicanum*. (Source: Steve and Janice Lucas)

Nahuatl (Zepeda and White 2008). This is known as *huacalxochitl/huacalxōchitl* (huacal flower) in the *Codex Cruz-Badianus* (Emmart 1940; Cruz and Badiano 1991; Alcántara Rojas 2008; Gates 2000; Clayton et al. 2009) or *huacalazochitl* (Zepeda and White 2008). Bown (1988) writes, in general, of the Araceae: "*Most of the species of Araceae which are used internally for bronchial problems contain saponins, soap-like glycosides which increase the permeability of membranes to assist in the absorption of minerals but also irritate the mucous membranes and make it more effective to cough up phlegm and other unwanted substances in the lungs and bronchial passages.*"

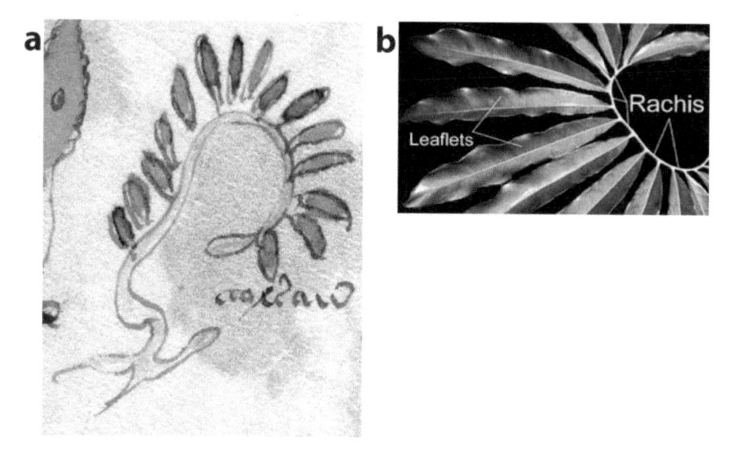

Fig. 4.8 *Philodendron* sp.: (**a**) folio 100r#7; (**b**) *P. goeldii*. (Source: Steve and Janice Lucas)

Folio 100r#7. *Philodendron* sp (Fig. 4.8) Phytomorph #7 on folio 100r (Fig. 4.8a) appears to be the leaf or stem of an aroid, probably a species of *Philodendron*, but the crudeness of the drawing cannot confirm whether this is a stem or compound leaf. If the latter, it may be a crude representation of the pedately compound leaf of the Mexican species *Philodendron goeldii* G. M. Barroso (Fig. 4.8b).

Asteraceae

Folio 53r. *Ambrosia* sp., cf. *A. ambrosioides* (Fig. 4.9) The phytomorph has composite heads borne on a raceme (Fig. 4.9a). The leaves are brown to green and sharply serrated. The roots are brown and segmented. This matches the variability of *Ambrosia ambrosioides* (Delpino) W. W. Payne, which is native from Arizona and California to Mexico (Fig. 4.9b, c).

Folio 93r. *Helianthus annuus* (Fig. 4.10) The phytomorph has a large single yellow asteroid head borne on a stout, thick stem (Fig. 4.10a). The leaves are green, alternate, ovate-lanceolate, acute, and whole. The petioles are short, with lines drawn down along the stem, possibly indicating a clasping base of the petiole or ridges along the stem. The roots are multiple, primary, and unbranched. The best match is sunflower, *Helianthus annuus* L. (Fig. 4.10b, c). This identification was first made by Hugh O'Neill (1944), plant taxonomist and curator of the LCU Herbarium, who confirmed the determination with six botanists.

Lincoln Taiz, emeritus plant physiologist at the University of California, Santa Cruz, confirms the resemblance (Taiz and Taiz 2011), whereas *Helianthus* authorities Robert Bye (personal communication 2014), distinguished ethnobotanist at Universidad Nacional Autónoma de México (UNAM) in Mexico City, and Billie

Fig. 4.9 *Ambrosia* sp.: (**a**) folio 53r; (**b**, **c**) flowers, inflorescences, and leaves of *A. ambrosioides*. (Sources: T. Beth Kinsey and Cabeza Prieta, http://cabezaprieta.org/plantpage.php?id=1196 respectively)

Turner (personal communication 2014), one of the world's leading experts on Mexican Asteraceae and former curator of the herbarium (TEX) of the University of Texas at Austin, also confirm this identification. Sunflower researcher Jessica Barb (personal communication 2015) of Iowa State University notes that inbred lines of sunflowers have very short petioles (Fig. 4.10c) and that leaf variation is quite high.

This is not *Helianthus tuberosus* L., Jerusalem artichoke, as proposed by Kennedy and Churchill (2006). They compare a European medieval "wild version" and conclude that it "*would possess a small heard, non-conical calyx, larger petals, and a more divided stem.*" Considering a cultivated selection from New Spain in the sixteenth century, these restrictions do not apply and *H. annuus* is a better fit.

The preponderance of evidence points to Mexico as the center of domestication for sunflowers (Harter et al. 2004; Heiser 2008; Lentz et al. 2008a, b; Rieseberg and Burke 2008; Bye et al. 2009; Blackman et al. 2011; Moody and Rieseberg 2012). In Mexico, names in period literature for *H. annuus* are *chilamacatl* (Sahagún 1963), *chimalacatl* or *chimalacaxochitl* (Hernández 1942; Sahagún 1963), and *chimalatl peruina* (Hernández et al. 1651) (all are Nahuatl names). Additional Nahuatl names

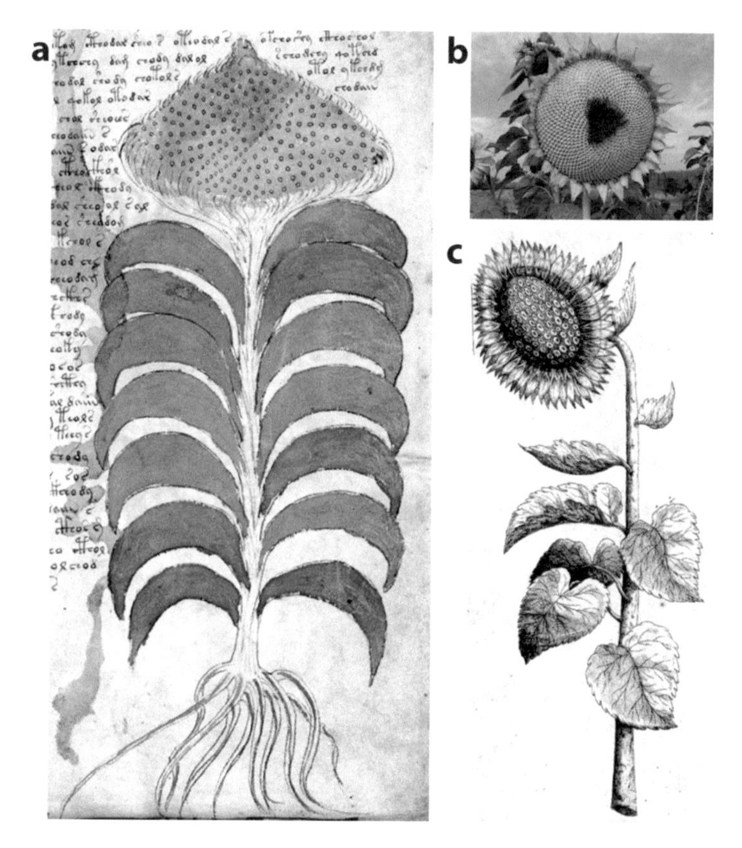

Fig. 4.10 *Helianthus annuus*: (**a**) folio 93r; (**b**) seedhead of *H. annuus*. (Source: Dave Fenwick www.aphotoflora.com); (**c**) *Solis flor Peruuinus* of Emanuel Sweerts of 1612 (Bleiler 1976), showing short petioles toward the apex

are *acahualli* (Ramírez and Alcocer 1902; Dressler 1953) and *chimalxochitl* (Zepeda and White 2008).

Folio 13r. *Petasites* sp., cf. *P. frigidus* var. *palmatus* (Fig. 4.11) Based on the asterid inflorescence, large cleft orbicular leaves, and relatively large root system, folio 13r (Fig. 4.11a) is probably a *Petasites* sp. The closest match might be *P. frigidus* (L.) Fr. var. *palmatus* (Aiton) Cronquist, the western sweet-coltsfoot (Fig. 4.11b). This is native to North America from Canada to California. *Petasites* sp. are used as anti-asthmatics, antispasmodics, and expectorants, and in salve or poultice form (Bayer et al. 2006).

Folio 2r. *Lactuca graminifolia* (Fig. 4.12) Folio 2r shows a plant that has linear leaves in groups and an asterid inflorescence. The bracts are prominently tipped in a brownish color, whereas the petals of the flowers are serrate and off-white. A match might be *Lactuca graminifolia* Michx., which has linear leaves, often in groups, and bracts tipped in purple. The petals are serrate and vary from a very light blue-pink to white. This is native to Mexico, north to New Jersey, and south to Guatemala.

Fig. 4.11 *Petasites* sp.: (**a**) folio 13r; (**b**) *P. frigidus* var. *palmatus*. (Source: Ben Legler)

Fig. 4.12 *Lactuca graminifolia*: (**a**) folio 2r; (**b**) *L. graminifolia* flower illustrating serrated petals and purple-tipped bracts. (Source: http://swbiodiversity.org/seinet/taxa/index.php?taxon=2786); (**c**) *L. graminifolia* illustrating tufted foliage that sometimes occurs. (Source: http://hasbrouck.asu.edu/imglib/seinet/ASU/ASU0021/ASU0021496.jpg)

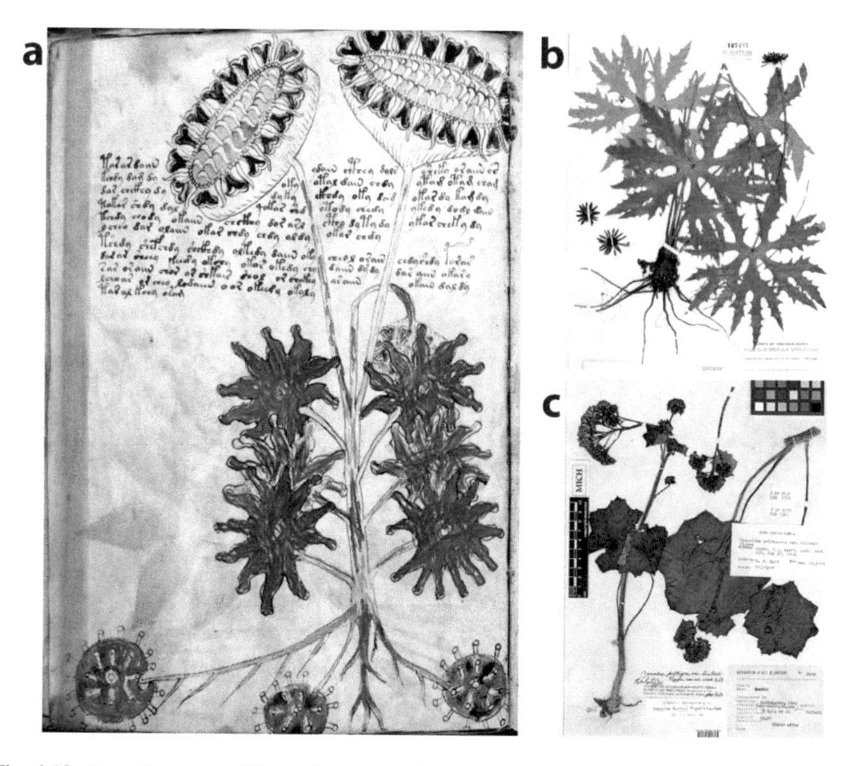

Fig. 4.13 *Psacalium* sp. + *Pippenalia* sp.: (**a**) folio 33v; (**b**) herbarium sheet of *Pippenalia delphinifolia* (ASU0029020). (Source: Arizona State University Herbarium); (**c**) herbarium sheet of *Psacalium peltigerum* var. *hintonii* R. W. Pippen (MICH1107637). (Source: University of Michigan Library Digital Collections. University of Michigan Herbarium Vascular Plant Type Collection with Specimen Images)

Folio 33v. *Psacalium* sp? *Pippenalia* sp.? (Fig. 4.13) This phytomorph (Fig. 4.13a) has lobed, peltate leaves and fleshy, round, subterranean tubers. The inflorescence is characteristic of the tribe Heliantheae, and the "achenes" or cypselae are round and naked, a rare feature in the Asteraceae family. This illustration is a conundrum. The leaves and tubers suggest *Psacalium* sp., possibly *P. peltigerum* (B.L. Rob. & Seaton) Rydb. (Fig. 4.13b), but the large flower suggests a *Pippenalia* sp., possibly *P. delphinifolia* (Rydb.) McVaugh (Fig. 4.13c). Is this a hybrid phytomorph, i.e., did the artist paint a combined image based on two species mixed together? This most definitely is not *Helianthus annuus* L., as proposed by Jorge Stolfi (Kennedy and Churchill 2006), a species that does not have tubers and has different leaves (see *Helianthus annuus,* mentioned above).

Folio 40v. *Smallanthus* sp. (Fig. 4.14) This folio contains two phytomorphs of the same plant, vegetative and flowering (Fig. 4.14a). Although definitely a member of the Asteraceae, the genus is less obvious. With bluish petals, reddish involucre,

Fig. 4.14 *Smallanthus* sp.: (**a**) folio 40v; (**b**) inflorescence *S. sonchifollius* (yacón). (Source: Rob Hille, https://upload.wikimedia.org/wikipedia/commons/4/49/Smallanthus_sonchifolius.P.jpg); (**c**) tuberous roots of *S. sonchifollius* (yacón). (Source: NusHub, http://www.cropsforthefuture.org)

palmately compound leaves, and tuberous roots, this seems to fit a *Smallanthus* sp. It resembles somewhat the leading *Smallanthus* species cultivated today, the edible yacón (*S. sonchifolius* (Po) H. Rob.), which is native to western South America (Fig. 4.14b, c).

Boraginaceae

Folio 47v. *Cynoglossum grande* (Fig. 4.15) This phytomorph has terminal blue flowers of six to seven petals with a raised white center, prominent cauline leaves, broadly elliptic basal leaves, and broad, branched, brown roots (Fig. 4.15a). This matches the variability of *Cynoglossum grande* Douglas ex Lehm., except that this species has only five petals and the cauline leaves are smaller and closer to the base (Fig. 4.15b). This species is native from British Columbia to California.

Folio 56r. *Phacelia campanularia* (Fig. 4.16) With blue flowers in a scorpioid cyme, dentate leaves, and overlapping leaf-like basal scales, the phytomorph (Fig. 4.16a) is a good match for *Phacelia campanularia* A. Gray, California bluebell (Fig. 4.16b, c), a California native.

Folio 39v. *Phacelia crenulata* (Fig. 4.17) This has bluish flowers in a cyme with deeply pinnately lobed green leaves on broad, brown, branched roots (Fig. 4.17a).

Fig. 4.15 *Cynoglossum grande*: (**a**) folio 47v; (**b**) *C. grande*. (Source: Eugene Zelenko, https://commons.wikimedia.org/wiki/File:Cynoglossum_grande-1.jpg)

Fig. 4.16 *Phacelia campanularia*: (**a**) folio 56r; (**b**) inflorescence of *P. campanularia*. (Source: Chez Brungraber; (**c**) plant of *P. campanularia*. (Source: George Williams)

This closely matches *Phacelia crenulata* Torr. (Fig. 4.17b), which is native from Colorado to Mexico.

Folio 51v. *Phacelia integrifolia* (Fig. 4.18) The phytomorph has blue flowers on a scorpioid cyme and narrowly elliptic crenate leaves that curl (Fig. 4.18a). The roots are brown and branched. This matches fairly well the variability of *Phacelia integrifolia* Torr. (Fig. 4.18b, c), which is native from Utah and Kansas, and south to Mexico.

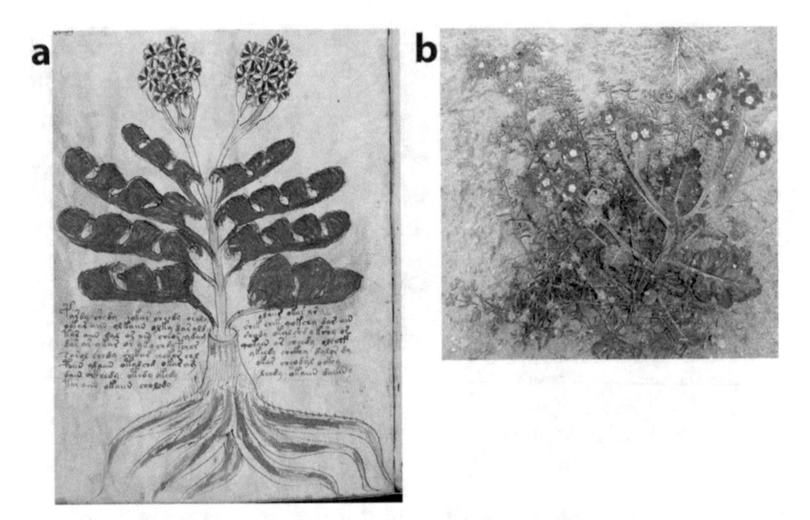

Fig. 4.17 *Phacelia crenulata*: (**a**) folio 39v; (**b**) *P. crenulata*. (Source: Tom Chester, http://tchester.org/sd/plants/floras/borrego_mtn.html)

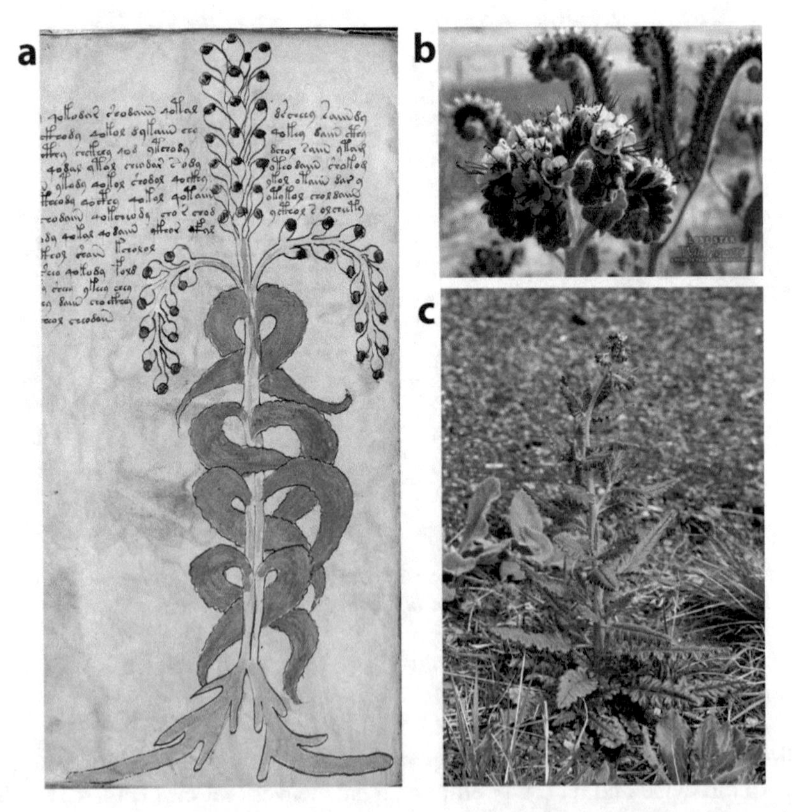

Fig. 4.18 *Phacelia integrifolia*: (**a**) folio 51v; (**b**) inflorescence of *P. integrifolia.* (Source: Nieland and Finley 2009, copyright 2009, published by Texas Tech University Press); (**c**) shoot of *P. integrifolia.* (Source: Russell Kleinman)

Fig. 4.19 *Wigandia urens*: (**a**) folio 26r; (**b**) *W. urens*. (Source: Dick Culbert, https://commons.wikimedia.org/wiki/File:Wigandia_caracasana_(9361123018).jpg)

Folio 26r. *Wigandia urens* (Fig. 4.19) This bears what can be interpreted as bluish flowers on a scorpioid cyme with leaves that are green, crenate, and obtuse (Fig. 4.19a). This matches *Wigandia urens* (Ruis & Pav.) Kunth very well (Fig. 4.19b), a shrub found from Mexico and south to Peru. This is called *chichicaztle* (Díaz 1976) and also matches *patlāhua-ctzītzicāztli* (wide/broad nettle) on folio 47r in the *Codex Cruz-Badianus* (Emmart 1940; Díaz 1976; Cruz and Badiano 1991; Gates 2000; Clayton et al. 2009; de Ávila Blomberg 2012).

Brassicaceae

Folio 90v. *Caulanthus heterophyllus* (Fig. 4.20) This phytomorph may be *Caulanthus heterophyllus* (Nutt.) Payson, San Diego wild cabbage, or San Diego jewelflower (Fig. 4.20a). The flowers of *C. heterophyllus* are four-petaled (white with a purple streak down the center), with four protruding dark purple anthers (Fig. 4.20b). The leaves vary from dentate to lobed, but are typically clasping, not petiolate (Fig. 4.20c). This annual species is native to California and Baja California (Al-Shehbaz 2012).

Fig. 4.20 *Caulanthus heterophyllus*: (**a**) folio 90v; (**b**) inflorescence of *C. heterophyllus*. (Source: National Park Service, http://www.nps.gov/media/photo/gallery.htm?id=04B6C1FE-155D-4519-3E69D153D09B72DA); (**c**) shoots of *C. heterophyllus*. (Source: Anthony J. Valois)

Cactaceae

Folio 100r#8. *Opuntia* sp., cf. *O. ficus-indica* (Fig. 4.21) Phytomorph #8 on folio 100r has the shape of a prickly pear cactus pad or fruit with areoles bearing leaf primordia and tiny fruits on the top edge, i.e., *Opuntia* sp., possibly *Opuntia ficus-indica* (L.) Mill., *O. megacantha* Salm-Dyck, or *O. streptacantha* Lem. (Dressler 1953). This is called *nochtli* and *tlatoc nochtli/tla-tōc-nōchtli* in the *Codex Cruz-Badianus* (Emmart 1940; Dressler 1953; Cruz and Badiano 1991; Gates 2000; Zepeda and White 2008; Clayton et al. 2009; de Ávila Blomberg 2012). Today, *Opuntia ficus-indica* is widely cultivated, but apparently native to central Mexico. *Nopalea cochenillifera* (L.) Salm-Dyck is also widely cultivated for the insect that is the source for cochineal (Standley 1920–1926: 863).

Caryophyllaceae

Folio 24r. *Silene* Sp., cf. *S. menziesii* Infected with *Microbotryum violaceum* (Fig. 4.22) This is probably a *Silene* sp., but the crudeness of the image prevents accurate designation of a species. This phytomorph may be based, in part, on *Silene menziesii* Hook., Menzie's catchfly. This variable species is native from Alaska

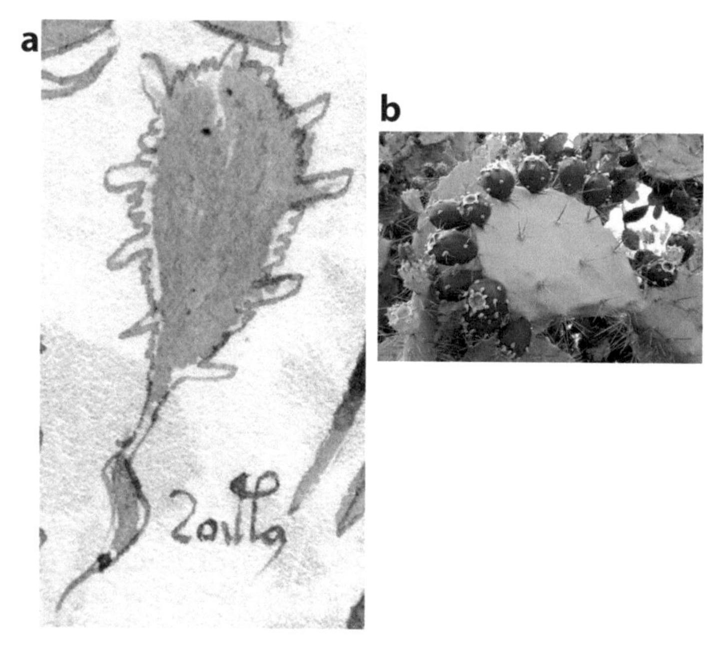

Fig. 4.21 *Opuntia* sp.: (**a**) folio 100r#8; (**b**) cladode (pad) of *O. ficus-indica* with fruits. (Source: Atozxyz, https://en.wikipedia.org/wiki/File:Prickly_pear_cactus_beed.jpg)

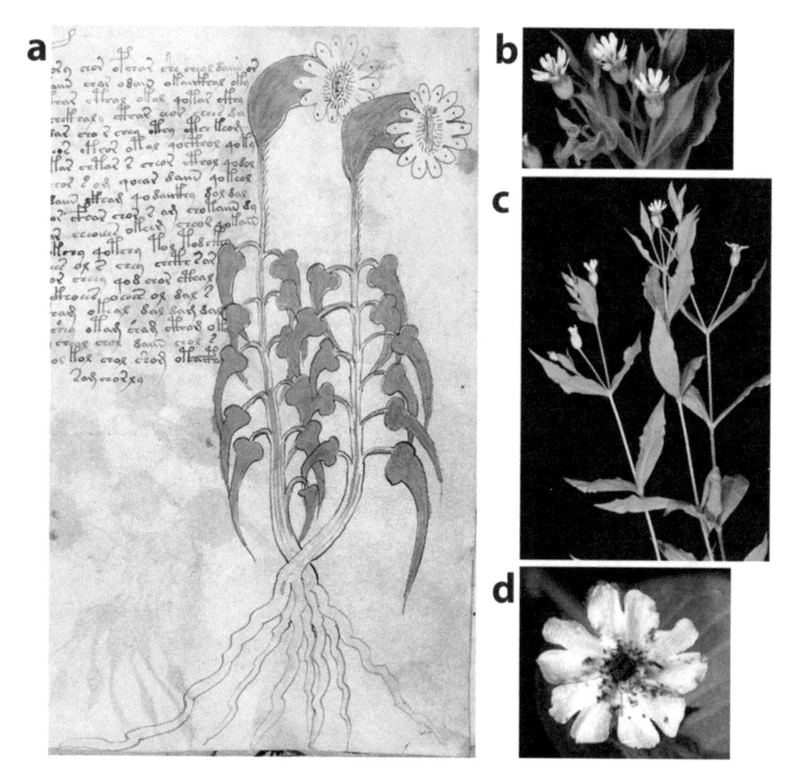

Fig. 4.22 *Silene* sp.: (**a**) folio 24r; (**b**, **c**) stems and inflorescence of *S. menziesii*. (Source: Robert L. Carr, http://web.ewu.edu/ewflora/Caryophyllaceae/Silene%20menziesii.html); (**d**) flower infected with *Microbotryum violaceum*. (Source: Michael Hood)

to California and New Mexico (Morton 2005). The white flowers are a good match, even showing the typical infection with the fungus *Microbotryum violaceum* (Pers.) G. Deml & Oberw., another smut fungus, which turns the anthers purple. However, the leaves are shown as hastate, and *C. menziesii* has attenuated leaf bases. Are the leaves another case of the disparity between reality and portrayal, or is there another species of *Silene* that is a better match for the illustration?

Convolvulaceae

Folios 1v and 101v(2)#4. *Ipomoea arborescens* (Fig. 4.23) The phytomorph shows a single leafy shoot with a single terminal flower bud arising from a thick caudex (Fig. 4.23a). The flower bud has whitish sepals and brownish petals. The leaves are alternate, cordate, petiolate, and green on the adaxial surface and tan on the abaxial surface. Coarse roots emanate from the basal caudex. This phytomorph is repeated on folio 101v(2)#4 (Fig. 4.23e) of the Pharmaceutical section. This illustration is overwhelmingly similar in style and substance to *xiuhamolli*/*xiuhhamolli* (soap plant) found in the 1553 Cruz and Badiano (1991) [plate 11 of Emmart (1940) and folio 9r, Fig. 4.11 (Fig. 4.23b) in Gates (2000), and Clayton et al. (2009)], which has been identified as *Ipomoea murucoides* Roem. & Schult. Reko (1947). Bye and Linares (2013) also identify this as an *Ipomoea* sp. The flower bud of *I. murucoides* (Fig. 4.23c) is similar.

Phytomorphs in the *Codex Cruz-Badianus* and in the *Voynich Codex* have large, broad, gray to whitish basal caudices with ridged bark. However, leaves in the

Fig. 4.23 *Ipomoea arborescens*: (**a**) folio 1v; (**b**) *I. murucoides* from *Codex Cruz-Badianus*, Fig. 4.11; (**c**) bud of *I. murucoides*. (Source: Kevin C. Nixon); (**d**); flower, fruits, and leaves of *I. arborescens*. (Source: Tony Rodd, https://www.flickr.com/photos/tony_rodd/532873040); (**e**) folio 101v(2)#4

Voynich Codex are cordate rather than attenuate, as observed in the *Cruz-Badianus Codex*. The phytomorph in the *Voynich Codex* must then be *Ipomoea arborescens* (Humb. & Bonpl. ex Willd.) G. Don (Fig. 4.23d), found from northern to southern Mexico (Standley 1920–1926: 1205) and commonly known in Nahuatl as *quauhtzahuatl* (Ocaranza 2011).

Additional botanical characteristics of both species are discussed by Standley (1920–1926) and McPherson (1981). Curiously, McPherson described the bases of the leaves of *I. murucoides* as truncate, whereas Standley described the bases as rounded or obtuse, but all herbarium sheets of this species that we have seen would be better described as cuneate. The leaves of *I. murucoides* are described by McPherson as variously pubescent, whereas the leaves of *I. arborescens* are usually tomentose, especially on the lower surface, rendering the abaxial surface gray-green and the adaxial surface green. The tomentose abaxial surface often turns brownish green upon drying, which is similar to that of the phytomorph in the *Voynich Codex*. Additional names for *I. arborescens* include: *palo blanco* (Sonora, Sinaloa); *palo del muerto, casahuate, quauhzahuatl, casahuate blanco* (Morelos); *palo santo* (Sonora); *palo bobo* (Morelos, El Salvador); and *tutumushte, siete pellejos,* and *siete camisas* (El Salvador) (Standley 1920–1926: 1205). In English, it is known as morning glory tree. The ashes of the arborescent *Ipomoea* species, *I. murucoides* and *I. arborescens* are used to prepare soap and also in hair and skin care (Batres et al. 2012; Standley 1920–1926: 1205).

Folio 57r. *Ipomoea nil* (Fig. 4.24) The phytomorph has a terminal dark blue flower with a white edge, acute petals, and elongated calyx lobes. The leaves are lobed peltate on an herbaceous bush. The roots are brown and branched (Fig. 4.24a). This may match the variability of *Ipomoea nil* (L.) Roth (Fig. 4.24b, c). This is native

Fig. 4.24 *Ipomoea nil*: (**a**) folio 57r; (**b**, **c**) flowers and leaves of *I. nil*. (Sources: Rare and Exotic Seeds and ghost32writer.com respectively)

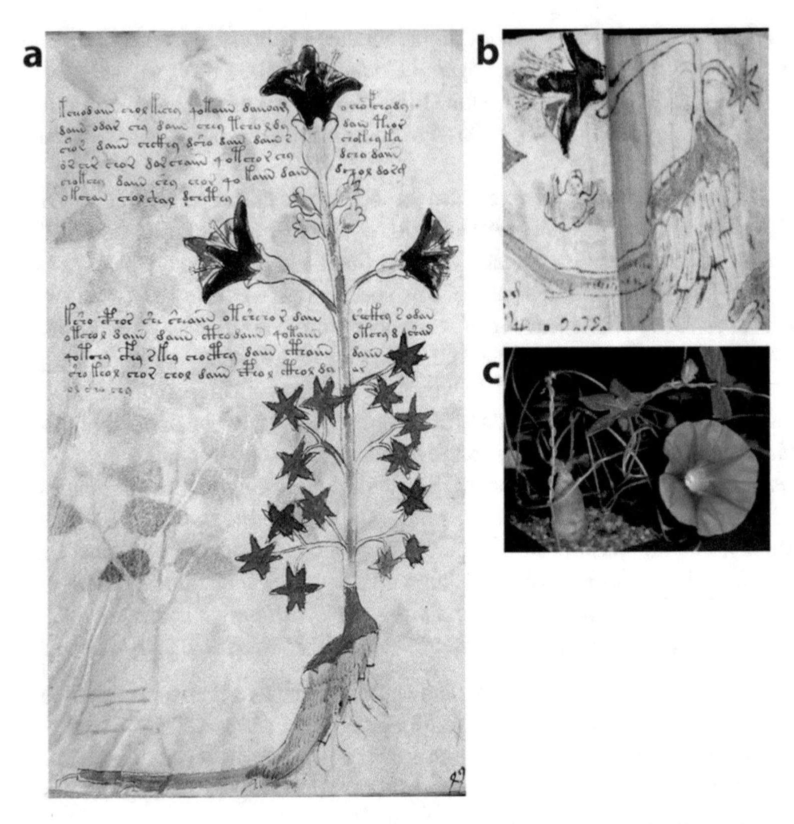

Fig. 4.25 *Ipomoea pubescens*: (**a**) folio 32v; (**b**) folio 101v(3)#2; (**c**) flower, leaf, and rhizome of *I. pubescens*. (Source: Apostolou Starvos)

from northern Mexico to Argentina, and is extremely variable, from a vine to a herbaceous bush, with floral colors from blue to pink to white, but often with a distinctive white edge. The leaves often are hastate, but vary to palmately lobed.

Folios 32v and 101v(3)#2. *Ipomoea pubescens* (Fig. 4.25) The blue flowers, deeply lobed leaves, and tuberous roots (Fig. 4.25a, b) are all good fits for most probably *Ipomoea pubescens* Lam., silky morning glory (Fig. 4.25c). The vine is native from Arizona and New Mexico to Mexico, and in Bolivia, Peru, and Argentina. This phytomorph is also repeated on folio 101v(3)#2 of the *Voynich Codex*. Species of *Ipomoea* are known for their resin glycosides and used to counter several diseases (Pereda-Miranda et al. 2010; Batres et al. 2012; Meira et al. 2012).

Dioscoreaceae

Folio 17v. *Dioscorea composita* (Fig. 4.26) Most probably this phytomorph (Fig. 4.26a) is *Dioscorea composita* Hemsl., barbasco (Fig. 4.26b). This is native from southern Mexico to Costa Rica. The roots (Fig. 4.26c) are quite often

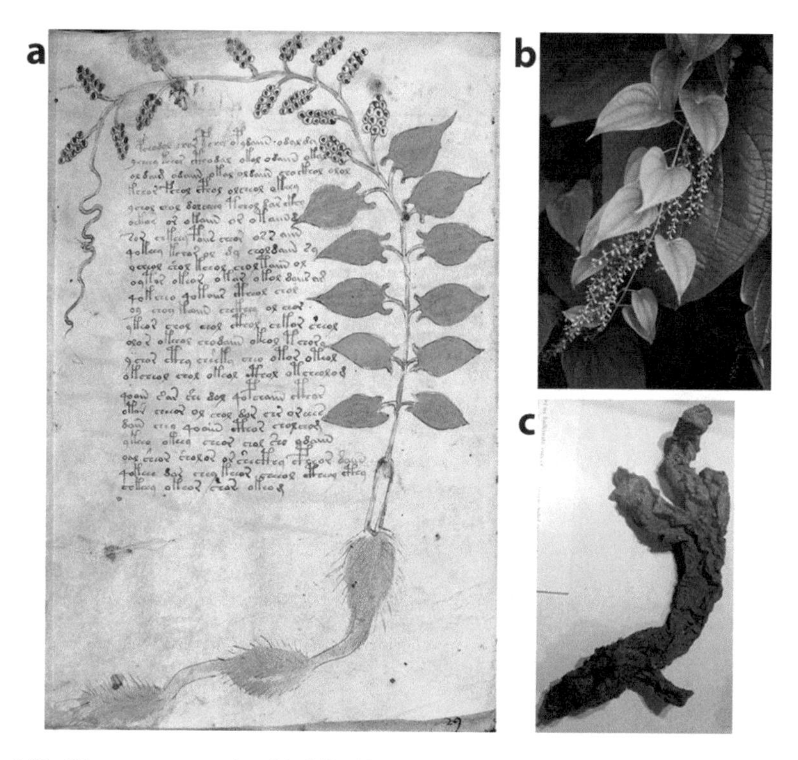

Fig. 4.26 *Dioscorea composita*: (**a**) folio 17v; (**b**) leaves and inflorescences of *D. composite*. (Source: Abisaí García Mendoza); (**c**) segmented roots of *D. composite*. (Source: Ryan Somma, https://commons.wikimedia.org/wiki/File:Dioscorea_composita.jpg)

segmented, as shown in the *Voynich Codex*, and a major source of diosgenin. The flowers (yellow when fresh, but rust-colored upon drying, and borne on a vine) fit rather well, but the phytomorph is shown with leaves more hastate than *D. composita* normally exhibits.

Folio 96v. *Dioscorea mexicana* (Fig. 4.27) The rust-colored flowers and sagittate leaves of the phytomorph (Fig. 4.27a) fit rather well for *Dioscorea mexicana* Schweidw., Mexican yam (Fig. 4.27b, c). This vine is native from northern to southern Mexico to Panama and is a source of diosgenin.

Folio 99r#28. *Dioscorea* sp., cf. *D. remotiflora* (Fig. 4.28) The 28th phytomorph on folio 99r (Fig. 4.28a) is most probably *Dioscorea remotiflora* Kunth (Fig. 4.28b), which is native from northern to southern Mexico. The large, dark root is paddle- or bat-like (Figs. 4.28c, d). The rust-colored flowers and cordate leaves on a vine also match.

Fig. 4.27 *Dioscorea mexicana*: (**a**) folio 96v; (**b**) inflorescence of *D. mexicana*. (Source: Michael Charters); (**c**) rhizome of *D. mexicana*. (Source: The Smithsonian, https://smithsoniangardens.wordpress.com/2013/04/11/i-yam-not-a-tortoise-but-a-plant)

Euphorbiaceae

Folio 6v. *Cnidoscolus texanus* (Fig. 4.29) The palmately compound leaves and trichomes on the fruit (Fig. 4.29a) match a *Cnidoscolus* sp. Both *C. chayamansa* McVaugh and *C. aconitifolius* (Mill.) I.M. Johnst. are called chaya and are widely cultivated from Mexico to Nicaragua, and the leaves are matches (Ross-Ibarra and Molina-Cruz 2002). However, these cultivated species have relatively smooth fruits, and a closer correspondence would be *C. texanus* (Müll. Arg.) Small (Fig. 4.29b, c), with fruits that are coated with trichomes.

Folio 21r. *Euphorbia thymifolia* (Fig. 4.30) The spreading growth pattern, tiny green to reddish leaves, and reddish axillary flowers (Fig. 4.30a) were good fits for *Euphorbia thymifolia* L. (*Chamaesyce thymifolia* (L.) Millsp.; Fig. 4.30b). It is native to the tropics in Africa, Asia, and the Americas (Florida to Argentina). The leaves, seeds, and fresh juice of the whole plant are used in worm infections, bowel complaints, and therapeutically in many more diseases (Mali and Panchal 2013).

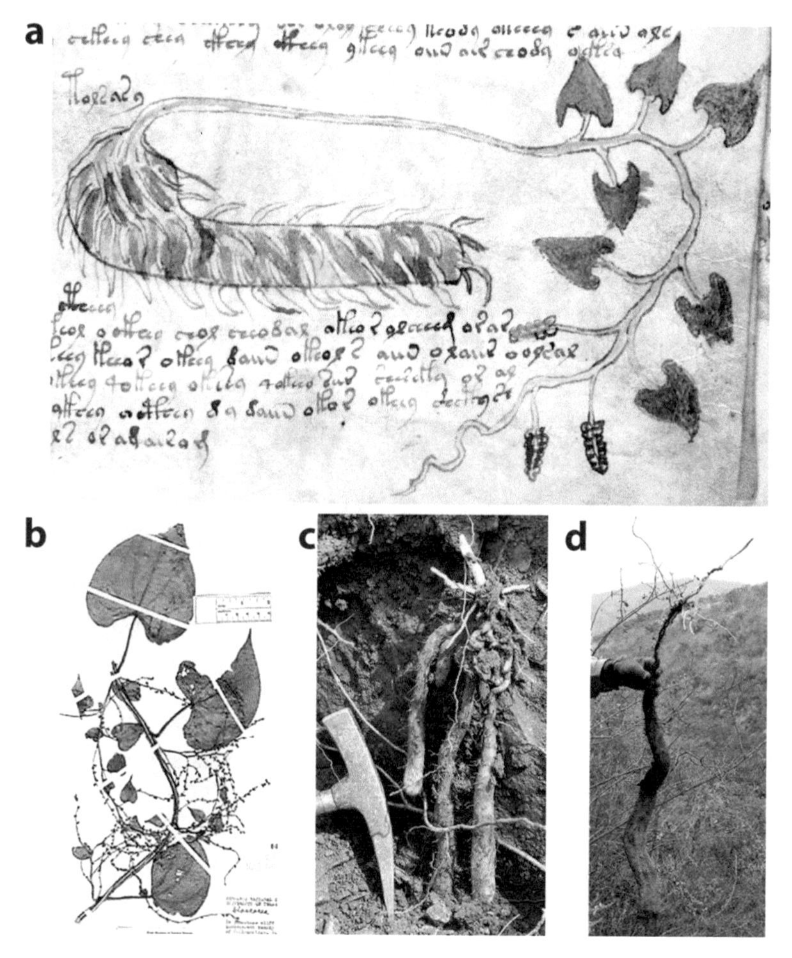

Fig. 4.28 *Dioscorea remotiflora*: (**a**) folio 99r#28; (**b**) herbarium sheet of *D. remotiflora* (F1405679). (Source: The Field Museum; (**c**, **d**) bat- or paddle-like roots of *D. remotiflora*. (Source: Ignacio Garcia Ruiz)

Folio 5v. *Jatropha cathartica* (Fig. 4.31) The appropriate identification for this phytomorph (Fig. 4.31a) is probably *Jatropha cathartica* Terán & Berland., *jicamilla* (Fig. 4.31b). The palmately compound dentate leaves, red flowers, and tuberous roots are similar. It is native from Texas to northern Mexico. As the name implies, this is cathartic and poisonous. Another similar species is *J. podagrica* Hook., native from southern Mexico to Nicaragua, but the leaflets are typically broader and not as deeply cut as *J. cathartica.*

Folio 93v. *Manihot rubricaulis* (Fig. 4.32) The stout, thickened roots, palmately compound leaves, and reddish fruits (Fig. 4.32a) all fit the genus *Manihot*. This phytomorph is most probably *Manihot rubricaulis* I. M. Johnst. (Fig. 4.32b) from

Fig. 4.29 *Cnidoscolus* sp.: (**a**) folio 6v; (**b**) fruits of *C. texanus* with trichomes. (Source: Carl Fabre, Lady Bird Johnson Wildflower Center); (**c**) leaves and flower of *C. texanus*. (Source: Tiana Rehman)

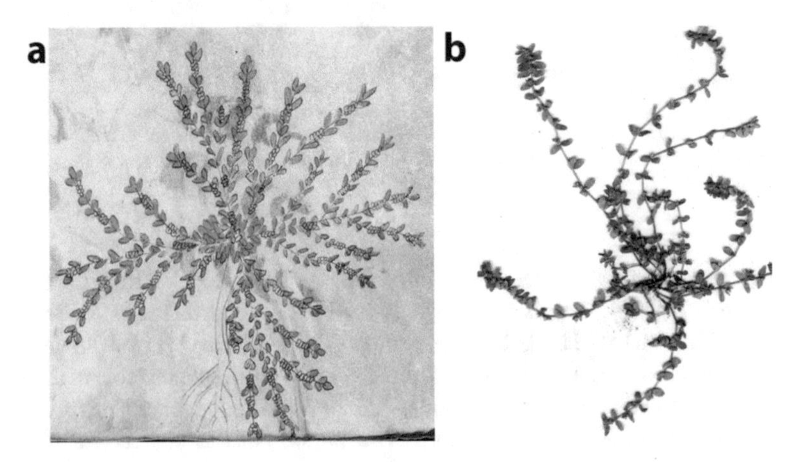

Fig. 4.30 *Euphorbia thymifolia*: (**a**) folio 21r; (**b**) *E. thymifolia*. (Source: Forest and Kim Starr, http://tropical.theferns.info/image.php?id=Euphorbia+thymifolia)

Fig. 4.31 *Jatropha cathartica*: (**a**) folio 5v; (**b**) *J. cathartica*. (Source: Frank Vincentz, https://en.wikipedia.org/wiki/Jatropha_cathartica#/media/File:Jatropha_cathartica1_ies.jpg)

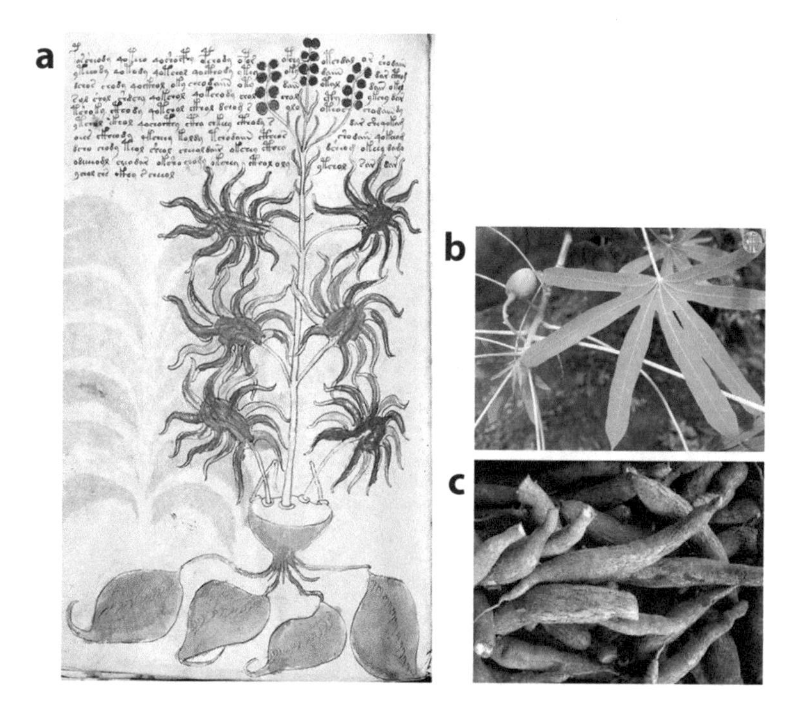

Fig. 4.32 *Manihot rubricaulis*: (**a**) folio 93v; (**b**) leaf of *M. rubricaulis*. (Source: Sky Jacobs, wildsonora.com); (**c**) tubers of *M. esculenta*. (Source: David Monniaux, https://en.wikipedia.org/wiki/Cassava#/media/File:Manihot_esculenta_dsc07325.jpg)

northern Mexico. This close relative to the cassava, *M. esculenta* Crantz, has thinner, more deeply lobed leaves, and also bears tuberous roots (Fig. 4.32c) (Hancock 2012). *Manihot rubricaulis* is perhaps illustrated in folio 43v of the *Codex Cruz-Badianus* as *yamanquipatlis* (gentle or weak medicine) (Emmart 1940; Cruz and Badiano 1991; Gates 2000; Clayton et al. 2009).

Fabaceae

Folio 88r#11. *Lupinus* sp., cf. *L. montanus* (Fig. 4.33) Phytomorph #11 on folio 88r (Fig. 4.33a) displays stylized compound peltate leaves and callus-like, nitrogen-fixing root nodules on one side of the roots. Many members of the Fabaceae have nitrogen-fixing nodules, but none has leaves exactly as depicted. Although the image is crudely rendered, it may be *Lupinus montanus* Humb., Bonpl. & Kunth (Fig. 4.33b) of Mexico and Central America. The compound, peltate leaves and soft, callus-like, nitrogen-fixing root nodules on one side of the roots (Fig. 4.33c) are all typical of this species. This lupine is noted to contain alkaloids (Dunn and Harmon 1977; Ruiz-López et al. 2010).

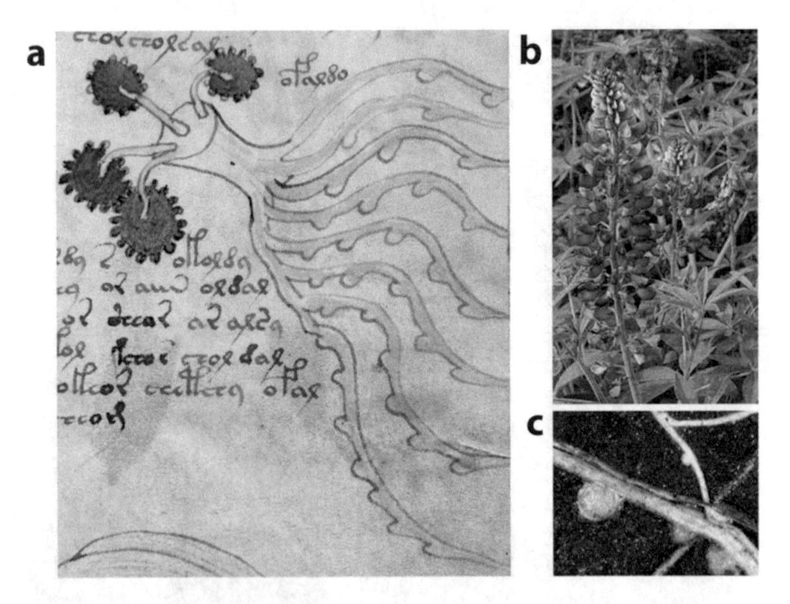

Fig. 4.33 *Lupinus* sp.; (**a**) folio 88r#7; (**b**) *L. montanus*. (Source: Harry Douwes); (**c**) root nodules on *L. montanus*. (Source: Wilderness and Backcountry Site Restoration Guide, USDA/US Forest Service)

Gesneriaceae

Folio 55r. *Diastema hispidum* (Fig. 4.34) This phytomorph has six petals, white and bluish, with a long corolla (Fig. 4.34a). The leaves are green and deeply lobed. Multiple stems arise from a rhizomatous base with many brown roots. Although this may be a species of *Geranium*, the swollen fruits seem incongruous with this genus. A better match might be *Diastema hispidum* (DC.) Fritsch. (Fig. 4.34b, c), which is native from Nicaragua to Peru.

Grossulariaceae

Folio 23r. *Ribes malvaceum* (Fig. 4.35) This phytomorph (Fig. 4.35a) is most probably *Ribes malvaceum* Sm., chaparral currant (Fig. 4.35b, c). This woody, stoloniferous shrub has purple-magenta flowers and palmately lobed leaves, and is native from California to Baja California Norte, Mexico (Standley 1920–1926: 316).

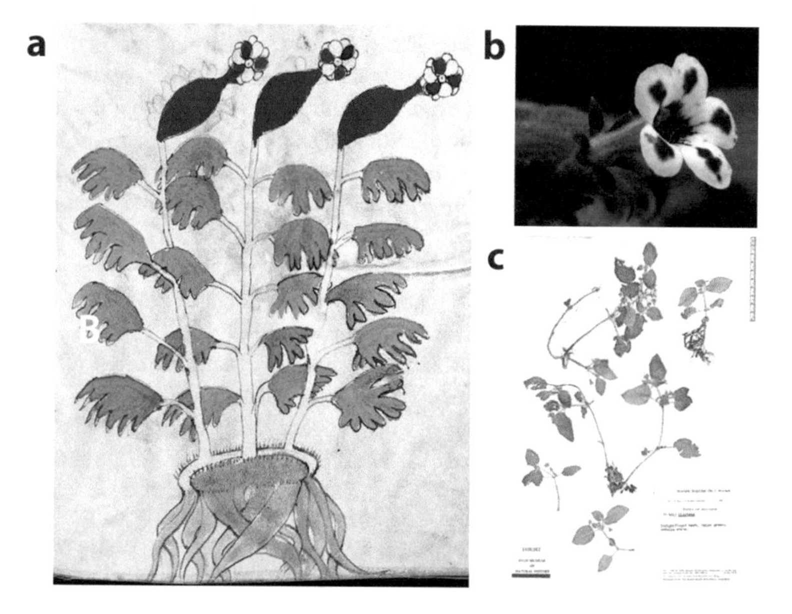

Fig. 4.34 *Diastema hispidum:* (**a**) folio 55r; (**b**) flower of *D. hispidum*. (Source: Leslie Brothers); (**c**) herbarium sheet of *D. hispidum* (F1836367) showing thin rhizomatous roots. (Source: The Field Museum)

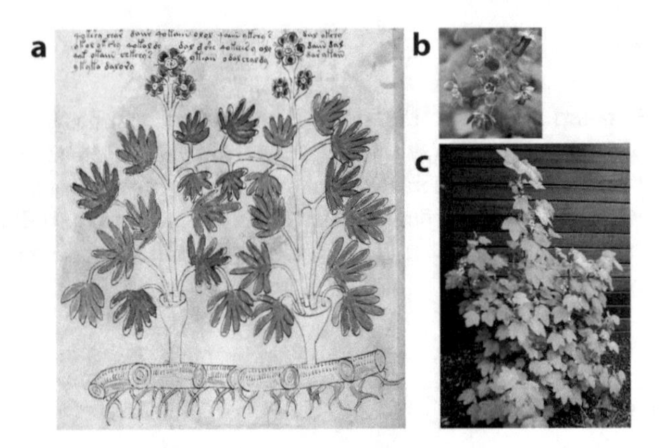

Fig. 4.35 *Ribes malvaceum*: (**a**) folio 23r; (**b**) flower of *R. malvaceum*. (Source: Stan Shebs, https://commons.wikimedia.org/wiki/File:Ribes_malvaceum_var_veridifolium_2.jpg); (**c**) shoots of *R. malvaceum*. (Source: Toedrifter, https://commons.wikimedia.org/wiki/File:Ribesmalvaceum12-2.jpg)

Lamiaceae

Folio 45v. *Hyptis albida* (Fig. 4.36) The gray leaves, blue flowers, and stout root (Fig. 4.36a) are all good fits for *Hyptis albida* Kunth (Fig. 4.36b, c). This shrub is native to Guanajuato and San Luis Potosí in central Mexico, and the Mexican states of Sonora, Chihuahua, and Guerrero. Standley (1920–1926: 1275) relates that: "*The leaves are sometimes used for flavoring food. In Sinaloa they are employed as a remedy for ear-ache, and in Guerrero a decoction of the plant is used in fomentations to relieve rheumatic pains.*"

Folio 32r. *Ocimum campechianum* (*O. micranthum*) (Fig. 4.37) This phytomorph (Fig. 4.37a) is probably *Ocimum campechianum* Mill. (*O. micranthum* Willd.) (Fig. 4.37b). This suffrutescent annual basil is native from Florida to Argentina. In Mexico, it is found from Sinaloa to Tamaulipas, Yucatán, and Colima. The terminal inflorescence, bluish flowers, and ovate leaves are good fits (Standley 1920–1926: 1272; Standley and Williams 1973: 269). Standley (1920–1926: 1272) relates: "*In El Salvador bunches of the leaves of this plant are put in the ears as a remedy for earache.*"

Folio 45r. *Salvia cacaliifolia* (Fig. 4.38) The blue flowers in a tripartite inflorescence and distantly dentate deltoid-hastate leaves (Fig. 4.38a) are characteristic of *Salvia cacaliifolia* Benth. (Fig. 4.38b). It is native from Mexico (Chiapas) to Guatemala and Honduras (Standley and Williams 1973: 278).

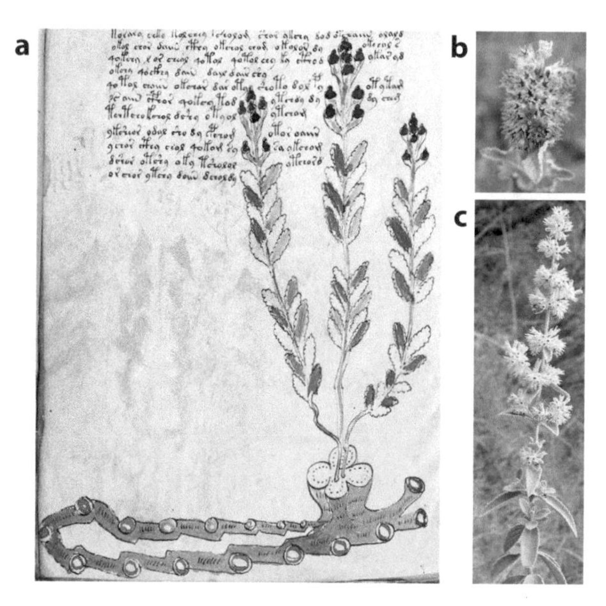

Fig. 4.36 *Hyptis albida*: (**a**) folio 45v; (**b**, **c**) inflorescence of . (Sources: Benjamin T. Wilder and Jim Conrad respectively)

Folio 100r#5. *Scutellaria mexicana* (Fig. 4.39) Phytomorph #5 on folio 100r (Fig. 4.39a) shows three flowers that closely resemble *Scutellaria mexicana* (Torr.) A.J. Paton (*Salazaria mexicana* Torr.) (Fig. 4.39b, c). This species also seems to match the description of *tenamaznanapoloa* (which could possibly mean "carrying triplets") of Hernández et al. (1651: 129) (alias *tenamazton* or *tlalamatl*). This shrub, native from Utah to Mexico (Baja California, Chihuahua, and Coahuila), bears inflated bladder-like calyces that vary in color, depending upon maturity, from green to white to magenta, with a dark blue and white corolla emerging from them (Standley 1920–1926: 1271).

Malvaceae

Folio 102r#11. *Chiranthodendron pentadactylon* (Fig. 4.40) Phytomorph #11 (Fig. 4.40a) is very curious and looks more like a very dark, blue-black flag than a possible plant part. However, this is often what the five-parted stamens of *Chiranthodendron pentadactylon* Larreat. (*C. platanoides* Bonpl.), the hand-flower, look like when pressed and dried. When fresh, the stamens are a brilliant vermillion (Fig. 4.40b), but they turn blue-black when improperly dried and/or aged (Fig. 4.40c), and the five-parted, hand-like stamens can assume a flag-like shape when pressed.

Fig. 4.37 *Ocimum campechianum*: (**a**) folio 32r; (**b**) inflorescence and leaves of *O. campechianum*. (Source: Roger L. Hammer)

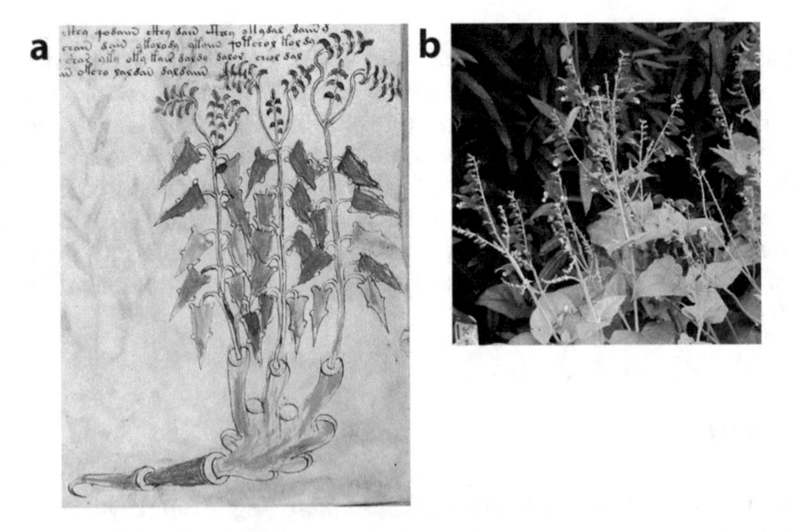

Fig. 4.38 *Salvia cacaliifolia*: (**a**) folio 45r; (**b**) inflorescence and leaves of *S. cacaliifolia.* (Source: Ashwood Nurseries Ltd)

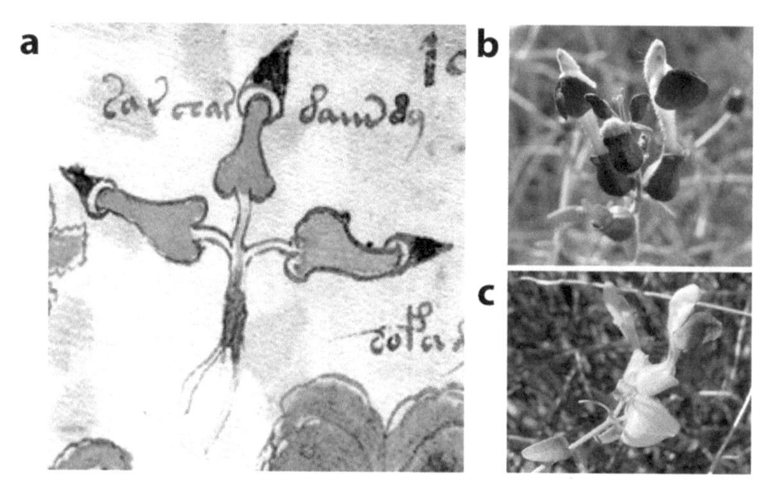

Fig. 4.39 *Scutellaria mexicana*: (**a**) folio 100r#5; (**b**, **c**) inflorescence of *S. mexicana*. (Source: Michael L. Charters)

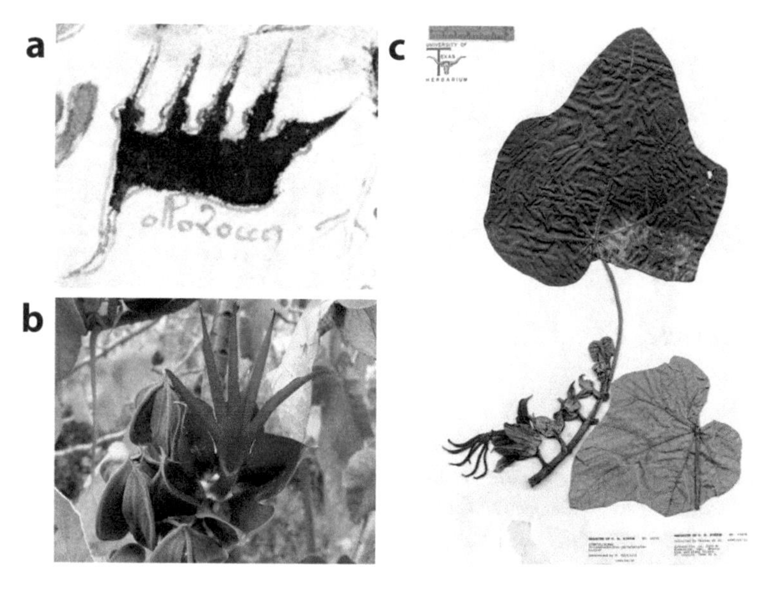

Fig. 4.40 *Chiranthodendron pentadactylon*: (**a**) folio 102r#7; (**b**) five-parted stamens of *C. pentadactylon*. (Source: Jan Conayne); (**c**) dried shoot of *C. pentadactylon* showing leaves and flower with protruding stamens. (Source: José Luis Villaseñor Ríos)

This species typically grows in wet areas in the mountains of Oaxaca and Guatemala, but is widely planted in the Valley of Mexico (Standley 1920–1926). It is called *macpalxochi quahuitl* in Hernández et al. (1651: 383, 459). Additional Nahuatl names are *mapasúchil, mapilxochitl*, and *teyacua* (Díaz 1976). *Mapasúchil* is derived from the Nahuatl *macpal-xochitl*, or "hand flower" (Standley 1920–1926).

Fig. 4.41 *Calathea* sp.: (**a**) folio 42v; (**b**) inflorescence and leaves of *C. loeseneri*. (Source: Milan Kořínek)

Marantaceae

Folio 42v. *Calathea* sp., cf. *C. loeseneri* (Fig. 4.41) The phytomorph inflorescence (Fig. 4.41a) is a crude representation of a *Calathea* sp., probably allied to *C. loeseneri* J.F. Macbr. (Fig. 4.41b), which yields a blue dye. Many species of *Calathea* were recently transferred to the genus *Goeppertia*, and a synonym of this species is now *G. loeseneri* (J.F. Macbr.) Borschs. & Suárez (Borschenius et al. 2012). The crudeness of the illustration, coupled with inadequate surveys of the genus *Calathea/Goeppertia* in Mexico, inhibit precise identification.

Menyanthaceae

Folio 2v. *Nymphoides aquatica* (Fig. 4.42) The rounded notched leaves with white flowers and thick rhizome (Fig. 4.42a) closely resemble *Nymphoides aquatica* (J. F. Gmel.) Kuntze (Fig. 4.42b). This aquatic plant with floating orbicular leaves

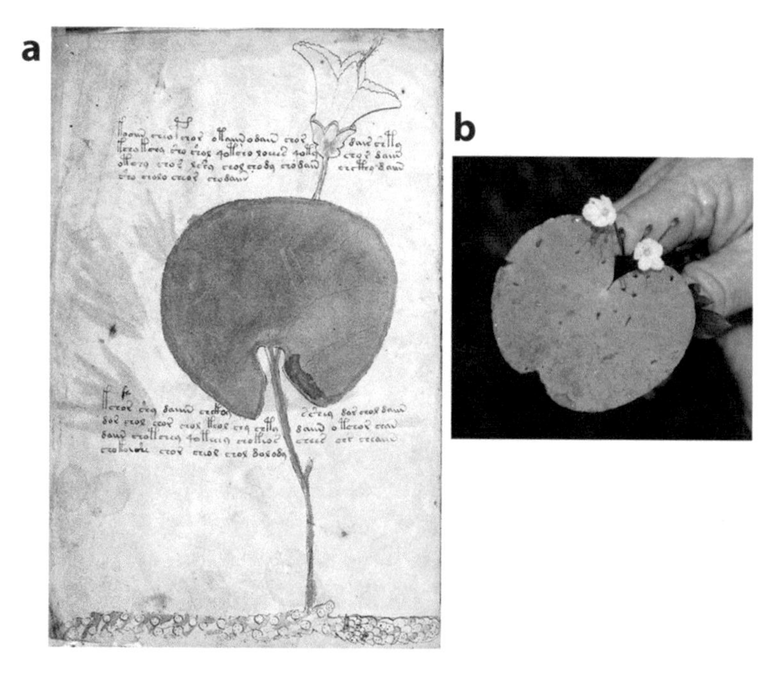

Fig. 4.42 *Nymphoides aquatica*: (**a**) folio 2v; (**b**) *N. aquatica*. (Source: Center for Aquatic and Invasive plants, University of Florida, Gainesville)

and white flowers with five petals is native to North America, from New Jersey to Texas. The horizontal rhizome bears thick, unbranched, adventitious roots that in young plants resemble a bunch of bananas (Richards et al. 2010), earning it the popular name the "banana plant." Another possibility, because the flower petals are illustrated with a crenate margin, is *N. indica* (L.) Kuntze, which has fringed petals and is native not only to Mexico, but also to Asia, Africa, and Australia.

Moraceae

Folio 36v. ***Dorstenia contrajerva*** **(Fig. 4.43)** The inflorescence (Fig. 4.43a), which looks like a fig split open, is quite distinct and probably matches a *Dorstenia* sp., likely the highly variable *D. contrajerva* L. (Fig. 4.43b). The leaves for this species vary "*in spirals, rosulate or spaced; lamina broadly ovate to cordiform to subhastate, pinnately to subpalmately or subpedately, variously lobed to parted with 3–8 lobes at each side or subentire*" (Berg 2001). This is native from Mexico to Peru. The Nahuatl name is either *tozpatli* or *tuzpatli* (Díaz 1976).

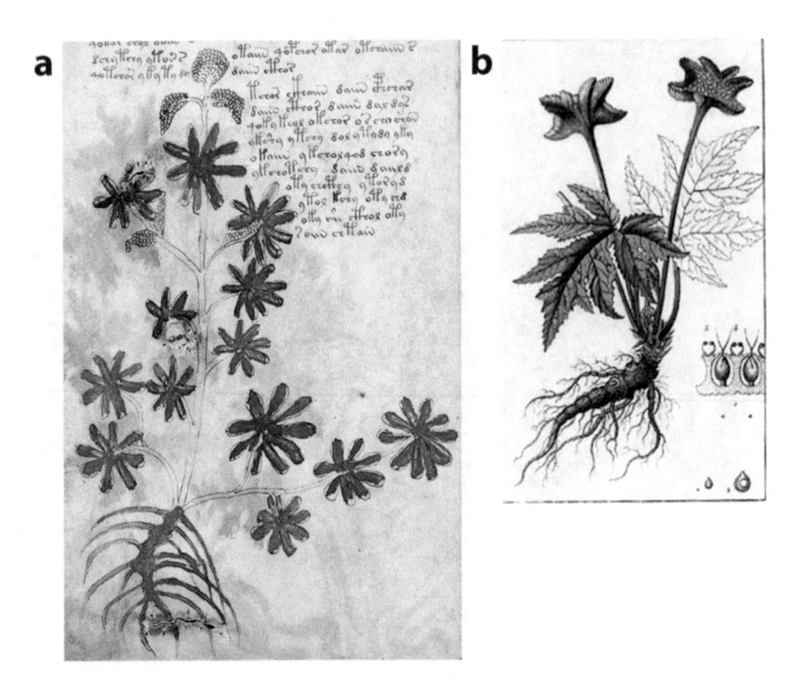

Fig. 4.43 *Dorstenia contrajerva*: (**a**) folio 36v; (**b**) botanical image of *D. contrajerva* by Pierre Turpin from Chaumeton (1830: 131)

Nyctaginaceae

Folio 33r. ***Allionia incarnata*** **(Fig. 4.44)** This has many petals united into a greenish corolla, backed by a swollen, whitish calyx, with sagittate, green leaves, and swollen, branched brown roots (Fig. 4.44a). It matches the wide variability of *Allionia incarnata* L. (Fig. 4.44b, c), commonly called "trailing four o'clock" or "trailing windmills," and known in Spanish as *hierba de la hormiga* (ant herb) or *hierba del golpe* (wound herb). This is native from Utah to Mexico. Curiously, the ends of the two main roots in the *Voynich Codex* phytomorph have a face. The roots of *Allionia incarnata* do bear bumps with indentations that could be interpreted as tiny faces.

Onagraceae

Folio 51r. ***Fuchsia thymifolia*** **(Fig. 4.45)** The phytomorph has four red petals and four pale sepals backed by a corolla tube and a swollen ovary. The leaves are green and deeply serrated and the roots are brown and tuberous (Fig. 4.45a). This may be *Fuchsia thymifolia* Kunth (Fig. 4.45b), which is native from Mexico to Guatemala.

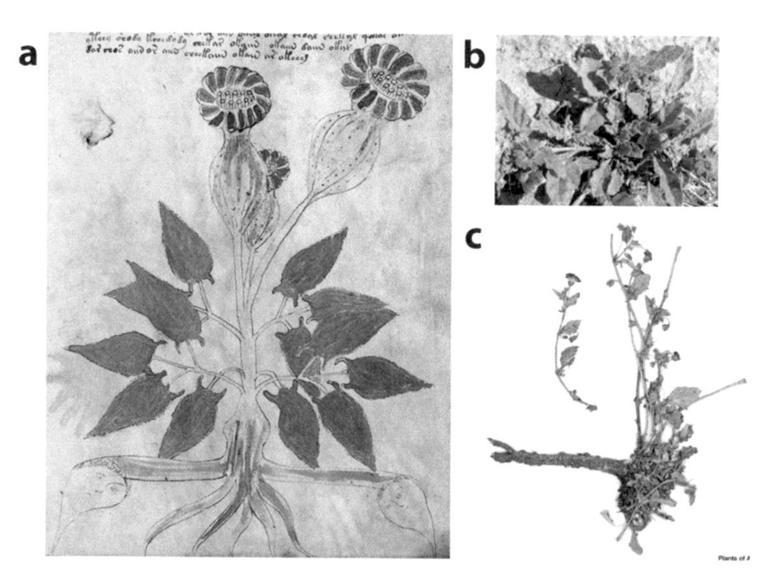

Fig. 4.44 *Allionia incarnata*: (**a**) folio 33r; (**b**) inflorescence and leaves of *A. incarnate*. (Source: Campbell and Lynn Loughmiller, Lady Bird Johnson Wildflower Center); (**c**) herbarium specimen of *A. incarnata* (DES00067664) showing the swollen, knobby roots. (Source: Desert Botanical Garden Herbarium Collection, http://intermountainbiota.org/portal/collections/individual/index.php?occid=2430103)

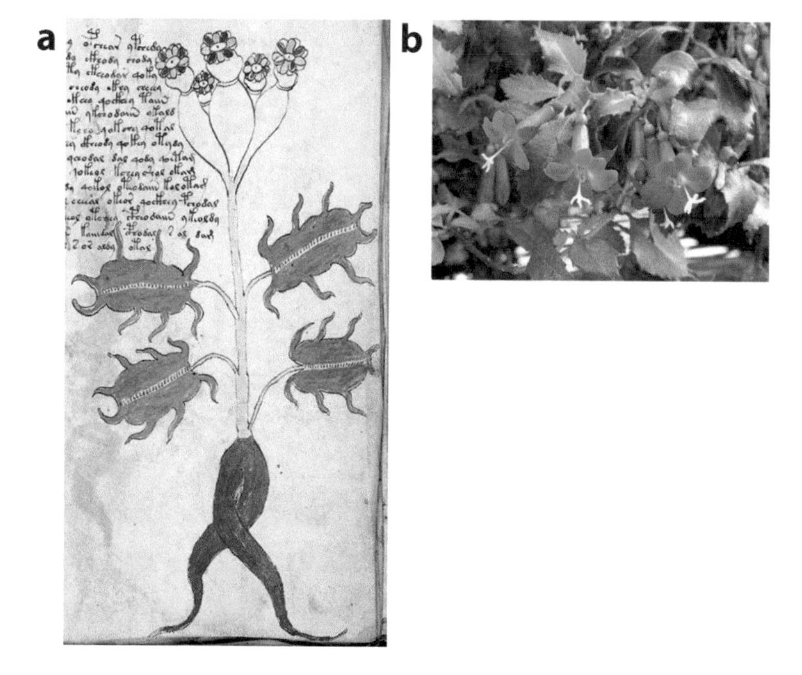

Fig. 4.45 *Fuchsia thymifolia*: (**a**) folio 51r; (**b**) leaves and flowers of *F. thymifolia*. (Source: Todd Boland)

Passifloraceae

Folio 23v. *Passiflora* Subgenus *Decaloba*, cf. *P. morifolia* (Fig. 4.46) From the flower alone, this is definitely a *Passiflora* sp. of subgenus *Decaloba* (Fig. 4.46a). *Passiflora* is primarily a New World genus (a few species also occur in Australia and Southeast Asia, but not Europe). The prominent corona with filaments of the genus *Passiflora* is very distinctive and cannot be confused with any other genus. The paired petiolar glands in the upper third of the leaf, blue tints in the flower, and dentate leaves that are deeply cordate only seem to match the variability of *P. morifolia* Mast. (Killip 1938) (Fig. 4.46b, c), although the artist has made the leaves slightly more orbicular than they normally occur in mature foliage. However, young plants, i.e., root suckers, often exhibit juvenile leaves that are entire orbicular.

Penthoraceae

Folio 30v. *Penthorum sedoides* (Fig. 4.47) The cymose inflorescence, dentate leaves (Fig. 4.47a) and stolons match *Penthorum sedoides* L. (Figs. 4.47b, c) quite well. This is a New World species native from Canada to Texas. The artist, though, has apparently illustrated this as a very early bud (or glossed over the details of the flowers) because the prominent pistils emerge later, and are very obvious in fruit, often turning rosy.

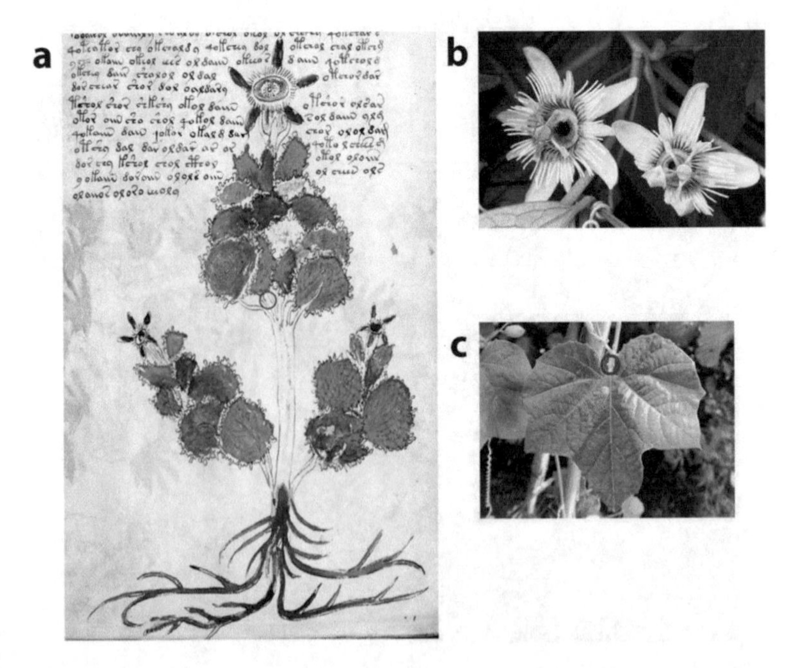

Fig. 4.46 *Passiflora* sp.: (**a**) folio 23v; (**b**) flower of *P. morifolia*. (Source: Anke and Ralf Schlosser); (**c**) leaf showing petiolar gland of *P. morifolia*. (Source: Hans B., https://commons.wikimedia.org/wiki/File:Passiflora_morifolia3.jpg)

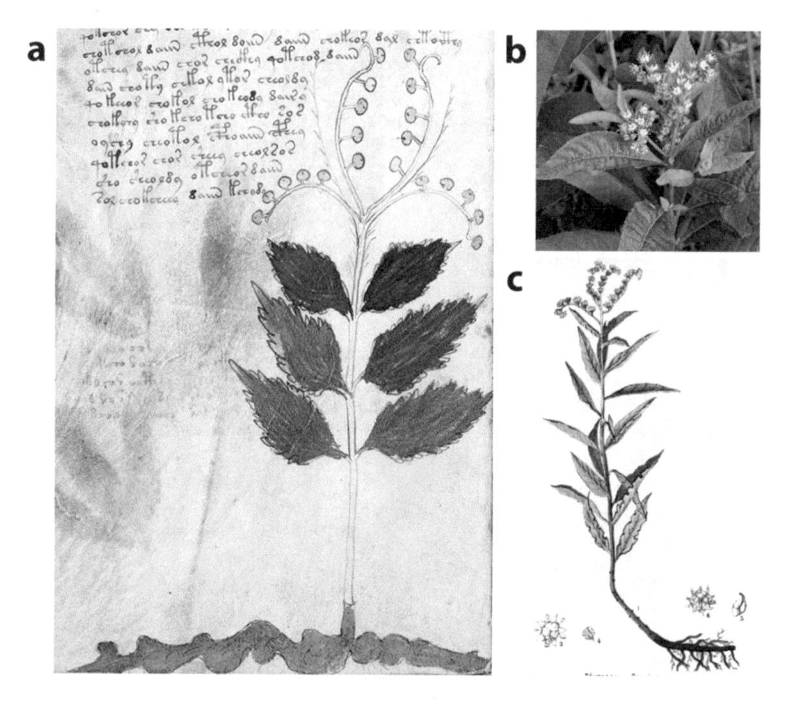

Fig. 4.47 *Penthorum sedoides*: (**a**) folio 30v; (**b**) inflorescence and leaves of *P. sedoides*. (Source: Fritz Flohr Reynolds, https://commons.wikimedia.org/wiki/File:Penthorum_sedoides_-_Ditch_Stonecrop.jpg); (**c**) botanical painting of *P. sedoides* from Millspaugh (1892 1: t. 57)

Polemoniaceae

Folio 4v. *Cobaea* sp., cf. *C. biaurita* (Fig. 4.48) With a basally woody stem, pinnately compound elliptic leaves, campanulate corolla, segmented calyx, and exserted style (Fig. 4.48a), the best match would be a *Cobaea* sp., probably *C. biaurita* Standl. (Figs. 4.48b, c), which is closely related to the cultivated *C. scandens* Cav. This vine is native to Chiapas and Oaxaca, Mexico, and has elliptic leaflets with acute to acuminate apices and flowers that emerge cream-colored, but later mature to purple (Standley 1914; Prather 1999).

Ranunculaceae

Folio 95r. *Actaea rubra* f. *neglecta* (Fig. 4.49) The crenate, pinnately compound leaves and noticeably white, globose fruits in a raceme (Fig. 4.49a) definitely fit an *Actaea* sp., probably the white-fruited *Actaea rubra* (Aiton) Willd. f. *neglecta* (Gillman) B. L. Rob (Fig. 4.49b). *Actaea rubra* is native to Eurasia and North America, from Canada to New Mexico, but f. *neglecta* is more common in North America (Compton et al. 1998). As the common name baneberry would indicate, this is poisonous.

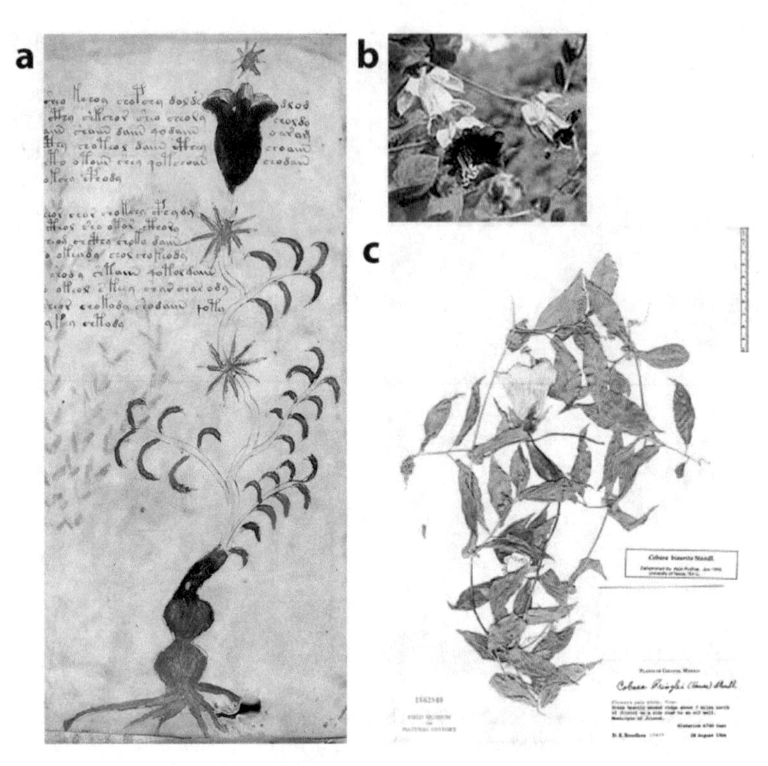

Fig. 4.48 *Cobaea* sp.: (**a**) folio 4v; (**b**) flower of *C. scandens*. (Source: Michael Wolf, https://commons.wikimedia.org/wiki/File:Cobaea_scandens_03.jpg); (**c**) herbarium sheet of *C. biaurita* (F1662840). (Source: The Field Museum)

Fig. 4.49 *Actaea rubra* f. *neglecta*: (**a**) folio 95r; (**b**) fruits and leaves of *Actaea rubra* f. *neglecta*. (Source: Donald Cameron)

Fig. 4.50 *Anemone patens:* (**a**) folio 52r; (**b**, **c**) flowers and leaves of *Anemone patens*. (Sources: Tom Koerner, U.S. Fish and Wildlife Service and John Richards respectively)

Folio 52r. *Anemone patens* (Fig. 4.50) This has a terminal, pubescent, blue flower with many linear bracts. The leaves are deeply laciniate and the roots are brown, long, and tuberous (Fig. 4.50a). It matches the variability of *Anemone patens* L., pasqueflower (Fig. 4.50b, c), which is circumboreal, south to Texas and New Mexico.

Folio 29v. *Anemone tuberosa* (Fig. 4.51) This has blue-green, hairy flower buds with a multi-petaled, bluish corolla. The basal leaves are deeply divided and the roots are tuberous (Fig. 4.51a). It matches *Anemone tuberosa* Rydb., desert anemone (Fig. 4.51b), which is native from Utah to Mexico.

Saxifragaceae

Folio 49r. *Lithophragma affine* (Fig. 4.52) This has blue flowers with fringed petals and red calyces. The leaves are numerous and borne tightly together on long petioles, and the roots are tuberous (Fig. 4.52a). It matches the variability of *Lithophragma affine* A. Gray (Fig. 4.52b, c), which is native from Oregon to Baja California, and intergrades with *L. parviflorum* (Hook.) Torr. & A. Gray (Park and Elvander 2012).

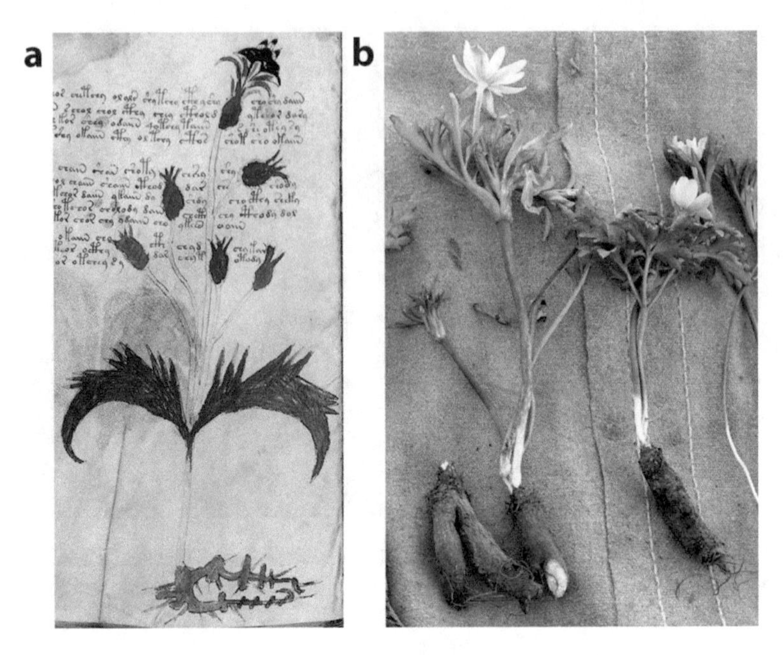

Fig. 4.51 *Anemone tuberosa*: (**a**) folio 29v; (**b**) *A. tuberosa*. (Source: 7Song)

Fig. 4.52 *Lithophragma affine*: (**a**) folio 49r; (**b**) inflorescence of *L. affine*. (Source: Doug Zimmerman, dazimmerman.com); (**c**) leaves of *L. affine*. (Source: Rebecca Snyder, www.fallbrooksource.com)

Solanaceae

1. Folio 101r#3 and Folio 101v(1)#2. *Capsicum annuum* (Fig. 4.53) Hugh O'Neill (1944) identified a *Capsicum* sp. in the *Voynich Codex*. However, in his text, he stated folio 101v, but the illustration provided was from folio 101r. Fortunately, we agree that the shape and color of the fruits of both phytomorphs (Fig. 4.53a, b) agree with the genus *Capsicum*. Phytomorph #3 on folio 101r has erect green fruits with depressed stem attachments and a forked primary root. Phytomorph #2 on folio 101v has pendant red fruits and a forked primary root. Both fall within the wide variation of *C. annuum* L. (Fig. 4.53c, d). The common red capsicum pepper originated in Mesoamerica, but was introduced to Eurasia and Africa by the early sixteenth century.

Fig. 4.53 *Capsicum annuum*: (**a**) folio 101r#3; (**b**) folio 101v(1)#2; (**c**) green fruit of *C. annuum*. (Source: Consell Comarca Baix Empordà, https://commons.wikimedia.org/wiki/File:Bitxos_de_girona.jpg); (**d**) red fruit of *C. annuum*. (Source: Simon Feiertag)

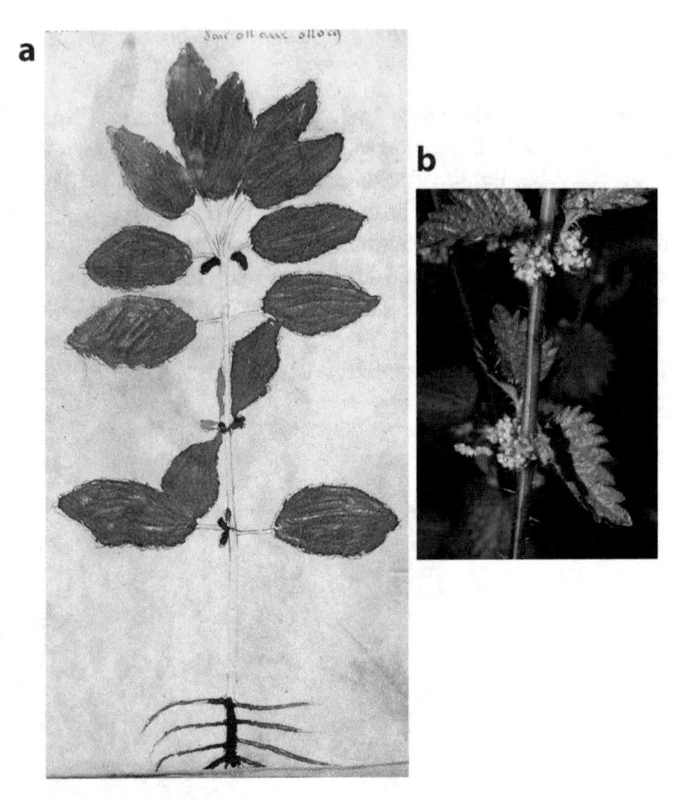

Fig. 4.54 *Urtica* sp.: (**a**) folio 25r; (**b**) *U. chamaedryoides*. (Source: Steve Baskauf, http://bioimages.vanderbilt.edu/)

Urticaceae

Folio 25r. *Urtica* sp., cf. *U. chamaedryoides* (Fig. 4.54) This phytomorph (Fig. 4.54a) was first postulated by Hugh O'Neill (1944) to be a member of the Urticaceae or nettle family. The best match, because of the dentate, lanceolate leaves and reddish inflorescences, seems to be *Urtica chamaedryoides* Pursh, the heart-leaf nettle (Fig. 4.54b). It is native from Canada to northern Mexico. *Urtica* and the closely related genus *Urera* also occur in the *Codex Cruz-Badianus* (Emmart 1940; Cruz and Badiano 1991; Gates 2000; Clayton et al. 2009) and Hernández et al. (1651).

Valerianaceae

1. Folio 65r. *Valeriana albonervata* (Fig. 4.55) The palmate- or cleft-lobed leaves, inflorescence, and napiform to fusiform, often forked taproots (Fig. 4.55a) are a good match for *Valeriana albonervata* B.L. Rob. (Fig. 4.55b). This is native to the Sierra Madre of Mexico (Meyer 1951).

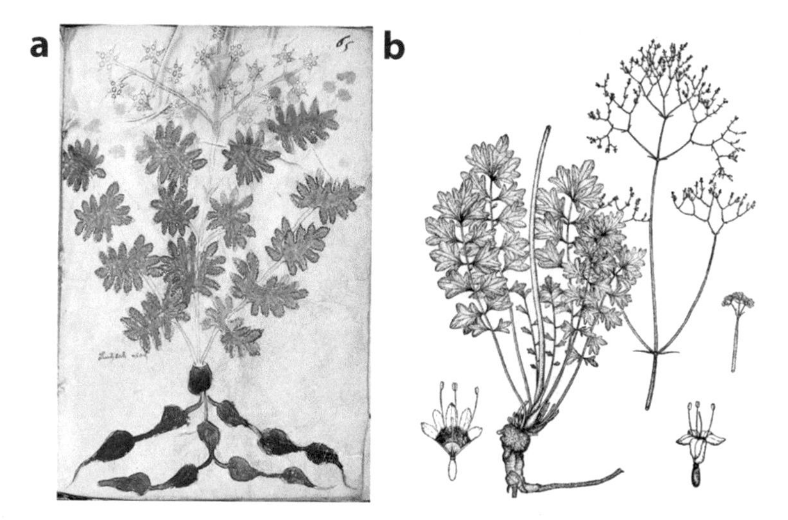

Fig. 4.55 *Valeriana albonervata*: (**a**) folio 65r; (**b**) botanical drawing of *V. albonervata* from Meyer (1951)

Verbenaceae

1. Folio 94r. ***Duranta erecta*** **(*D. repens*) (Fig. 4.56)** This phytomorph has terminal black fruits with a white terminal knob, arranged in a corymb or umbel. The leaves are obcordate and crenate. The roots are thick, brown, and branched (Fig. 4.56a). Except for the shape of the inflorescence (Fig. 4.56b), this matches perfectly a dried specimen of the repent form of *Duranta erecta* L. (*D. repens* L.), golden dewdrop (Fig. 4.56c). This shrub is native from Florida and Texas, south to Argentina. The fruits are golden yellow when fresh and born on a panicle, but when dried they often quickly fall apart and turn black. Presumably because the color of the fruits matches that of a dried specimen, the sample was conceivably fragmented when the artist prepared this illustration, and thus the artist was likely unaware of the color of the mature fruit on a living branch.

Violaceae

Folio 9v. ***Viola bicolor*** **(*V. rafinesquii*) (Fig. 4.57)** This phytomorph (Fig. 4.57a) clearly shows linear terminal, stipular lobes, as in the North American native *V. bicolor* Pursh (*V. rafinesquii* Greene; Fig. 4.57b), not spatulate as in the Eurasian *V. tricolor* L. Also, this phytomorph matches the blue flowers of *V. bicolor*, not the tricolored ones of *V. tricolor*. *V. bicolor* flowers are uniformly cream to blue, whereas those of *V. tricolor* usually have two purple upper petals and three cream to yellow lower petals. *Viola bicolor* is native from New Jersey to Texas, west to Arizona, with

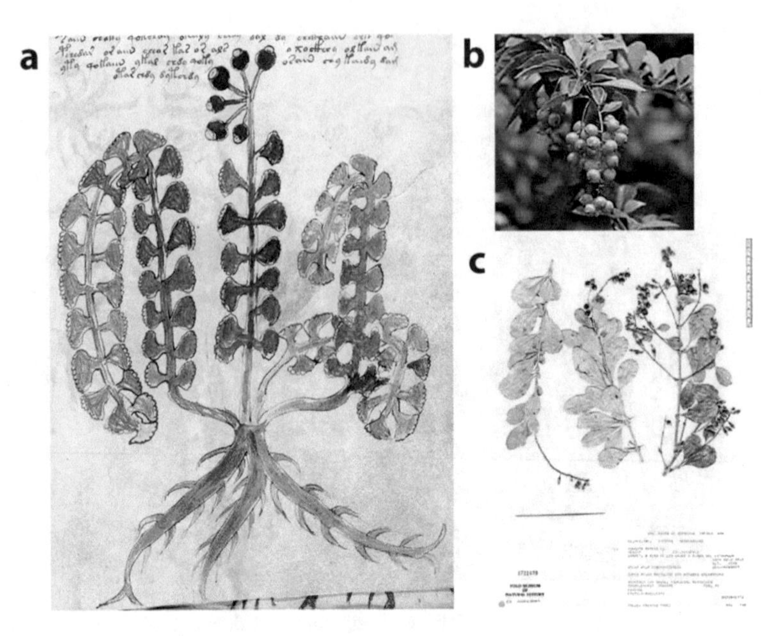

Fig. 4.56 *Duranta erecta*: (**a**) folio 94r; (**b**) fruit cluster of *D. erecta*. (Source: Hari Krishnan, https://commons.wikimedia.org/wiki/File:Duranta_erecta_fruits.jpg); (**c**) herbarium sheet of *D. erecta* (F1721033). (Source: The Field Museum)

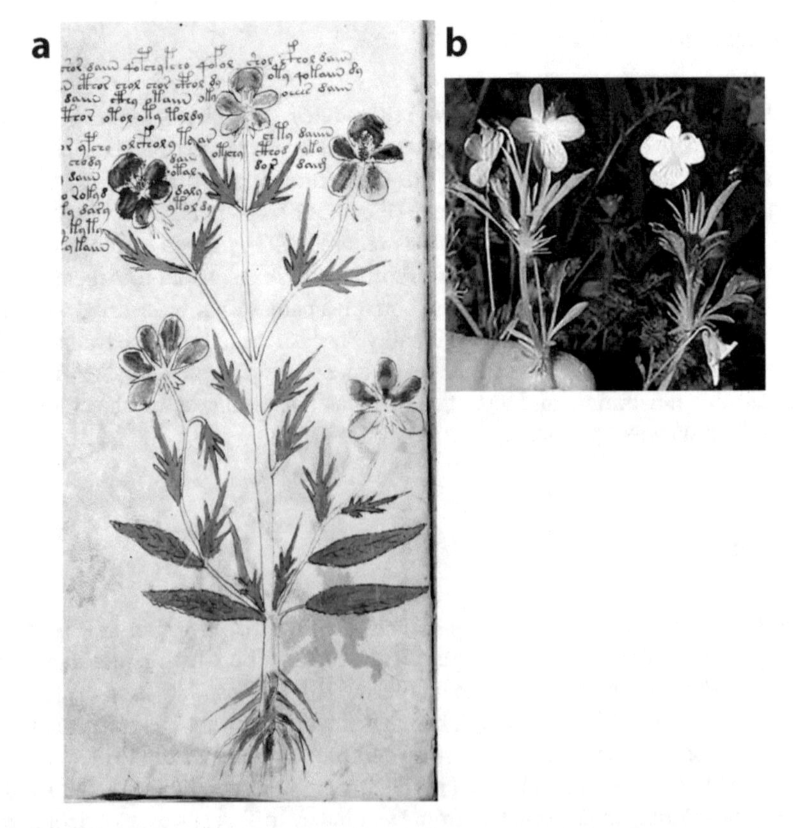

Fig. 4.57 *Viola bicolor*: (**a**) folio 9v; (**b**) flowers and leaves of *V. bicolor*. (Source: Daniel Reed)

a center of diversity in eastern Texas (Shinners 1961; Russell 1965). This was also proposed as a species of *Viola* (Kennedy and Churchill 2006).

The delineation of *V. bicolor* as native to North America and not introduced from elsewhere was only first fully elucidated by Shinners in 1961. Before 1961, *V. bicolor* was considered doubtfully native and also classified as a variety of the Eurasian *V. kitaibeliana* Schult. [var. *rafinesquii* (Greene) Fernald] and often confused with *V. tricolor* and *V. arvensis* Murray (Fernald 1938). The North American *V. bicolor*, besides being seasonally dimorphic with fertile and cleistogamous flowers, differs from the other three related species of Eurasia by "*roundish, almost entire basal leaves, and by pectinate, palmately divided stipules. The petals of its open flowers are twice as long as the sepals, slightly shorter than the petals of V. tricolor, but longer than those of V. arvensis and V. kitaibeliana*" (Clausen et al. 1964). Thus, because *V. bicolor* has only been known as a North American endemic since 1961, any attempt to propose a forgery before 1912 (Barlow 1986) would have to explain this anomaly.

Mineral Identification

Folio 102r#4. Boleite (Fig. 4.58) This image includes a cubic (isometric), blue mineral resembling a blue bouillon cube (Fig. 4.58a). This can only be boleite ($KPb_{26}Ag_9Cu_{24}Cl_{62}(OH)_{48}$; Fig. 4.58b). Boleite crystals are cubic, typically measure 2–8 mm on the sides, and occasionally up to 2 cm. Boleite is closely related to pseudoboleite and cumengeite, but the crystal structures are slightly different (isometric for boleite, tetragonal square tablets for pseudoboleite, and tetragonal pyramids for cumengeite).

Boleite occurs naturally in the oxidized zone of lead-copper deposits, but also in smelter slags immersed in and leached by sea water. The only sources of large crystals of this quality and quantity are three closely related mines in Baja California Sur, Mexico (principally the mine at the El Boleo deposit at Santa Rosalia).

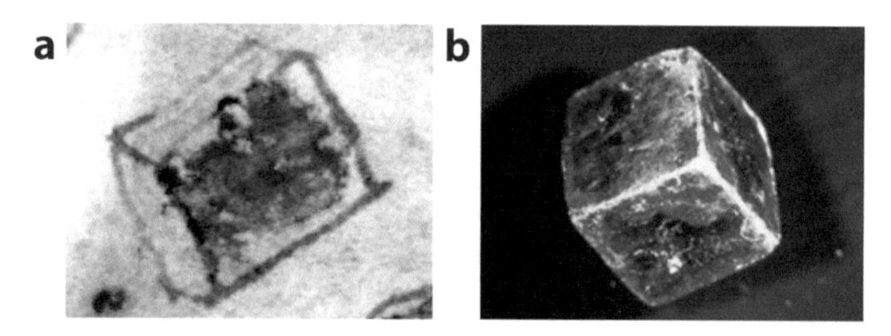

Fig. 4.58 Boleite: (**a**) cube from folio 102r#4; (**b**) boleite mineral. (Source: http://www.minfind.com/mineral-54337.html)

Typically, at El Boleo, crystals of boleite occur embedded in atacamite, the probable source of the green pigment (in the *Voynich Codex*). Boleite crystals also occur, more rarely, in the copper mines of Zacaecas, Mexico; Chile; Broken Hill, in Yancowinna County, Australia; and the St. Anthony deposit in Tiger, Arizona. The mineral also occurs as trace deposits in Arizona, California, Montana, Nevada, and Washington. Boleite has been reported as trace deposits in Eurasia and is more commonly found in slag localities, such as France, Germany, Greece, Iran, Italy, Russia, Spain, and the UK (Cooper and Hawthorne 2000; Mallard and Cumenge 1891; Mottana et al. 1978; Ralph and Chau 2012). In summary, the identification of this geomorph is critical. Boleite most likely came from the New World or, less likely, from Australia. An Eurasian origin is unlikely. Copper compounds have been used historically to treat pulmonary and skin diseases and parasitic infections, such as schistosomiasis and bilharzia (Copper Development Association 1971).

Conclusion

Sixty phytomorphs representing 57 plant species were identified from the Herbal and Pharmaceutical sections. In addition, the geomorph boleite was identified in the Pharmaceutical section. Of the 131 phytomorphs in the Herbal section, 123 had inflorescences or fruits, of which 46 (37.4%) were identified. In contrast to the 228 phytomorphs in the Pharmaceutical section, only 23 showed flowers or fruits (with most lacking key botanical characteristics), and 9 (39.1%) were identified. Thus, of the 146 phytomorphs in the *Voynich Codex* with inflorescences or fruits, 55 (37.7%) were identified. All phytomorphs identified were from colonial New Spain, primarily with a native range in the twenty-first century from Texas, west to California and south to Honduras. Asteraceae was the most common family and at least seven more phytomorphs have been tentatively identified from this family, but are not included here (folios 18r, 34r, 35r, 46r, 50r, 53v, and 54r).

No indigenous European, Asian, African, Australian, or South American plants have been identified other than circumboreal species (e.g., *Actaea rubra*). Some of the plant families, such as Cactaceae, and genera, such as *Passiflora*, were primarily native to the New World. In addition, the animals and minerals identified in the *Voynich Codex* were primarily from colonial New Spain (Tucker and Talbert 2013).

The plants identified include a fern and a gymnosperm, but the remainder were angiosperms, including dicotyledonous and monocotyledonous, herbaceous, and woody plants. Although a number of food plants, including the genera *Agave*, *Capsicum*, *Helianthus*, *Ipomoea*, *Opuntia*, and *Smallanthus*, were described, most of the plants appear to have medicinal value, indicating that the *Voynich Codex* was largely a medicinal herbal, although it was possible that maize may have been included in the missing herbal folios. The Aztec culture was rich in the knowledge of medicinal plants and had many botanical gardens that pre-dated those in Europe (Nuttall 1925; Granziera 2001, 2005).

Many of the genera identified herein can also be found in sixteenth century Aztec herbals, including the *Codex Cruz-Badianus* (Emmart 1940; Cruz and Badiano 1991; Gates 2000; Clayton et al. 2009), Book 11 of Sahagún (1963), and the collections of Francisco Hernández (Hernández et al. 1651; Hernández 1942). Thus, the plant identification alone is evidence that the *Voynich Codex* is from colonial New Spain, probably from the sixteenth century.

The accuracy of the drawings of some phytomorphs, such as folio 9v (*Viola bicolor*), versus the broad strokes of others, including folio 16v (*Eryngium heterophyllum*), suggest that more than one artist might have been involved, or more likely that both fresh and dried specimens were used as models. The mixed nature of some of the phytomorphs (e.g., folios 9r, 15v, and 33v), the flatness of many phytomorphs (e.g., folio 100r#4, *Agave* sp.), and the discoloration that could result from drying (e.g., folio 100r#11, *Chiranthodendron pentadactylon*) would point to the use of dried specimens. The *curanderos/curanderas* (folk healers) in Mexico today most often deal with dried herbs, not fresh ones, as this is the easiest means of preserving materials for future use.

Correct identification of the phytomorphs today is further rendered difficult by an extremely limited palette of colors. Thus, the red pigment seems to have been used for hues, shades, and tints from pink to dark red and from purple to orange. Furthermore, not only do vegetative pigments shift in color with age, but mineral pigments may also shift, and these changes are increased with humidity, heat, and light (Feller 1986; Eastaugh et al. 2008; Finlay 2014).

The *Voynich Codex* utilizes a number of other uncommon iconographic techniques:

1. A flat, two-dimensional representation of plants reminiscent of pressed specimens, e.g., folio 100r#4.
2. A foreshortening of large plants in which young shoots appear "grafted or inserted" upon older bases, e.g., folios 1v, 13r, 16v, 23r, 26r, and 45r (Kennedy and Churchill 2006).
3. Anthropomorphic faces among the roots, e.g., folio 33r.
4. Reptiles and amphibians among the roots and leaves, e.g., folios 25v and 49r.
5. Enlarged organs out of proportion to the rest of the plant, e.g., folio 40v.
6. A mixture of accurate botanical details versus crude representations, e.g., folio 9v versus folio 16v.

These methods of plant illustration were not those of the native Nahua of pre-conquest New Spain, so what were their origins? Fray Motolinía, one of the 12 Franciscan priests who accompanied Hernán Cortés, remarked that the Nahua were extremely talented at copying Latin and Greek manuscripts, so much so that the original and the copy were indistinguishable (Motolinía 1951). The Spanish friars routinely used European images of Biblical figures as inspiration for Nahua *tlacuiloque* (indigenous artists) (Camelo Arredondo et al. 1964; Morrill 2014). Furthermore, the library of El Colegio de Santa Cruz de Tlatelolco (in present-day Mexico City; also called Colegio de Santa Cruz) had an eclectic collection of books, including 51 from Lyon, France; 51 from Paris; 35 from Venice, Italy; 22 from Salamanca, Spain; 20 from Antwerp, Belgium; 19 from Basel, Switzerland; 13 from Mexico; 11 from

Cologne, Germany; 19 from Alcala de Henares, Spain; and 22 from Santiago de Tlatelolco, Mexico (Gravier 2011). Fray Juan de Zumárraga, the bishop of New Spain, created the Colegio for the Nahua elite after the Spanish conquest.

The style of plant illustrations in the Herbal section of the *Voynich Codex*, such as individual, naturalistic drawings interspersed with text, was common in ancient Dioscoridean illustrated herbals, such as the *Juliana Anicia Codex* of 512 (Collins 2000; Janick and Hummer 2012) and were continuously recopied for a millennium. Naturalism was reintroduced in the herbals of the Secreta Saliterniana, an Arab-influenced school based in Salerno, which also contained many of the iconographic features of the *Voynich Codex*. These Salerno-inspired herbals ranged in time from the *Tractatus de Herbis* (British Library, Ms. Egerton 747) of the first third of the fourteenth century, to the many variations of the *Livre des Simples Médecines* of the end of the fourteenth century to the first half of the fifteenth century (Collins 2000). The Salerno school had a great impact on European herbals before the conquest of New Spain. There may have been European herbals in the well-stocked library of the Colegio de Santa Cruz, which influenced the *Voynich Codex*, in addition to other herbals executed by native *tlacuiloque* (indigenous artists). However, caution must be exercised until antecedent herbals can be identified (Emmart 1940; Peterson 1988).

The phytomorph identifications of plants as indigenous to the New World is conclusive evidence that the *Voynich Codex* must be a post-Columbian, Mesoamerican manuscript. There is no other possible explanation. The misidentification of the *Voynich Codex* as a European manuscript of the fifteenth century does not withstand botanical scrutiny. We do not claim 100% accuracy as to species, but there is no doubt that the analysis confirms O'Neill's identification (in 1944) of the presence of New World plants in the *Voynich Codex*. The wide variety of phytomorphs suggests that the vast Aztec botanical gardens, in addition to the extensive trade in dried herbs, were among sources for many of the plants.

Literature Cited

Alcántara Rojas, B. 2008. *Nepapan Xochitl*: The power of flowers in the works of Sahagún. In *Colors between two worlds: The Florentine codex of Bernardino de Sahagún*, ed. L.A. Waldman, 106–132. Florence: Villa I Tatti.

Al-Shehbaz, I.A. 2012. Brassicaceae. In *The Jepson manual. Vascular plants of California*, ed. B.G. Baldwin, D.H. Goldman, D.J. Keil, R. Patterson, T.J. Bosatti, and D.H. Wilken, 2nd ed., 512–577. Berkeley: University California Press.

Barlow, M. 1986. The Voynich manuscript—By Voynich? *Cryptologia* 10: 210–216.

Bayer, R.J., A.L. Bogle, and D.M. Cherniawsky. 2006. *Petasites*. In *Flora of North America North of Mexico*, ed. Flora of North America Editorial Committee, 635–637. Vol. 20. Magnoliophyta: Asteridae, Part 7: Asteraceae, Part 2. New York: Oxford University Press.

Berg, C.C. 2001. Moreae, Artocarpeae, and *Dorstenia* (Moraceae). *Flora Neotropica* 83: 1–346.

Blackman, B.K., M. Scascitelli, N.C. Kane, H.H. Luton, D.A. Rasmussen, R.A. Bye, D. Lentz, and L.H. Rieseberg. 2011. Sunflower domestication alleles support single domestication center in eastern North America. *Proceedings of the National Academy of Sciences of the United States of America* 108: 14360–14365.

Bleiler, E.F., ed. 1976. *Early floral engravings: All 110 plates from the 1612 "florilegium" by Emanuel Sweerts*. New York: Dover Publications.
Borschenius, F., L.S. Suárez Suárez, and L.M. Prince. 2012. Molecular phylogeny and redefined generic limits of *Calathea* (Marantaceae). *Systematic Botany* 37: 620–635.
Bown, D. 1988. *Aroids, plants of the arum family*. Portland: Timber Press.
Bye, R.A., and E. Linares. 2013. Códice de la Cruz-Badiano: Medicine préhispánica. Primera parte. *Arqueología Mexicana* 50: 7–91.
Bye, R.A., E. Linares, and D.L. Lentz. 2009. México: Centro de origen de la domesticación del girasol. *Tip Revista Especializada en Ciencias Quimico-Biológicas* 12: 5–12.
Camelo Arredondo, R., J. Gurría Lacroix, and C. Reyes Valerio. 1964. Juan Gerson, Tlacuilo de Tecamachalco. In *Departamento de Monomuntos Coloniales*. México: Instituto Nacional de Anthropolgia e Historia.
Chaumeton, F.P. 1830. *Flore médicale*. Vol. 3. Paris: C.L.F. Panckoucke.
Clausen, J., R.B. Channell, and U. Nur. 1964. *Viola rafinesquii,* the only Melanium violet native to North America. *Rhodora* 66: 32–46.
Clayton, M., L. Guerrini, and A. de Avila. 2009. *Flora: The Aztec herbal*. London: Royal Collection Enterprises.
Collins, M. 2000. *Medieval herbals: The illustrative traditions*. London: The British Library.
Compton, J.A., A. Culham, and S.L. Jury. 1998. Reclassification of *Actaea* to include *Cimicifuga* and *Souliea* (Ranunculaceae): Phylogeny inferred from morphology, nrDNA ITS, and cpDNA *trn*L-F sequence variation. *Taxon* 47: 593–634.
Cooper, M.A., and F.C. Hawthorne. 2000. Boleite: Resolution of the formula $KPb_{26}Ag_9Cu_{24}Cl_{62}(OH)_{48}$. *The Canadian Mineralogist* 8: 801–808.
Copper Development Association. 1971. *Uses of copper compounds. CDA technical note TN11*. Hemel Hempstead: Copper Development Association http://www.copperinfo.co.uk/copper-compounds/downloads/tn11-uses-of-copper-compounds.pdf.
De Ávila Blomberg, A. 2012. Yerba del coyote, veneno del perro: La evidencia lexica para identificar plantas en el Códice de la Cruz Badiano. *Acta Botánica Mexicana* (100): 489–526.
De Batres, L.D.P., C.A.B. Alfaro, and J. Ghaemghami. 2012. Mesoamerica aesthetics: Horticultural plants in hair and skin care. *Chronica Horticulturae* 50 (2): 12–15.
De la Cruz, M., and J. Badianus. 1991. *Libellus de medicinalibus indorum herbis*. México: Fondo de Cultura Económica.
De Motolinia, T. 1951. Motolinia's history of the Indians of new Spain. Trans. F.B. Steck. Washington, DC: Academy of American Franciscan History.
De Sahagún B. 1963. *Florentine Codex. General history of the things of New Spain. Book 11—Earthly things*. Trans. C.E. Dibble and A.J.O. Anderson. Salt Lake City: University of Utah Press.
Díaz, J.L. 1976. Índice y sinonimia de las plantas medicinales de MéxicoInstituto Mexicano para el Estudio de las Plantas, México.
Dressler, R.L. 1953. The pre-Columbian cultivated plants of Mexico. *Botanical Museum Leaflets* 16: 115–172.
Dunn, D.B., and W.E. Harmon. 1977. The *Lupinus montanus* complex of Mexico and Central America. *Annals of the Missouri Botanical Garden* 64: 340–365.
Eastaugh, N., V. Walsh, T. Chaplin, and R. Siddall. 2008. *Pigment compendium: A dictionary and optical microscopy of historical pigments*. London: Routledge.
Emmart, E.W. 1940. *The Badianus manuscript*. Baltimore: Johns Hopkins Press.
Farfán, J.A.F., and J.G.R. Elferink. 2010. *Ethnobotany and Aztec sexuality*. Muenchen: Lincom Europa.
Feller, R.L., ed. 1986. *Artists' pigments. A handbook of their history and characteristics*. Vol. 1. Washington, DC: National Gallery of Art.
Fernald, M.L. 1938. Noteworthy plants of southeastern Virginia. *Rhodora* 40: 434–459.
Finlay, V. 2014. *The brilliant history of color in art*. Los Angeles: J. Paul Getty Museum.

Friedman, E.S. 1962. 'The most mysterious MS' still an enigma. *Washington Post* 5 August: E1, E5.
Gates, W. 2000. *An Aztec herbal*. New York: Dover Publication (First publ. Maya Society. 1939. Baltimore).
González Stuart, A. 2004. Plants used in Mexican traditional medicine. http://www.herbalsafety.utep.edu/documents/Plants%20Used%20in%20Mexican%20Traditional%20Medicine-July%2004.pdf.
Granziera, P. 2001. Concept of the garden in pre-Hispanic Mexico. *Garden History* 29: 185–213.
———. 2005. Huaxtepec: The sacred garden of an Aztec emperor. *Landscape Research* 30: 81–107.
Gravier, M.G. 2011. Sahagún's codex and book design in the indigenous context. In *Colors between two worlds*, ed. G. Wolf and J. Connors, 156–197. Milan: Florentine Codex of Bernardino de Sahagún. Officina Libraria.
Hancock, J.F. 2012. *Plant evolution and the origin of crop species*. Wallingford: CABI.
Harter, A., K.A. Gardner, D. Falush, D.L. Lentz, R.A. Bye, and L.H. Rieseberg. 2004. Origin of extant domesticated sunflowers in eastern North America. *Nature* 430: 201–205.
Heiser, C.B. 2008. How old is the sunflower in Mexico? *Proceedings of the National Academy of Sciences of the United states of America* 105: E48.
Hernández, F. 1942. Historia de las Plantas de Nueva España. Ed. I. Ochoterena. Imprenta Universitaria. Mexico.
Hernández, F., F. Celsi, F. Colonna, B. Deversini, J. Faber, J. Greuter, V. Mascardi, N. A. Recchi, and J. Terentius. 1651. *Rerum medicatum Novae Hispaniae Thesarus, seu, Plantarum animalium mioneralium Mexicanorum historia*. Vitalis Mascardi, Romae.
Hough, W. 1908. The pulque of Mexico. *Proceedings of the United States National Museum* 33: 577–592.
Janick, J., and K.E. Hummer. 2012. The 1500th anniversary 512–2012 of the Juliana Anicia codex: An illustrated Diorcoridean recension. *Chronica Horticulturae* 52 (3): 9–156.
Kennedy, G., and R. Churchill. 2006. *The Voynich manuscript*. Rochester: Inner Traditions.
Killip, E.P. 1938. The American species of Passifloraceae. *Fieldiana* 19: 1–613.
Lentz, D.L., M.D. Pohl, J.L. Alvarado, S. Tarighat, and R. Bye. 2008a. Sunflower (*Helianthus annuus* L.) as a pre-Columbian domesticate in Mexico. *Proceedings of the National Academy of Sciences United States of America* 105: 6232–6237.
Lentz, D.L., M.D. Pohl, and R. Bye. 2008b. Reply to Rieseberg and Burke, Heiser, Brown, and Smith: Molecular, linguistic, and archaeological evidence for domesticated sunflower in pre-Columbian Mesoamerica. *Proceedings National Academy of Sciences of the United states of America* 105: E59–E50.
Mali, P.Y., and S.S. Panchal. 2013. A review on phyto-pharmacological potentials of *Euphorbia thymifolia* L. *Ancient Science Life* 32: 165–172.
Mallard, E., and E. Cumenge. 1891. Sur une nouvelle espèce minérale, la Boléite. *Bulletin de la Societe Francaise de Mineralogie et de Cristallographie* 14: 283–293.
Mathias, M.E., L. Constance. 1941. A synopsis of North American species of *Eryngium*. *The American Midland Naturalist* 24: 361–387.
McPherson, G. 1981. Studies in *Ipomoea* (Convolvulaceae) I. The Arborescens group. *Annals of the Missouri Botanical Garden* 68: 527–545.
Meira, M., E.P. da Silva, J.M. David, and J.P. David. 2012. Review of the genus *Ipomoea:* Traditional uses, chemistry and biological activities. *Revista Brasileira de Farmacognosia* 22: 682–713.
Meyer, F.C. 1951. *Valeriana* in North America and the West Indies (Valerianaceae). *Annals of the Missouri Botanical Garden* 38: 377–503.
Mickel, J.T., and A.R. Smith. 2004. *The pteridophytes of Mexico*. Garden: New York Bot.
Millspaugh, C.F. 1892. *American medicinal plants*. Philadelphia: J.C. Yorston & Co.
Moody, M.L., and L.H. Rieseberg. 2012. Sorting through the chaff, nDNA gene trees for phylogenetic inference and hybrid identification of annual sunflowers (*Helianthus* sect. *Helianthus*). *Molecular Phylogenetics Evolution* 64: 145–155.

Morrill, P.C. 2014. *The casa del Deán: New World imagery in a sixteenth century Mexican mural cycle*. Austin: University Texas Press.
Morton, J.K. 2005. *Silene*. In *Flora of North America*, ed. Flora of North America Editorial Committee, 168–216. Vol. 5. Magnoliophyta: Caryophyllidae, part 2. New York: Oxford University Press.
Mottana, A., R. Crespi, and G. Liboro. 1978. In *Guide to rocks and minerals*, ed. M. Prinz, G. Harlow, and J. Peters. New York: Simon & Schuster.
Navarrete, A., D. Niño, B. Reyes, C. Sixtos, E. Aguirre, and E. Estrada. 1990. On the hypocholesteremic effect of *Eryngium heterophyllum*. *Fitoterapia* 61: 183–184.
Nieland, LaShara, and Willa Finley. 2009. *Lone Star Wildflowers*, 164–165. Lubbock: Texas Tech University Press.
Nutttall, Z. 1925. The gardens of ancient Mexico. *Annual Report of the Board of Regents of the Smithsonian Institution* 1923: 453–464.
O'Neill, H.T. 1940. The sedges of the Yucatan peninsula. *Carnegie Institute Washington Publication* 522: 247–322.
O'Neill, H. 1944. Botanical observations on the Voynich MS. *Speculum* 19: 126.
Ocaranza, F. 2011. *Historia de la medicina en México*. 2nd ed. Cauhtémoc: Cien de México.
Park, M.S., and P.E. Elvander. 2012. Saxifragaceae. In *The Jepson manual. Vascular plants of California*, ed. B.G. Baldwin, D.H. Goldman, D.J. Keil, R. Patterson, T.J. Bosatti, and D.H. Wilken, 2nd ed., 1234–1244. Berkeley: University of California Press.
Pereda-Miranda, R., D. Rosas-Ramírez, and J. Castañeda-Gómez. 2010. Resin glycosides from the morning glory family. *Fortschritte der Chemie Organischer Naturstoffe* 92: 77–153.
Peterson, J.F. 1988. The Florentine codex imagery and the colonial tlacuilo. In *The work of Bernardino de Sahagun, pioneer ethnographer of sixteenth-century Aztec Mexico*, ed. J. Klor de Alva, H.B. Nicholson, and E. Quiñones Keber, 273–293. Austin: Univeristy of Texas Press.
Plant List. 2013. Version 1.1. http://www.theplantlist.org/.
Prather, A. 1999. Systematics of *Cobaea* (Polemoniaceae). *Systematic Botanical Monographs* 57: 1–81.
Ralph, J., and I. Chau. 2012. Boleite. http://www.mindat.org/min-712.html.
Ramírez, J., and G.V. Alcocer. 1902. *Sinonimia vulgar y científica de las plantas mexicanas*. Mexico: Oficina Tipográfica de la Secretaría del Fomento.
Reko, B.P. 1947. Nombres bótanicos del manuscrito Badiano. *Boletín de la Sociedad Botánica de México* 5: 23–43.
Richards, J.H., M. Dow, and T. Troxler. 2010. Modeling *Nymphoides* architecture: A morphological analysis of *Nymphoides aquatica* (Menyanthaceae). *American Journal of Botany* 97: 1761–1771.
Rieseberg, L., and J.M. Burke. 2008. Molecular evidence and the origin of the domesticated sunflower. *Proceedings of the National Academy of Sciences of the United States of America* 105: E46.
Ross-Ibarra, J., and A. Molina-Cruz. 2002. The ethnobotany of chaya (*Cnidoscolus aconitifolius* ssp. *aconitifolius* Breckon): A nutritious Maya vegetable. *Economic Botany* 56: 350–365.
Ruiz-López, M.A., P.M. Garcia-López, R. Rodríguez, and J.F. Zamora Natera. 2010. Mexican wild lupines as a source of quinolizidine alkaloids of economic potential. *Polibotánica* 29: 159–164.
Russell, N.H. 1965. Violets (*Viola*) of central and eastern United States: An introductory survey. *Sida* 2: 1–113.
Sherwood, E., and E. Sherwood. 2008. The Voynich botanical plants. http://www.edithsherwood.com/voynich_botanical_plants/index.php. 26 July 2012.
Shinners, L.H. 1961. *Viola rafinesquii*: Nomenclature and native status. *Rhodora* 63: 327–335.
Stafleu, F. A., and R. S. Cowan. 1979. *Taxonomic literature: A selective guide to botanical publications and collections, with dates, commentaries and types*. Vol. 2, H-Le. Utrecht: Bohn, Scheltema & Holkema.
Standley, P.C. 1914. A revision of the genus *Cobaea*. *Contributions from the United States National Herbarium* 17: 448–458.
———. 1920–1926. Trees and shrubs of Mexico. *Contributions from the United States National Herbarium* 23: 1–1721.

Standley, P.C., and L.O. Williams. 1973. Labiatae, mint family. In *Flora of Guatemala*, eds. P.C. Standley and L.O. Williams, Part IX, no. 3. *Fieldiana* 24: 237–317.

Taiz, L., and S.L. Taiz. 2011. The biological section of the Voynich manuscript: A textbook of medieval plant physiology. *Chronica Horticulturae* 51 (2): 9–23.

Tucker, A.O., and R.H. Talbert. 2013. A preliminary analysis of the botany, zoology, and mineralogy of the Voynich manuscript. *HerbalGram* 100: 70–85.

Tucker, A.O., and J. Janick. 2016. Identification of phytomorphs in the Voynich codex. *Horticultural Reviews* 44: 1–64.

Tucker, A.O., M.E. Poston, and H.H. Iltis. 1989. History of the LCU herbarium, 1895–1986. *Taxon* 38: 196–203.

USDA, ARS. 2015. National Genetics Resources Program. Germplasm resources information network – GRIN [online database]. National Germplasm Resources Laboratory, Beltsville, Maryland. http://www.ars-grin.gov/cgi-bin/npgs/html/taxgenform.pl.

Velinska, E. 2013. The Voynich Manuscript: Plant id list. ellievelinska.blogspot.com/2013/07/the-voynich-manuscript-plant-id-list.html. 5 June 2017.

Zepeda, G.C., and L.O. White. 2008. Herbolaria y pintura mural: Plantas medicinales en los murals del Convento del Divino Salvador de Malinalco, Estado de México. *Polibotanica* 25: 173–199.

Chapter 5
Phytomorphs in the Pharmaceutical Section: The Rosetta Stone of the Voynich Codex

Arthur O. Tucker and Jules Janick

The Decipherment of Lost Languages

The language of the *Voynich Codex* (referred to here as Voynichese) has defied decipherment since it was brought to public attention in 1912. The text as shown in Fig. 2.6 of Chap. 2 appeared to be a combination of an unknown alphabet or syllabary in an unknown language by an unknown author (Brumbaugh 1978; Kennedy and Churchill 2006). In 1950, William F. Friedman, renowned cryptologist of the US National Security Agency, extensively analyzed the text and indicated that it has all of the qualities of a real language and is not a cipher (Tiltman 1967; Reeds 1995). Rugg (2004a, b) claimed that overlaying a sixteenth century Cardan Grille "proves" that the *Voynich Codex* text was written in gibberish. Rugg was supported by Schinner (2007), but disproven by Hermes (2012). Meanwhile, modern computer analyses have found that the text is, indeed, in a language, perhaps more than one, supporting Friedman's initial assessment. Antoine Casanova (1999; not cited by Rugg) stated: "*We regard the manuscript as one single text or as a conglomerate of cryptograms endowed with six separate alphabets*." Landini (2001) stated: [By] "*spectral analysis … findings shown here favor the natural language theory*." Montemurro and Zanette (2013) found a real language sequence, not gibberish. They found semantic word networks and other statistical features. This was supported by Amancio et al. (2013), who stated that the text of the *Voynich Codex* is mostly compatible with natural languages and incompatible with random texts.

The approach to decipherment of any lost language is twofold:

1. The transliteration of the symbols into either an alphabet (letters that can be arranged to form sounds, usually by combining consonants and vowels) or a syllabary (in which the symbols represent syllables directly) with Latin letters.
2. The translation of the resulting words into a recognizable language.

J. Janick, A. O. Tucker, *Unraveling the Voynich Codex*, Fascinating Life Sciences, https://doi.org/10.1007/978-3-319-77294-3_5

In the past, the transliteration of lost languages was accomplished via proper names. Then, translation into a modern language was achieved via extant ancient languages. If a multilingual text were available, this provided additional intersecting evidence. A primary example was the decipherment of Egyptian hieroglyphics and hieratic on the Rosetta Stone, a stone stele found in 1799 near Rosetta, bearing parallel inscriptions in Egyptian hieroglyphics, demotic script, and ancient Greek. Hieroglyphs were used for more formal declarations, whereas hieratic was the written form used for secular texts, such as letters, business documents, poems, and stories. Hieratic was also the text in the cartouches. A third script, demotic or "popular," was used for mundane, daily purposes. Initially, Thomas Young (1773–1829), the British polymath, transliterated the royal name Ptolemy from the Rosetta Stone, but it was insufficient for a full decipherment. Meanwhile, Jean-François Champollion (1790–1832) visited Abu-Simbel, Egypt, and provided the additional royal name of Rameses, and then, coupled with his knowledge of Coptic, finally deciphered the Rosetta Stone (Jacq 1987; Champollion 2009; Robinson 2012).

Sumerian and Akkadian (Babylonian and Assyrian) were transliterated by many researchers finding the names of kings, and then translated with ancient Persian languages (Cottrell 1972; Ceram 1986; Gordon 1987). Linear B of Crete was transliterated as proto-Greek, or Mycenean, by Michael Ventris (1922–1956) after discovering additional material on mainland Greece and the names of kings. He was helped in the grammar of Linear B by Alice Kober (1906–1950), in conjunction with the Greek scholar John Chadwick (1920–1988) (Robinson 2002; Fox 2013). Mayan logograms were first deciphered by finding the names of cities, ruling dynasties, and gods, with preliminary efforts by Heinrich Berlin (1915–1988), but Yuri Knorosov (1922–1999) discovered the keys to decipherment, and this was further refined by Michael Coe (b. 1929) (Coe 2012). The *Popol Vuh* (also known as the Mayan Bible), along with surviving languages in Yucatan, provided additional materials for decipherment. Note that in all these decipherments, the transliteration of names required a body or "corpus" of literature. One name on a few fragments of text was insufficient.

The discovery of proper names attached to some plants in the Pharmaceutical section of the *Voynich Codex* shows some words to be orthographic variants of languages spoken in sixteenth century New Spain. The objective of the current chapter is to decode the symbols of the *Voynich Codex* as a first step in decipherment. This chapter is based on the seminal paper of Arthur O. Tucker and Rexford H. Talbert (2013). The Voynichese symbols were decoded by associating the labels attached to Pharmaceutical section plants with the names of plants in Taino, Spanish, or Nahuatl. These labeled phytomorphs may be considered the "Rosetta Stone" of the *Voynich Codex*. Although the decoded symbols helped to translate a number of words, including cities and cognates, the main text defied translation and Voynichese may be a mixed synthetic language or a mixed Nahuatl lingua franca of commerce (Dakin 1981).

Deciphering the Language Symbols of the Voynich Codex

The Voynichese Symbols

Voynichese contains about 32 strange characters or symbols (Knight 2009), but there has been no scholarly consensus on the number or value of the various symbols (see Chap. 11, Table 11.1). Because of the difficulty of typing them on a computer keyboard, Zandbergen () transposed them to about 77 keystrokes in a transcription font called the Extensible Voynich Alphabet (or EVA), which was formerly called the European Voynich Alphabet. Although it allows easy keyboarding of the Voynichese characters, this transliteration has no basis for pronunciation or meaning. Thus, in this presentation, to avoid confusion, Voynichese symbols have been retained (Table 5.1).

Table 5.1 Voynichese characters with proposed Latin equivalent letters or syllables. Voynichese character counts and percentage frequencies (out of 160,602 total characters) were calculated from Knight (2009). Selected seventeenth century Nahuatl incantations of a *ticitl* (doctor or seer) from Alcarón (1987, Sixth Treatise, Chapters 1–24) were typed and percentage frequencies were then calculated to obtain an approximation of early Nahuatl as a comparison (9451 total characters)

Voynichese character	Proposed Latin equivalent[a]	EVA	Voynichese character count and frequency	Count and frequency seventeenth century classical Nahuatl
o	ā	o	25,468 = 15.9%	ā = 609 = 6.4%
c	a	e	20,227 = 12.6%	a = 753 = 8.0%
9	i/y	y	17,655 = 11.0%	i = 1318 + y = 160 = 1478 = 15.6%
[illegible]	} tl	k	16,020 = 10.0%	tl = 443 = 4.7%
[illegible]		t		
a	o	a	14,281 = 8.9%	o = 685 = 7.2%
8	ch	d	12,973 = 8.1%	ch = 125 = 1.3%
[illegible]	m	ch	11,008 = 6.9%	m = 285 = 3.0%
[illegible]	câ	l	10,471 = 6.5%	ca = 180 = 1.9%
[illegible]	e	r	6,716 = 4.2%	e = 449 = 4.8%
[illegible]	qu/kw	q	5,423 = 3.4%	qu = 214 = 2.3%
[illegible]	ts/tz	Sh	4,501 = 2.8%	tz = 113 = 1.2%
[illegible]	ll	iin	4,076 = 2.5%	ll = 53 = 0.6%
2	n	s	2,886 = 1.8%	n = 741 = 7.8%
[illegible]	} cu	cKh	1,858 = 1.2%	cu = 85 = 0.9%
[illegible]		cTh		
[illegible]	} hu/gu	f	1,844 = 1.1%	hu = 306 + uh = 155 = 461 = 4.9%
[illegible]		p		
[illegible]	l	in	1,752 = 1.1%	l = 710 = 7.5%

(continued)

Table 5.1 (continued)

Voynichese character	Proposed Latin equivalent[a]	EVA	Voynichese character count and frequency	Count and frequency seventeenth century classical Nahuatl
	yâ/hâ	m	1,046 = 0.7%	ya = 13 + ha = 2 = 15 = 0.2%
	c/k	ir	591 = 0.4%	c = 496 = 4.9%
	?	x	524 = 0.3% (listed by Knight as *?)	
	sh/x	i	316 = 0.2%	x = 206 = 2.2%
	} p	cFh	291 = 0.2%	p = 148 = 1.6%
		cPh		
	z/ç	n	157 = 0.1%	z = 351 = 3.7%
	t	iiin	156 = 0.1%	t = 545 = 5.8%
	?	iir	148 = 0.1%	
	h	g	96 = 0.1%	h = 306 = 3.2%
	?	im	52 = <0.1%	
	?	il	31 = <0.1%	
	?	iim	17 = <0.1%	
	?	iil	14 = <0.1%	
	?	iiil	2 = <0.1%	
	?	iiim	1 = <0.1%	
	?	iiir	1 = <0.1%	
^		v		

EVA Extensible Voynich Alphabet
[a]Decipherment not underlined is tentative

Clemens (2016:xiii), in an essay on the *Voynich Codex*, has stated that, "*the letters are altogether different from any known exemplar. At present we know of no other scripts on which the scribe might have drawn as a model.*" However, some symbols in the *Voynich Codex* show similarities to letters found in sixteenth century codices from New Spain (Tucker and Talbert 2013; Comegys 2013) particularly the *Codex Osuna* (Valderrama 1600; Chávez Orozco 1947). For example, the four most unique symbols in the *Voynich Codex* (, 2, 8, and) show similarities to letters in the *Codex Osuna* written in Spanish in courtesan script, the handwriting of sixteenth century New Spain, along with some scribal abbreviations of the era (Fig. 5.1). The symbol is similar to a scribal abbreviation for "de," but we propose that this should be pronounced "tl" in the *Voynich Codex* (Fig. 5.1a). Throughout the *Codex Osuna*, the Nahuatl "h" (Fig. 5.2b) is written as a large, conspicuous backward version of that found in the *Voynich Codex* 2, with the new pronunciation of "n" (Fig. 5.1b). In folio 475-13v of the *Codex Osuna*, the florid Spanish signatures (Fig. 5.1c) have several inspirations for the in the *Voynich Codex* with the new

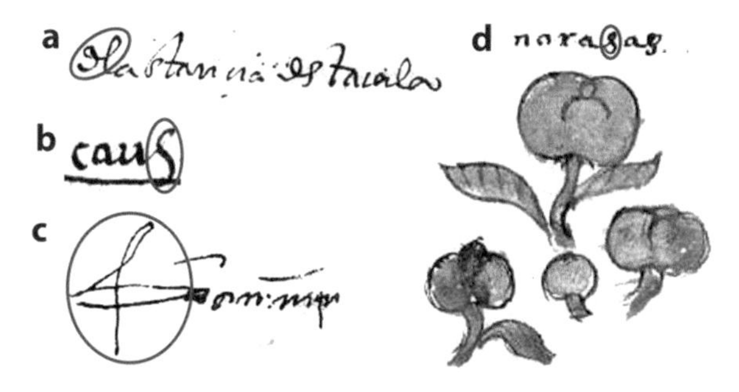

Fig. 5.1 Calligraphy in the *Codex Osuna* (Chávez Orozco 1947) with corresponding symbols from the *Voynich Codex*. (**a**) Scribal abbreviation for "de" merged with "l" from the *Codex Osuna* matching the Voynichese character ʮ (*dela stancia destacalco,* folio 474-12v); (**b**) "h" from the *Codex Osuna* when reversed matches the Voynichese 2 (*cauh*, folio 469-7r); (**c**) a florid Spanish signature; (**d**) "s" sound in the word *narasas* (*Codex Osuna*, folio 474-12v) resembles the Voynichese character 8, which is pronounced "ch"

Fig. 5.2 Cherokee syllabary created by Sequoyah (also known as George Gist) in 1821 by modifying letters from Latin, Greek, and Cyrillic that he had encountered

pronunciation of "hu/gu." In folio 474-12v of the *Codex Osuna*, the "s" sound in the word *narasas* accompanying the picture of orange fruits (Fig. 5.1d) resembles the 8 in the *Voynich Codex*, which has the pronunciation "ch."

Mary E. D'Imperio (1978/1979) briefly discussed brachygraphy, the history of shorthand, and its relation to the *Voynich Codex*. Although prescient, D'Imperio did not pursue this further and investigate scribal abbreviations of sixteenth century Spain and New Spain. Also, she did not mention that this shorthand is as personal as handwriting and can indicate the school or even individual scribe. A number of scribal abbreviations and calligraphy of the era, in addition to alchemic symbols, show remarkable similarity to the Voynichese symbols (Table 5.2). For example, ʮ, one of the most enigmatic Voynichese symbols, may be derived from a scribal abbreviation for "de" common in sixteenth century Spain and New Spain. The similarity of these symbols indicates that the author of the *Voynich Codex* did not have to invent them, but merely to borrow or to modify calligraphy from other documents in New Spain. Some of the Voynichese symbols (e.g., *4*, 8, 9, ʮ, ʯ) are also found in alchemy (Gessmann 1922).

Table 5.2 Possible sources of *Voynich Codex* symbols

Voynichese symbol	Source	Example	References
a	Calligraphy	a	Muñoz y Rivero (1917:58)
o	Calligraphy	o	Muñoz y Rivero (1917:58)
c	Calligraphy	c	Muñoz y Rivero (1917:58)
[illegible]	Calligraphy	[illegible]	Muñoz y Rivero (1917:58)
9	Scribal abbreviation	9	Muñoz y Rivero (1917:92)
[illegible]	Scribal abbreviation	[illegible]	Ayala (1990:58/117,113)
	Alchemic symbols	[illegible]	Gessmann (1922)
8	Scribal abbreviation	8	Muñoz y Rivero (1917:95)
[illegible]	Scribal abbreviation	[illegible]	Muñoz y Rivero (1917:58)
[illegible]	Scribal abbreviation	[illegible]	Muñoz y Rivero (1917:95)
2	Scribal abbreviation	2	Muñoz y Rivero (1917:95)
[illegible]	Scribal abbreviation	[illegible]	Ayala (1990:57/49)

Although some of the symbols of Voynichese appear unique, they are similar in concept to the Cherokee syllabary created by Sequoyah (also known as George Gist) in 1821 by modifying letters from Latin, Greek, and Cyrillic that he had encountered (Fig. 5.2). The resulting Cherokee syllabary did not carry along the pronunciations from the original language, and many new characters were created that had no direct correspondence with existing languages, but were obviously inspired by them.

Associations of Phytomorphs with Plant Names

We hypothesize that the symbols of the *Voynich Codex* might have made up a syllabary/alphabet from letters borrowed from contemporary post-conquest Mesoamerican manuscripts, such as the *Codex Osuna*, and scribal abbreviations. In six folios (88r-89v and 99v-102v, containing 20 pages including foldouts) of the Pharmaceutical section, apothecary jars, roots, shoots, and one mineral are portrayed. There are at least 228 images of plants or plant parts, of which 188 are accompanied by names (see Chap. 2, Fig. 2.2, and Appendix E). The approach was to match proper names of identified phytomorphs with the names of plants in dialects of Nahuatl (Tucker and Talbert 2013).

The first plant name deciphered (from folio 100r#8; Table 5.3) was identified as *Opuntia ficus-indica* (Cactaceae) or a related species. It shows what is obviously a cactus pad (cladode) and is labeled [illegible]. In Nahuatl, this could either be *nopal,* the

Table 5.3 Critical phytomorphs whose proper names allowed the preliminary development of a Voynichese syllabary/alphabet

Code, folio, & plantNo. of *Voynich Codex*	**Voynichese**	**Phytomorph**	**Transliteration**	**Cognate, translation, and identification**
100r#8	=n =ā =sh/x =tl =i/y Red letters are new decipherments		nāshtli	*nochtli* (Nahuatl) = fruit ("*tuna*") of the prickly pear cactus or the plant bearing the fruit, most probably rooted cladode of *Opuntia ficus-indica*
100r#4	=m =ā =gu/hu =o =e =i/y		māguoey	*maguey* (Spanish from Taino mid-16th century), most probably *Agave atrovirens* Source: Sahagún
88r#12	=ā =gu/hu =o =câ =ch =ā		guocâchā	*agua* (Spanish) = water + *cacha* (Nahuatl) = callus, most probably a lupine (*Lupinus montanus*)

(continued)

Table 5.3 (continued)

Code, folio, and plant no. of *Voynich Codex*	Voynichese	Phytomorph	Transliteration	Cognate, translation, and identification
99v#10	[glyph] [glyph]=ch [glyph]=o [glyph]=ll [glyph] =a		cholla	*cholla* (Spanish) = skull, cactus unknown, perhaps root of *Cylindropuntia* sp.
100r#1	[glyph] [glyph]=m [glyph]=ā [glyph]=n [glyph]=o [glyph]=e o=ā [glyph]=ts/tz o=ā [glyph]=câ	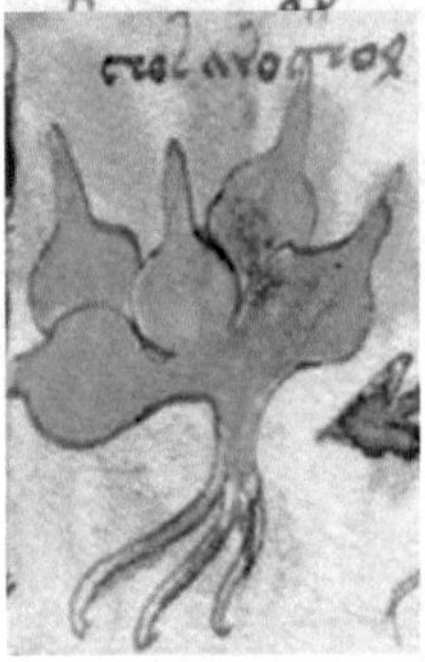	mānoetzācâ	*mano* (Spanish) = hand + *tzacua* (Nahuatl) = to close, enclose unknown, perhaps an *Agave* sp.?
100r#7	[glyph] [glyph]=m [glyph]=ā [glyph]=câ [glyph]=n [glyph]=o [glyph]=l		mācanol	*macana* (Taino) = obsidian & wooden sword similar to the Aztec *macuahuitl* (Nahuatl)  most probably *Philodendron goeldii* 

(continued)

Table 5.3 (continued)

Code, folio, and plant no. of *Voynich Codex*	Voynichese	Phytomorph	Transliteration	Cognate, translation, and identification
100r#14	[glyph] [glyph]=ā [glyph]=câ [glyph]=m [glyph]=a [glyph]=ā [glyph]=yâ/hâ		ācâmaāya	*acamaya* (Nahuatl) = crab or crayfish, most probably fruit of *Gonolobus chloranthus*
100r#5	[glyph] [glyph] [glyph] [glyph]=n [glyph]=o [glyph]=a [glyph]=m [glyph]=o [glyph]=e [glyph]=ch [glyph]=o [glyph]=ll [glyph]=ch [glyph]=i/y		noe moe chollchi	*cholla* (Spanish) = skull + *chi* (Nahuatl) = root word for owl most probably three flowers of bladder sage or paperbag bush (*Scutellaria mexicana*)

name of the pad, or *nochtli*, the name of the fruit ("*tuna*" in Spanish) or plant with fruit (Díaz 1976). Nine extensions, which are clearly incipient fruits, are shown attached to the pad; thus, *nochtli* would seem to be the most appropriate. Coincidentally, *Opuntia ficus-indica* is often propagated by planting the cladodes in the ground, and the fruits are harvested green, later turning red (see Chap. 4, Fig. 4.21b). Later, with more transliterations, this was refined to *nāshtli,* an orthographic variant or a phonetic spelling of *nochtli*. This transliteration provided the sound of five Voynichese letters: [glyph] = n, [glyph] = ā, [glyph] = sh, [glyph] = tl, and [glyph] = i or y.

The second deciphered name was *Agave atrovirens* (Asparagaceae/Agavaceae). On folio 100r#4 what is obviously a young agave with serrated or spiny leaves, folded as pressed, is most probably *Agave atrovirens* (the sixteenth century source of pulque and tequila) and is labeled [glyph]. Putative Nahuatl names that have been applied to species of *Agave* include: *acametl, amole/amolli, macoztic metl,*

metl, metl coztli/mecoztli, mexcalmetl, nequametl, quametl, quetzalichtli, teohamatl/teoamatl, teometl, theometl, tlacametl, tlacoamatl/tlacametl, and *xolometl* (Hernández et al. 1651; Hernández 1942; Reko 1947; Dressler 1953; Sahagún 1963; Díaz 1976; Ocaranza 2011). However, none of these fits the word length properly. One word that entered sixteenth century Latin American Spanish from the Taino was *maguey*, and this fits rather well as *māguoey*, again an orthographic variant or phonetic spelling. Thus, transliteration of *māguoey* ([Voynich glyph] = m, [Voynich glyph] = ā, [Voynich glyph] = gu/hu, [Voynich glyph] = o, [Voynich glyph] = e, [Voynich glyph] = i/y) preserves the transliteration of [Voynich glyph] and [Voynich glyph] and transliterates four new symbols: [Voynich glyph], [Voynich glyph], [Voynich glyph], and [Voynich glyph]. Note that both the Nahuatl, *nochtle,* and the Spanish loan word, *maguey,* fit the primary "*folk generic names following the same line of logic,*" presented by Tucker and Talbert (2013). Subsequent plant names that revealed new transliterated letters included the following:

Lupinus montanus (Fabaceae), a lupine on folio 88r#12 and labeled [Voynich glyphs], transliterated as *āguocâchā*, assuming [Voynich glyph] = câ and [Voynich glyph] = ch, and possibly derived from *agua* (Spanish), water + *cacha* (Nahuatl), callus, i.e., watery callus. In fact, the nitrogen-fixing nodules on one side of the roots of this lupine are callus-like and watery in texture. This would be an example of a descriptive phrase.

Folio 99v#100 has a root labeled [Voynich glyphs] and the new symbol was transliterated as [Voynich glyph] = ll, producing *cholla* (Spanish), skull or cactus. Thus, the phytomorph may be the root of a *Cylindropuntia* sp. (Cactaceae).

Agave sp. (Asparagaceae/Agavaceae), an agave on folio 100r#1 labeled [Voynich glyphs], *mānoeātzācâ,* may be derived from *mano* (Spanish), hand + *tzacua* (Nahuatl), to close or enclose, and this phytomorph does resemble a closed hand.

Philodendron goeldii (Araceae), a philodendron on folio 100r#7, is labeled [Voynich glyphs] and the transliteration of [Voynich glyph] = l produces *mācânol,* probably derived from *macana* (Taino), a wooden/obsidian sword. Indeed, the phytomorph does resemble the unique *macuahuitl* (Nahuatl term for wooden club) of the Aztecs.

Gonolobus chloranthus (Apocynaceae), an asclepiad on folio 100r#14, is labeled [Voynich glyphs], and the transliteration of the new symbol [Voynich glyph] = yâ/hâ yields *ācâmăyâ.* This was probably derived from *acamaya* (Nahuatl), crab or crayfish, and the knobby fruit does resemble the shell of a crab.

Scutellaria mexicana, Lamiaceae, bladder sage or paper bag bush, on folio 100r#5, is labeled [Voynich glyphs], transliterated as *noe, moe choll chi.* Note that this transliteration is based on previous decoding. The [Voynich glyphs] may be derived from *cholla* (Spanish), skull + *chi* (Nahuatl), the root word for owl. The flowers of bladder sage consist of a white to green to rosy calyx with a protruding dark blue corolla, somewhat resembling the head of an owl with its white skull and black beak. The words *noe, moe* have not yet been translated.

Development of a Syllabary/Alphabet

A tentative syllabary/alphabet (Table 5.4) was produced based on these initial decipherments shown in red in Table 5.3, in addition to others from plants, animals, a mineral boleite, and cities using similar procedures. Diacritical marks were added

Table 5.4 Selected words in Voynichese with cognates in Classical Nahuatl of sixteenth century New Spain

Folio	Voynichese symbols	Transliteration	Cognates and translation
1v, 86v		ācâ	*aca* (Nahuatl) = someone
99v		ācâchi	*Acachi* is located in the municipality of Urique in the Mexican state of Chihuahua
86v		ācân	*acan* (Nahuatl) = in any place, in any part
88r		āhu/guintlchocâ	*ahuitl* + *choca* (Nahuatl) = to weep for an aunt
86v		āhu/guoe	*ahuoj* (Nahuatl) = dry arroyo
88r		ānocâ	*Anoca* is located in the municipality of Techaluta de Montenegro in the Mexican state of Jalisco
86v		ātlā	*atla* (Nahuatl) = place of abundant water
86v, 1r		ātlācâ	*atlaca* (Nahuatl) = mean people
102r		ātlācuo	*atlacui* (Nahuatl) = to draw water
88r		ātlāematli	*atla* (Nahuatl) = place of abundant water + *matli* (Nahuatl) = animal front leg
99v		ātlami	*atlamica* (Nahuatl) = drowning
88v		ātlātlācâ	*atlatlac* (Nahuatl) = flooded (field)
88r, 99v		ātlātli	*atlatl* (Nahuatl) = spear-thrower
1r, 86v, 99r		ātli	*atli* (Nahuatl) = to drink water or cocoa
86v		āyâ/hâ	*aya* (Nahuatl) = not yet
1v		câācâ	*caca* (Nahuatl) = frog, toad
88v		chācâi	*chacah* (Nahuatl) = tip of a tree
86v		chālli	*chicalli* (Nahuatl) = spittle
1r, 1v		chocâ	*choca* (Nahuatl) = to weep, cry; for animals to make various sounds
100v		chocâni	*chocani* (Nahuatl) = a weeper
1r, 1v		cuācâ	*cuac* (Nahuatl) = at the end, at the top, after
1v		cuācui	*cui* (Nahuatl) = to take something or someone (see *qua*) + *cui* (Nahuatl) = to eat
1v		itlocâ	*itloc* (Nahuatl) = by, with against
1v		mā	*ma* (Nahuatl) = preceding the imperative
1r, 1v		mācâ	*maca* (Nahuatl) = singular of *macamo*, no (before the imperative)
1r, 1v		moe	*mo* (Nahuatl) C = negative particle on its own in questions expressing doubt
99r		nocâā	*noca* (Nahuatl) = while

(continued)

Table 5.4 (continued)

Folio	Voynichese symbols	Transliteration	Cognates and translation
1v		otlocânācâmai	*-itloc* (Nahuatl) = adjacent to, close to + *ana* (Nahuatl) = to seize + *cama* (Nahuatl) = let
86v		paãe	*pa* (Nahuatl) = color, dye, paint
1r		pācâ	*paca* (Nahuatl) = to wash, bathe, launder something
1v		qua	*qua* (Nahuatl) = biting, eating someone
86v		tlācâ	*tlaca* (Nahuatl) = men
1v		tlāchi	*tlachia* (Nahuatl) = to look, see, or observe from a watchtower
1r		tlāe	*tla* (Nahuatl) = something, if
1r, 1v, 86v		ts/tzācâ	*tzaca* (Nahuatl) = to enclose, lock up
100r		ts/tzācuāi	*tzacua* (Nahuatl) = close, to enclose

based on Classical Nahuatl. A comparison of the frequency of symbols found in the *Voynich Codex* based on a total of 160,602 characters (Knight 2009) with the frequency of 9451 characters from Nahuatl incantations (Alcarón 1987) in this table demonstrated that Voynichese is similar to or includes Classical Nahuatl, but is not entirely in this dialect. Yet, a number of words in the *Voynich Codex* have cognates in Classical Nahuatl (Table 5.3), providing evidence that Voynichese includes this language. Furthermore, the use of "tl" and "chi" or "chy" endings places this language in central or northern Mexico (Canger 1988; Lacadena 2008).

The transliteration of the proper names in historical sources shows the same phoneme transformations, such as "gu" versus "hu," "ts" versus "tz," and "x" versus "sh," depending upon the source and time. For example, Sir Francis Drake sacked a town in April 1579, listed in contemporary accounts as Guatulco in the Gulf of Tehuantepec, Mexico, which is now written as Huatulco. We also made the assumption that the "looped" variants (sometimes referred to as gallows symbols), such as versus , versus , versus , and versus , may not be separate characters, but either represent sloppy calligraphy or more likely could reflect differences in pronunciation.

The general guidelines for rendering spoken Nahuatl into written Spanish were formulated by Catholic friars of the sixteenth and seventeenth centuries in New Spain (Nueva España). However, if the *Voynich Codex* was written by a layperson and not proofread by a Spanish priest, the transcribing of the Nahuatl phonemes may not have followed these generalities, further compounding the difficulty of transliterating Voynichese.

Voynichese and Problems in Decipherment

If the *Voynich Codex* is from central New Spain in the mid-sixteenth century, then we should also expect sixteenth century Latin American Spanish nouns, as indicated in stage 2 of Nahuatl (Lockhart 1992). In addition, along with dialects of Nahuatl and sixteenth century Latin American Spanish, and the above indications of Taino and Mixtec, we have also found some words in the text in Huastec and Arabic. However, we have not been able to find most of the Voynichese text in any dictionary of languages of New Spain that have survived from the sixteenth century.

According to Okrent (2010), in the past nine centuries about 900 synthetic languages, or "conlang," have been created. These were one of three types:

1. Languages completely from scratch called à priori languages.
2. Languages that take most of their material from natural languages, called à posteriori.
3. Languages that contain elements of both, called mixed.

The purposes of conlang have varied over the centuries, but in past years the overriding reason was secrecy. At present, films and gaming require synthetic languages.

The original hypothesis was that Voynichese was an extinct dialect of Nahuatl, but the current hypothesis assumes that it is a mixed synthetic language, i.e., cognates of extant languages in a synthetic language, essentially agreeing with William Friedman (1950, quoted in Tiltman 1967) of a Nahuatl lingua franca (Dakin 1981). Clearly, linguistic expertise is needed at this point. We have identified many nouns with labeled names, including 26 villages, 288 nymphs, 63 stars, 23 apothecary jars, and 207 plants in the Pharmaceutical section, and believe that this is a promising lead to follow (see Appendices A to E).

Cognates of Voynichese in Classical Nahuatl

Besides the proper names attached to the plants, animals, minerals, and cities, a number of other words in the *Voynich Codex* have cognates in Classical Nahuatl, providing evidence that Voynichese is not Classical Nahuatl, but is related to a dialect of Nahuatl. These are listed in Table 5.4. The use of "tl" and "chi" (or "chy") endings places this dialect of Nahuatl in central or northern Mexico (Canger 1988; Lacadena 2008).

Orthography of Voynichese and Comparison with Nahuatl

Nahuatl is the language of the Nahua, including the Mexica, the chief tribe of the Triple Alliance, popularly known as the Aztecs, and their predecessors. A language of the Uto-Aztecan language family, Nahuatl, has been spoken in central

Mexico since the seventh century CE. The conquest introduced the Latin alphabet to written Nahuatl. The Tenochtitlan variety is known as Classical Nahuatl. With almost 30 recognized dialects of Nahuatl today, it is difficult to apply general orthographic principles that cover all of the dialects. Upon surveying Nahuatl dictionaries and grammar books (Robinson 1969; Bierhorst 1985; Karttunen 1983; Canger 1988; Lockhart 1990, 2001; Sullivan 1997; Herrera 2004; Lacadena 2008; Launey 2011; Siméon 2010; Walters et al. 2002; Whittaker 2009), we found that the syllabary/alphabet that we developed for Voynichese agrees with the basics.

Pre-conquest Nahuatl was ideographic, and attempts to transcribe the Nahuatl phonemes into Spanish evolved into some customary orthographic principles for Nahuatl today. Short vowels are written without diacritical marks ("a," "e," "y/i," "o") in transcribing Nahuatl into Spanish. The letter "u" by itself is not part of Nahuatl, but the vowel "o" is sometimes written as "u" when the pronunciation approaches that of Spanish and English. Vowel length is conventionally unrepresented, but when it is written, the convention is that of the macron ("ā"). For typographical reasons, the circumflex ("â") is also used. Double vowels ("aa") are not uncommon, but typically belong to different syllables.

The Nahuatl phonemes "p," "t," "l/r," "n," and "m" are written as the Spanish ("p," "t," "l," "r," "n," "m"). The letter "s" is not used in Classical Nahuatl, but "c" before "e" or "i" is pronounced like the English "s," and "z" is also pronounced like the English "s" ("ç" may be used sometimes in place of "z" in transliteration). The Spanish "tl" (as a consonant rather than a full syllable), "ts/tz," "ch," "x," "cu" (not a syllable), and "hu/uh" are attempts to approximate the Nahuatl phonemes (Spanish does not have a "w," but some modern linguists now use the English "w" instead of the Spanish "hu"). The Nahuatl phoneme "k" is written as a "c" when preceding a back vowel or consonant and "qu" when preceding a front vowel. "H" is an attempt to approximate the Nahuatl phoneme and actually represent a "silent" glottal stop. The Nahuatl double "ll" is not pronounced like the Spanish "y," but rather as a long "l." Sometimes this double "ll" is lost and written as one "l." In some dialects, it is pronounced as "hl" and typically written as "jl." The letter "x" is pronounced in English as "sh." Further pronunciations of Nahuatl in the International Phonetic Alphabet (IPA) are given in a number of sources (e.g., Anonymous 2017).

As mentioned before, these are the general guidelines for rendering spoken Nahuatl into written Spanish that were formulated by the Catholic priests of the sixteenth and seventeenth centuries in New Spain. However, if the *Voynich Codex* was written by an amateur and not proofread by a Spanish priest, then we have to accept that the transcribing of the Nahuatl phonemes may not have followed these generalities, further compounding the difficulty of transliterating Voynichese. Additionally, a Nahuatl lingua franca, a language of commerce that became extinct as Spanish arose in ascendency, would not be expected to follow rules (Dakin 1981).

SpanNahuatl

Lockhart (1992) outlines three phases in the development of Nahuatl in New Spain. Stage 1 encompassed the arrival of the Spaniards in 1519 to ca. 1540–1550 and was characterized by no change in Nahuatl. Stage 2 extended from 1540 to 1600 to the mid-seventeenth century and saw massive borrowing of Spanish nouns, but the Nahuatl language remained unaltered in other aspects. Stage 3, from about 1640–1650 until today, involved a deeper and broader Spanish influence.

In the *Voynich Codex* there are various nouns that are either Spanish or combine both Nahuatl and Spanish (what we have called SpanNahuatl). For example, folio 88r#12 (Table 5.3) includes the fusion of Spanish and Nahuatl words, such as *aguocâchā* (*agua*, Spanish for water + *cacha*, Nahuatl for callus). In other examples of geographical names, folio 86v provides [Voynichese word] (or Huoxeatl(o)capi) as an alternative name for Huejotzingo (of today; alternatively called Huexucinco, Huexutcinco, Huexotzinco, or Guaxocingo in period documents and maps). This is a combination of a cognate of the Nahuatl *huexotl* for "willow" and the Spanish *capi* for "Latin American capital." These examples of proper names provide evidence that the *Voynich Codex* falls within stage 2 of the development of the Nahuatl language, i.e., from 1540–1550 to 1640–1650.

An Alphabetical Sequence of Voynichese Symbols

Folio 57v displays four women's heads and shoulders, each with an arm extended in a circle surrounding a "rosette" form (Fig. 5.3). Surrounding the figures are four rings of text. What is unusual and may be a key to decipherment is that ring 2 from the top contains 17 symbols repeated four times in the same sequence. Most symbols are not found in the text, but nine of them, [Voynichese symbols: o, ꭗ, 8, ʔ, ſſ, ꭗ, ꝑ, ꝑꝑ, 9], are very frequent in the *Voynich Codex*. The other symbols, many of which are rare or absent in the Voynich, are ignored here. The assumption was made that the sequence contained important information and the first conjecture was that the sequence was alphabetical. To test this, the Voynichese symbols decoded by Tucker and Talbert (2013) were put in sequential order based on the Spanish alphabet and compared with the sequence of folio 57v (Table 5.4). We were immediately struck by the fact that the first four symbols, [Voynichese symbols: o, ꭗ, 8, ʔ], in the sequence (1, 2, 3, 4) of folio 57v had the same relative sequence as decoded symbols in the Spanish alphabet sequence (1, 3, 4, 6).

We noted two similar symbols, [Voynichese symbol] and [Voynichese symbol], which had been assumed to be variant forms of [Voynichese symbol], the difference being a result of sloppy calligraphy. Their presence together on folio 56v indicated that two separate symbols were intentional. They were out of sequence assuming that they had a "tl" sound. However, the Nahuatl pronunciation of "tl" was given as follows: *"no equivalent in English. Pronounce by touching the back of the teeth with the tip of the tongue and releasing the air*

Fig. 5.3 Folio 57v of the *Voynich Codex* considered a "key to decipherment" containing four repeated lists of symbols

laterally (side of the tongue). Do not pronounce as kettle, sprinkle, or metal" (Herrera 2004).

It seemed therefore that the [symbol] symbol should not be under "t," but under the letter "l" in the English alphabet. Furthermore, the Spanish alphabet contains two letter combinations with the "l" sound: "l" and "ll," and it was conjectured that [symbol] referred to the Spanish letter "ll" and [symbol] referred to the Spanish letter "l." However, two other symbols, [symbol] = l and [symbol] = ll, may have similar pronunciations.

Another problem was the symbol [symbol], which we had decoded as "i/y." Transferring [symbol] from "i" to "y" in the alphabet improved the sequence. There was one other problem: the symbol [symbol] (pronounced "e") appears twice, 5th and 12th on folio 57v. We have no explanation why this symbol was recorded twice.

It was conjectured that these nine symbols taken together with pronunciations as above were close to an alphabetical sequence (Table 5.4). It suggests that the author, when inventing the code, started with the Spanish alphabet in a sequence from A to

Table 5.5 Spanish alphabet and corresponding Voynichese character and sequence compared with sequence in folio 57v

Sequence	Spanish alphabet	Spanish pronunciation	Voynichese symbol based on plant labels	Voynichese symbol in folio 57v	Sequence in folio 57v	Comments
1	a	(a) amigo	o	o^z	1	
2	b	(be) bonita				
3	c	(ce) cereal	ɣ	ɣ	2	
4	ch	(ch) hache chocolate	8	8	3	
5	d	(de) dedo				
6	e	(e) Español	ʔ	ʔ	4, 12	
7	f	(efe) feo				
8	g	(ge) gato				
9	h	(hache) hormiga	ɣ̃	ɣ̃	8	
			ꟻ, ꟻ	ꟻ	9	
10	i	(i) iglesia				See 28
11	j	(jota) José				
12	k	(ka) kilo	ıʔ			
13	l	(ele) lobo	ꝉꝉ or ıɔ	ꝉꝉ	7	
14	ll	(elle) lluvia	ꝉꝉ		11	
15	m	(eme) mama	cz			
16	n	(ene) no	ʔ			
17	ñ	(eñe) ñoño				
18	o	(o) ojo				
19	p	(pe) pelo	ꟻz, ꟻz			
20	q	(cu) quemar	4			
21	r	(erre) ratón				
22	s	(ese) soso	ˎ			
23	t	(te) tocar	ıııɔ			
			ꝯz			
24	u	(u) uva				
25	v	(uve) vamos				
26	w	(uve doble) whisky				
27	x	(equis) xilófono				
28	y	(i griega) yate	9	9	15	
29	z	(zeta) zorro	ɔ			

Tentative symbols are underlined

[a]Note: ɕ (Voynichese) is pronounced as a short "a"

Z. If this is correct, it supports most of the symbol decipherment of Tucker and Talbert (2013). However, we were concerned with one issue. Many of the cities in Mexico begin with the letter "t," such as Tecamachalco, Tenochtitlan, Tepeaca, Texcoco, Tlatelolco, and Tlaxcala. It is unclear which Voynichese symbol would be pronounced "t" at the beginning of a word. As we assumed that circle 4 on folio 86v is Tlaxcala, and that circle 6 is Tecamachalco or possibly Tepeaca, we had thought that these words might be included in folio 86v, but we have not found them. According to our translation of Voynichese, the cities beginning with the letter "t" could be spelled as follows: Tecamachalco = [Voynichese symbols]; Tepeaca = [Voynichese symbols]; Tlaxcala = [Voynichese symbols] or [Voynichese symbols].

Conclusion

Many of the symbols in the *Voynich Codex* have been decoded based on the association of Voynichese-labeled names on some plants with their names in Nahuatl or Spanish. Despite our apparent success in decoding some of the Voynichese symbols and the decipherment of some plants, a mineral, an animal, and a city, most of the *Voynich Codex* text defies translation, suggesting that another language might be involved besides Classical Nahuatl. This may be a dialect or an extinct Mesoamerican language. Another possibility is that Voynichese is a mixed synthetic language or a Nahuatl lingua franca. This will be covered in more detail in Chap. 13.

Literature Cited

Amancio, D.R., E.G. Altmann, D. Rybski, O.N. Oliveira, and L. da F. Costa. 2013. Probing the statistical properties of unknown texts: Application to the Voynich manuscript. *PLoS One* 8 (7): e67310. https://doi.org/10.1371/journal.pone.0067310.

Anonymous. 2017. Help: IPA for Nahuatl. https://en.wikipedia.org/wiki/Help:IPA_for_Nahuatl. 17 May 2017.

Ayala, J.V. 1990. *Introduccion a la paleografia*. Fondo Editorial, Lima.

Bierhorst, J. 1985. *A Nahuatl-English dictionary and concordance to the Cantares Mexicanos with an analytic transcription and grammatical notes*. Stanford: Stanford University Press.

Brumbaugh, R.S. 1978. *The most mysterious manuscript*. Carbondale: Southern Illinois Press.

Canger, U. 1988. Nahuatl dialectology: A survey and some suggestions. *International Journal of American Linguistics* 54: 28–72.

Casanova, A. 1999. Méthodes d'analyse du langage crypté: Une contribution à l'étude du manuscript du Voynich. Docteur thesis. Université de Paris. http://voynich.free.fr/a_casanova_these_19mars1999.pdf. 18 January 2014.

Ceram, C.W. 1986. *Gods, graves, and scholars: The story of archaeology*. 2nd rev. ed. New York: VintageBooks.

Champollion, J.-F. 2009. *The code breaker's secret diaries*. Trans. M. Rynja. Gibson Square Books, London.

Chávez Orozco, L. 1947. *Codice Osuna*. Mexico D.F: Ediciones del Instituto Indigenista Interamericano.

Clemens, R., ed. 2016. *The Voynich manuscript*. New Haven: Yale University Press.
Coe, M. 2012. *Breaking the Maya code*. 3rd ed. New York: Thames & Hudson.
Comegys, J.D. 2013. The Voynich manuscript: Aztec herbal from New Spain. voynichms.com/wp-content/uploads/2014/03/VMSAztecHerbal-JDComegys.pdf. Accessed 5 June 2017.
Cottrell, L. 1972. *Reading the past: The story of deciphering ancient languages*. London: J. M. Dent.
Dakin, K. 1981. The characteristics of a Nahuatl *lingua franca*. In *Nahuatl studies in memory of Fernando Horcasitas*, ed. Frances Karttunen, 55–67. Austin: Texas Linguistic Forum 18.
de Alarcón, H.R. 1987. *Treatise on the heathen superstitions that today live among the Indians native to This New Spain, 1629*. Trans. & ed. J. R. Andrews and R. Hassig. Norman: University of Oklahoma Press.
de Sahagún, B. 1963. *Florentine Codex. General history of the things of New Spain. Book 11—Earthly things*. Trans. C.E. Dibble and A.J.O. Anderson. Salt Lake City: Univ. Utah.
de Valderrama, J. 1600. *Pintura del gobernador, alcaldes y regidores de México, Osuna Codex*. Mexico. http://www.theeuropeanlibrary.org/exhibition-reading-europe/detail.html?id=108151. 4 January 2013.
Díaz, J.L. 1976. *Índice y sinonimia de las plantas medicinales de México*. México: Instituto Mexicano para el Estudio de las Plantas.
Dressler, R.L. 1953. The Pre-Columbian cultivated plants of Mexico. *Botanical Museum Leaflets* 16: 115–172.
Fox, M. 2013. *The riddle of the labyrinth: The quest to crack an ancient code*. New York: HarperCollins.
Gessmann, G.W. 1922. *Die Geheimsymbolev de Alchymie. Arzneikunde und Astrologie des Mittelalters*. Berlin: Verlag von Karl Siegismund.
Gordon, C.H. 1987. *Forgotten scripts: Their ongoing discovery and decipherment*. Rev. ed. New York: Dorset Press.
Hermes, J. 2012. Textprozessierung—Design und Applikation. Ph.D. thesis, Universität zu Köln. http://kups.ub.uni-koeln.de/4561/. 6 March 2016.
Hernández, F. 1942. In *Historia de las Plantas de Nueva España*, ed. I. Ochoterena. Mexico: Imprenta Univ.
Hernández, F., F. Celsi, F. Colonna, B. Deversini, J. Faber, J. Greuter, V. Mascardi, N. A. Recchi, and J. Terentius. 1651. *Rerum medicatum Novae Hispaniae Thesarus, seu, Plantarum animalium mioneralium Mexicanorum historia*. Vitalis Mascardi, Romae.
Herrera, F. 2004. Hippocrene concise dictionary. In *Nahuatl-English, English-Nahuatl (Aztec)*. New York: Hippocrene Books.
Jacq, C. 1987. *Champollion the Egyptian*. Trans. G. Le Roy. London: Pocket Books.
Karttunen, F. 1983. *An analytical dictionary of Nahuatl*. Norman: University of Oklahoma Press.
Kennedy, G., and R. Churchill. 2006. *The Voynich manuscript*. Rochester: Inner Traditions (First published in 2004 by Orion Press, London).
Knight, K. 2009. *The Voynich manuscript*. MIT. http://www.isi.edu/natural-language/people/voynich.pdf. 1 May 2014.
Lacadena, A. 2008. Regional scribal traditions: Methological implications for the decipherment of Nahuatl writing. *PARI Journal* 8: 1–22.
Landini, G. 2001. Evidence of linguistic structure in the Voynich manuscript using spectral analysis. *Cryptologia* 25: 275–295.
Launey, M. 2011. *An introduction to Classical Nahuatl*. Trans. C. Mackay. New York: Cambridge University Press.
Lockhart, J. 1990. Postconquest Nahua writing and concepts viewed through Náhuatl writings. *Estudio de Cultura Náhuatl (México: UNAM Instituto de Investigaciones Históricas)* 20:91–116.
———. 1992. *The Nahuas after the conquest: A social and cultural history of the Indians of Central Mexico, sixteenth through eighteenth centuries*. Stanford: Stanford University Press.

———. 2001. *Nahuatl as written. Lessons in older written Nahuatl, with copious examples and texts*. Los Angeles: Stanford University Press.

Montemurro, M.A., and D.A. Zanette. 2013. Keywords and co-occurrence patterns in the Voynich manuscript: An information-theoretic analysis. *PLoS One* 8 (6): e66344. https://doi.org/10.1371/journal.pone.0066344.

Muñoz y Rivero, J. 1917. *Manual paleografía diplomática Española*. Madrid: Daniel Jorro.

Ocaranza, F. 2011. *Historia de la medicina en México*. 2nd ed. Cauhtémoc: Cien de México.

Okrent, A. 2010. *In the land of invented languages*. New York: Spiegel & Grau Trade Paperbacks.

Reeds, J. 1995. William F. Friedman's transcription of the Voynich manuscript. *Cryptologia* 19: 1–23.

Reko, B.P. 1947. Nombres bótanicos del manuscrito Badiano. *Boletín de la Sociedad Botánica de México* 5: 23–43.

Robinson, D.F., ed. 1969. *Aztec studies. I. Phonological and grammatical studies in modern Nahuatl dialects*. Norman: Univ. Oklahoma.

Robinson, A. 2002. *The man who deciphered linear B: The story of Michael Ventris*. New York: Thames & Hudson.

———. 2012. *Cracking the Egyptian code. The revolutionary life of Jean-François Champollion*. Oxford: Oxford University Press.

Rugg, G. 2004a. The mystery of the Voynich manuscript: New analysis of a famously cryptic medieval document suggests that it contains nothing but gibberish. *Scientific American* 291 (1): 104–109.

———. 2004b. An elegant hoax? A possible solution to the Voynich manuscript. *Cryptologia* 28: 31–46.

Schinner, A. 2007. The Voynich manuscript: Evidence of the Hoax hypothesis. *Cryptologia* 3: 95–107.

Siméon, R. 2010. *Diccionario de la lengua Nahuatl o Mexicana*. México: Siblo Veintiuno.

Sullivan, T.D. 1997. *Primeros Memoriales de Fray Bernardino de Sahagún. Paleography of Nahuatl text and English translation*. Norman: University of Oklahoma Press.

Tiltman, J.H. 1967. *The Voynich Manuscript: "The most mysterious manuscript in the world."* National Security Agency/Central Security Service, Fort George G. Meade, Maryland. DOCID:631091. [released 23 April 2002]. www.nsa.gov/public_info/_files/tech_journals/voynich_manuscript_mysterious.pdf. 5 June 2017.

Tucker, A.O., and R.H. Talbert. 2013. A preliminary analysis of the botany, zoology, and mineralogy of the Voynich manuscript. *Herbalgram* 100: 70–85.

Walters, J.C.W., M.M. de Wolgemuth, P.H. Pérez, E.P. Ramírez, and C.H. Upton. 2002. Dicconario Náhuatl de los municipios de Mecayapan y Tatahuicapan de Juárez, Veracruz. Instituto Lingüístico de Verano.

Whittaker, G. 2009. The principles of Nahuatl writing. *Göttinger Beiträge zur Sprachwissenschaft* 16: 47–81.

Zandbergen, R. 2004–2017. *The Voynich manuscript*. www.voynich.nu/. 6 June 2017.

Chapter 6
Zoomorph Identification

Elizabeth E. Flaherty, Arthur O. Tucker, and Jules Janick

Animal Images in the Voynich Codex

There are 21 different animal images (zoomorphs) in the various folios of the *Voynich Codex*: three are incidental sketches in the Herbal and Pharmaceutical sections (folios 25v, 49r, 101v); eight are zodiac animal signs in the Astrological section (folios 70r-73v), as discussed in Chap. 8; and seven are in the Balneological section, five of which are on a page with a pool (Fig. 6.1) and two are in the margins of other folios (78r, 80v). In addition, there are two birds – one nesting and one in flight (folio 86r) – and a sheep (folio 116v), scribbled on the back page. Most of the illustrations appear hastily drawn; yet, in many cases there are sufficient details to permit identification to the species level. The animals are discussed in sequence as invertebrates, fish, amphibians, reptiles, birds, and mammals. Some folios such as folio 101v are trifold; thus, the section of the folio number is indicated in parentheses (folio 101v(3) is the third section).

Zoomorph Identification

Invertebrates

Crayfish (Phylum: Arthropoda; Subphylum: Crustacea; Class: Malacostraca; Order Decapoda) An illustration associated with the zodiac, folio 71v(4) depicts two crayfish (Fig. 6.2a) identified by four pairs of walking legs (located along the cephalothorax), in addition to large chelipeds that end in grasping or pinching claws (Fig. 6.2b). Crayfish also possess smaller feeding appendages between the chelipeds and long antennae that extend from the head (Fig. 6.2b). Both structures are shown in the Voynich drawing (Fig. 6.2a). This image was tentatively identified

J. Janick, A. O. Tucker, *Unraveling the Voynich Codex*, Fascinating Life Sciences, https://doi.org/10.1007/978-3-319-77294-3_6

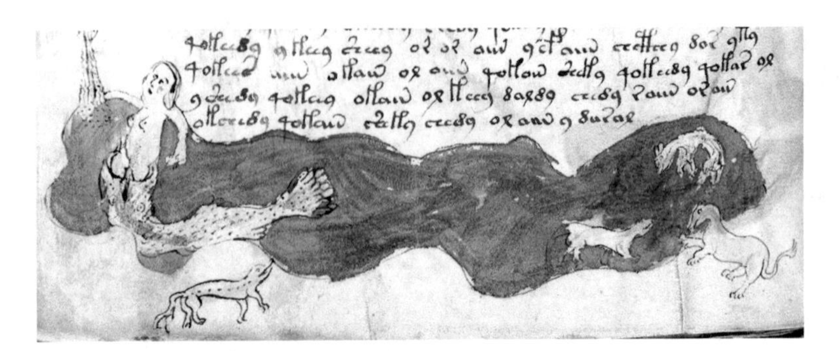

Fig. 6.1 Folio 79v from the *Voynich Codex* depicts a pool of water with five different animals (left to right): resembling an alligator gar (*Atractosteus spatula*) swallowing a woman, a lizard, an iguana (*Ctenosaura similis*), a paca (*Agouti paca*), and a coatimundi (*Nasua* sp.)

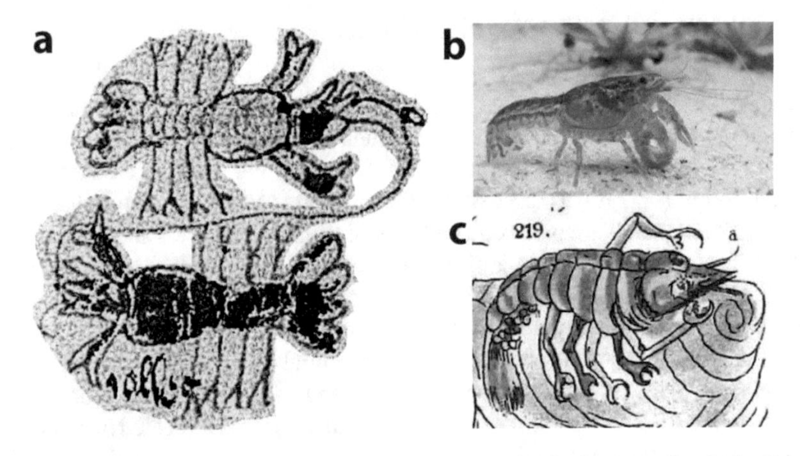

Fig. 6.2 (**a**) Two crayfish – one red and one white –are associated with the zodiac in the *Voynich Codex*, folio 71v(4). Words were added later under the red crayfish and are thought to spell *jollio*, which is similar to the Occitan month of *juhlet* or July. (**b**) A modern crayfish, *Cambarellus* spp., with morphological features similar to those shown in the Voynich image, such as long antennae, four walking legs, large front pincers, and tail structure. (Source: Wikimedia Commons). (**c**) Crayfish are also included in the *Florentine Codex*, another sixteenth century Mesoamerican manuscript

as the Mexican crayfish, *Cambarellus montezumae* Saussure, 1857, by Tucker and Talbert (2013), primarily based on its popularity as food and wide distribution. Mexico is the most crayfish-diverse country in the world, with more than 50 described species (Gutiérrez-Yurrita 2004), and many indigenous people consumed crayfish. Sahagún (1963: 64, Fig. 219a) included a drawing of a crayfish (*acocili*) in the *Florentine Codex* (Fig. 6.2c) and described it as a source of food: *"It is like the shrimp. Its head is like a grasshopper's. It is small, dark; it has legs. But when cooked, it is red, ruddy, hard, and firm. It is edible; it can be toasted, it can be cooked."* A crustacean figure also appears in Mayan art (Wyllie 2010), again supporting the observation that this was a popular, well-known animal in early Mexican cultures.

Fig. 6.3 (**a**) A jellyfish was drawn near the bottom of folio 78r. Two common jellyfish found on the coasts of Mexico are: (**b**) Pacific sea nettles (*Chrysaora fuscescens*). (Source: Ed Bierman); and (**c**) moon jellies (*Aurelia aurita*). Note that the tentacles in the Voynich image more closely resemble the longer tentacles of the Pacific sea nettle, whereas the dome-shaped bell resembles both species. (Source: Andreas Augstein)

Jellyfish (Phylum: Cnidaria; Class: Scyphozoa) The image on the bottom of folio 78r (Fig. 6.3a) clearly shows the bell and tentacles of a jellyfish (Fig. 6.3b, c). This tiny zoomorph is crudely rendered and impossible to identify to species level, but there are two possibilities: *Chrysaora fuscescens* Brandt, 1835, commonly known as the Pacific sea nettle or the West Coast sea nettle (Fig. 6.3b), or the moon jelly, *Aurelia aurita* Linneaus, 1758 (Fig. 6.3c), which is the most common jellyfish in the Gulf of Mexico. These cnidarians pulsate their bells for locomotion, and their tentacles contain stinging nematocysts used to capture prey. Jellyfish have a worldwide distribution and regularly wash ashore in most coastal environments, suggesting that observations by indigenous people would have been common.

Fish

Alligator Gar (Phylum: Chordata; Class: Actinopterygii; Order: Lepisosteiformes) The image on folio 70r (Fig. 6.4a) illustrates two alligator gar (*Atractosteus spatula* Lacepède, 1803). These large-bodied, torpedo-shaped, freshwater fish have long, alligator-like jaws that contain many sharp teeth used to capture prey. The artist who drew this detailed zoomorph included several key characteristics of the species: the upper jaw extends past the lower and is upturned; the bulb or bulge on the upper jaw contains olfactory organs (Echelle and Grande 2014) used to locate prey; the soft-rayed fins lack spines and appear lobed when out of water; and the heavy, diamond-shaped, interlocking scales (ganoid scales) serve as body armor (Echelle and Grande 2014, page 253).

These sizeable fish (Fig. 6.4c, d) can grow up to 3 m long (Echelle and Grande 2014), and females are generally larger than males (García de León et al. 2001). The largest reported alligator gar measured 3.05 m long and weighed 137 kg (McClane 1978: 179). Their distribution extends from the Mississippi River Basin, with St. Louis, Missouri, as the northern extent of the range, and south into northeastern Mexico. Although they are sport fish and were used as food by Native Americans

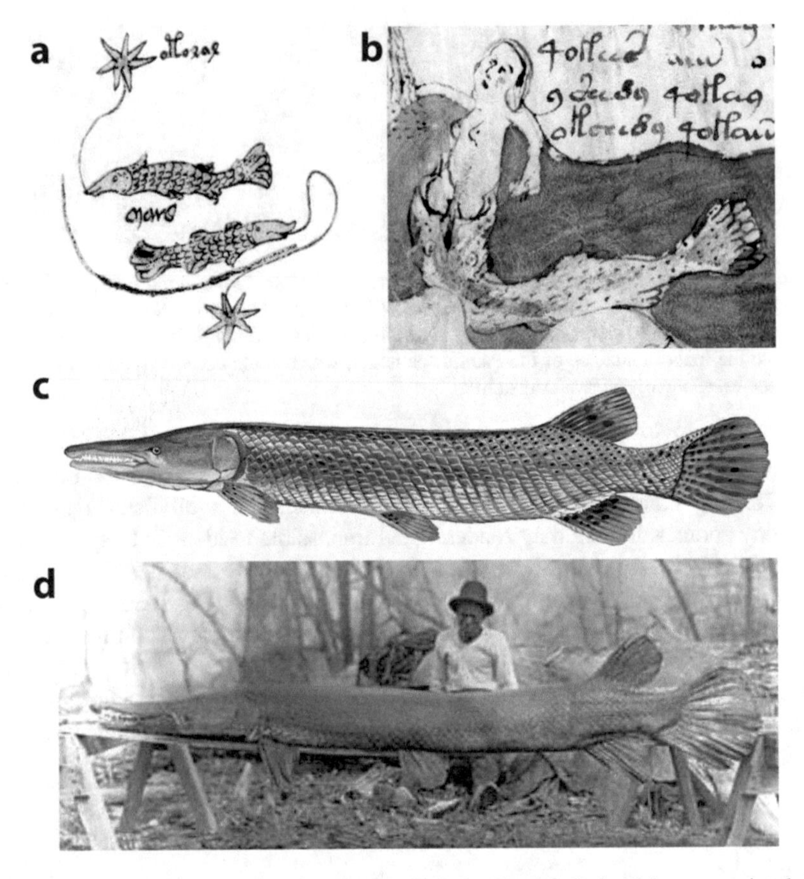

Fig. 6.4 Three alligator gars (*Atractosteus spatula*) in the *Voynich Codex*: (**a**) two associated with the zodiac and (**b**) one in a pool in folio 79v. The word *mars* (written between the two zodiac gars at a later date) refers to the Occitan word *març* (or March). The word [illegible] above the gar can be decoded as *ātlācâocâ*, according to Tucker and Talbert (2013). Based on similarities to Nahuatl, we suggest that *ātlācâocâ* might mean "still fished" (*ātlācâ* = fished, fisherman/spear throwers + *oc(a)* = still). The Voynich drawings include several key characteristics of the species, such as (**c**) soft-rayed fins (that appear lobed when the fish is out of the water) and ganoid scales, both of which are clearly represented in this zoological drawing. (Source: Duane Raver, US Fish and Wildlife Service). (**d**) Size and an elongated body are also key features. (Source: Wikimedia Commons)

(Scarnecchia 1992), alligator gar were feared by some to be a potential threat to humans because of their large size and role as a top predator in aquatic ecosystems. There was a report of an alligator gar attack on a 9-year-old girl, while she dangled her feet in Lake Pontchartrain in southeastern Louisiana (Scarnecchia 1992). This concern appears to have been shared by the artist, because folio 79v shows a woman in the jaws of an alligator gar (Fig. 6.4b). The zoomorph on folio 70r was also identified by Tucker and Talbert (2013) as the alligator gar. Dr. Michael Coe (personal communication 2013), a prominent Mayan epigrapher from Yale University, noted that he has caught, cooked, and eaten this fish at Olmec sites in Mexico.

Amphibians

Frog or Toad (Phylum: Chordata; Class: Amphibia; Order: Anura) Folio 101v(3) (Fig. 6.5a) depicts either a frog or toad with what appears to be the vine of the silky morning glory, *Ipomoea pubescens* Lamarck, 1791. The drawing shows the animal's muscular, long hind legs, which are used for jumping locomotion. Mexico has approximately 231 species of anurans (Flores-Villela and Canseco-Márquez 2004), which were likely frequently observed by local people. Indeed, eight different images of tadpoles, frogs, or toads were included in the *Florentine Codex* (Sahagún 1963, Figs. 213–217, 242–244). The size and crudeness of the zoomorph in the *Voynich Codex* prevent an accurate identification to species level, but the Mexican tree frog, *Smilisca baudinii* Duméril & Bibron, 1841 (Fig. 6.5b), is within the range of possibilities and often appears on vines in Mesoamerica.

Caecilian (Phylum: Chordata; Class: Amphibia; Order: Apoda) Two long, thin-bodied animals are hidden among the roots of what is probably the flowering plant, *Lithophragma affine* A. Gray, on folio 49r of the *Voynich Codex* (Fig. 6.6a). The animals appear to be caecilians, or legless amphibians. There are nine species

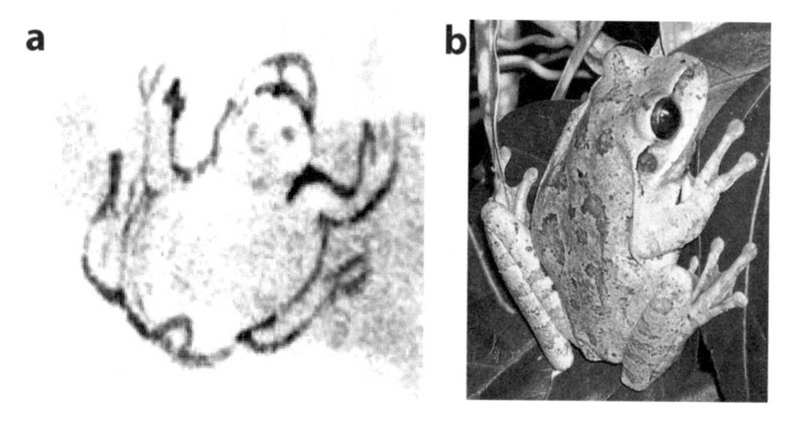

Fig. 6.5 (**a**) A crude drawing of an anuran, either a frog or toad, appears in folio 101v(3). The artist included the strong jumping hind legs and shorter front legs of these animals, which are visible in (**b**) the photo of a Mexican tree frog, *Smilisca baudinii*. (Source: A.H. Vega)

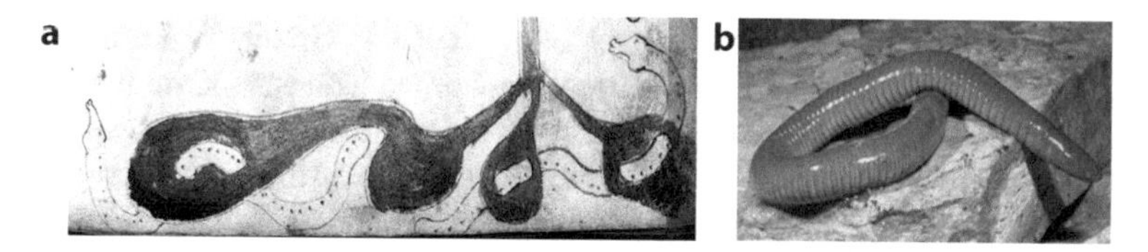

Fig. 6.6 (**a**) Intertwined within the roots of a plant (likely *Lithophragma affine* A. Gray) in folio 49r are two long, thin-bodied animals that appear to be caecilians, or legless amphibians. Both the *Voynich Codex* illustration and (**b**) a photo of *Dermophis mexicanus* (which is relatively common in Mexico) depict annular grooves, consistent with a fossorial lifestyle. (Source: Franco Andreone)

within three genera of caecilians that occur in Central America (Savage and Wake 2001), but they are also found in the Old World in Africa and Asia. Caecilians resemble large earthworms, not only because of their body shape, but also because of their segmented annular groves (Lee 2000: 48; Fig. 6.6b). The dashes along the side of the body in the drawing may be a rough depiction of these annular grooves (Fig. 6.6a). These animals eat invertebrates (e.g., earthworms, termites, orthopteran instars), lizards, and other small vertebrates (Lee 2000). Modern biologists acknowledge the difficulty in collecting these animals (Nussbaum and Wilkinson 1989) in some parts of their range, indicating a low probability of sightings by humans in the absence of concerted searching. Such rare sightings may have generated idealized stories heard by the artist and inspired the Voynich illustration. However, as Tucker and Talbert (2013) noted, this zoomorph may have been inspired by *Dermophis mexicanus* Duméril & Bibron, 1841. This terrestrial member of the subfamily Dermophiinae has a distribution ranging from southern Mexico to Columbia and is an adept fossorial (burrowing) amphibian, capable of living in a wide variety of soil types. Caecilians are rather common today in the compost of some coffee *fincas* (rural estates), where deep, organic soils attract earthworms, their primary food (Summers and O'Reilly 1997; Wake 2003). They are rather common in nature, and sometimes appear in the international pet trade.

In addition to the caecilians, Mesoamerica has many species of native earthworms (Phylum: Annelida; Class Oligochaeta). The family Megascolecidae includes a very large-bodied species (Fragoso et al. 1994: 71) that may have provided inspiration for the long-bodied, worm-like creatures in the Voynich illustration. Approximately 48 native species of earthworms occur in Mexico (Fragoso et al. 1994: 71), likely resulting in common interactions with people, and there is some evidence that the Aztecs ate earthworms (Keane 1908: 273). However, the distinct heads in the two creatures depicted in the illustration provide further evidence that it is not an earthworm. The Mexican mole lizard (*Bipes biporus* Stegner & Barbour, 1917) also appears segmented with annular grooves. Although this reptile lacks hind legs, it does have two short front legs. The absence of front legs in the illustration makes it an unlikely candidate as an inspiration for the drawing.

Reptiles

Horned Lizard (Phylum: Chordata; Class: Reptilia; Order: Squamata) The illustration on folio 25v (Fig. 6.7a) resembles a horned lizard based on the rounded abdomen, armor-like scales on the dorsum, horns projecting from the back of the head (Fig. 6.7b), and rounded, dorsoventrally flattened body that allows these lizards to heat quickly in the sun, avoid predation, and collect water droplets (Sherbrooke 2003: 10). All horned lizard species possess temporal and occipital spines that serve as protection against predation (Young et al. 2004); however, they vary in their arrangement, number, and size (Sherbrooke 2003: 13). Horned lizards also squirt blood from their orbital sinuses, which may have made them a species of particular interest to people within their distributional range.

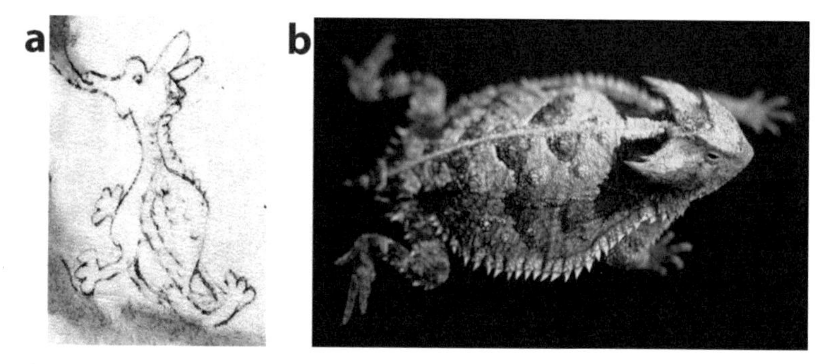

Fig. 6.7 (**a**) A horned lizard eating a plant in folio 25v of the *Voynich Codex*. The artist included the typical rounded abdomen, armor-like scales, and cranial horns projecting from the back of the head. (**b**) The two large cranial horns are characteristic of the Mexican horned lizard, *Phrynosoma taurus*. (Source: Wade Sherbrooke)

In total, there are 13 extant species of horned lizards in North America and their distribution extends south through Mexico to the Guatemalan border and north to southwestern Canada (Sherbrooke 2003: 10). Twelve species are found in Mexico in diverse habitats. Tucker and Talbert (2013) suggested that the illustration on folio 25v might be the Texas horned lizard, *Phrynosoma cornutum* Harlan, 1825, but a better fit might be the Mexican horned lizard, *P. taurus* Dugès, 1873 (Fig. 6.7b), which has the largest horns of any *Phrynosoma* species, with two projecting from the temporal region of the head and two spines above the eyes. It is endemic to the arid highlands of southern Mexico (Guerrero, Puebla, and Oaxaca). Only one other notable armored reptile occurs in the world, the thorny devil (*Moloch horridus* Gray, 1841), which is found in Australia (Pianka and Parker 1975). This zoomorph is not a fictional "dragon" (Kennedy and Churchill 2006), but rather matches a living species of lizard.

Other Lizards Lizards are also ubiquitous in Mexico and Central America and it is not surprising that the two additional images in the *Voynich Codex* (folio 79v) depict lizards. The lizard drawn on the bottom left of folio 79v (Fig. 6.8a) appears to have a row of dots extending horizontally from its head along the flank to near the tail. This marking is present on several Central American lizard species, including the Cozumel spiny lizard (*Sceloporus cozumelae* Jones, 1927), the rose-bellied lizard (*Sceloporus variabilis* Wiegmann, 1834; Fig. 6.8b), and the brown anole (*Anolis sagrei* Duméril & Bibron, 1837). However, identification of any one of these species is not possible based on the characteristics in the drawing. The image on the right middle of folio 79v (Fig. 6.9a) appears to be a more robust-bodied lizard than the other drawing and is most likely an iguana species. The black iguana (*Ctenosaura similis* Gray, 1831; Fig. 6.9c) is found primarily in Central America, whereas the green iguana (*Iguana iguana* Linneaus, 1758) has a larger range, which extends from southern Mexico to central South America. Humans have used both species as food (Lee 2000: 192), possibly making them important for inclusion in drawings and records. Lizards are also included in four images of the *Florentine Codex* (Sahagún 1963, Figs. 202–205, 213; Fig. 6.9b).

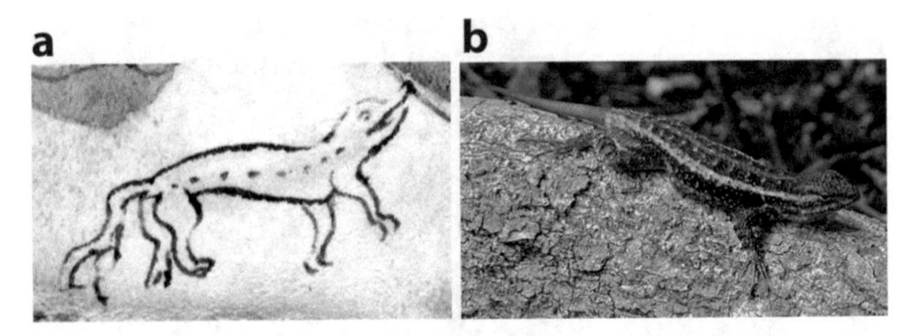

Fig. 6.8 (**a**) A lizard-like animal in folio 79v of the *Voynich Codex* has a row of spots extending horizontally from its head to tail. (**b**) This is a feature common in several species of Mexican lizards, including the rose-bellied lizard (*Sceloporus variabilis*). (Source: Bernard Dupont)

Fig. 6.9 (**a**) Another animal roughly sketched in the pool scene, folio 79v of the *Voynich Codex*, is a large lizard, possibly a black iguana (*Ctenosaura similis*). (**b**) An iguana was also included in the *Florentine Codex*. (**c**) Black iguanas are found in Mexico and Central America and are a robust lizard, similar to the animal in the Voynich image. Green iguanas (*Iguana iguana*) have larger spikes along the spine than the black iguana, and are lacking in the drawing. (Source: Ken Thomas)

Birds

Birds (Phylum: Chordata; Class: Aves) There are two images of flying birds on folio 86r: one nesting (Fig. 6.10a) and one flying (Fig. 6.10b). Both birds appear to be large bodied with speckled feathers across the breast. In both figures, the beak is relatively large compared with the body and is highly suggestive of a raptor species. Although many birds of prey in Central America have spots that appear across at least part of the body, the crested caracara (*Caracara plancus audubonii* Miller, 1777), shown in Fig. 6.10c, d, has notable spots on its breast during both the juvenile and adult stages. Its range is widespread throughout Mexico and Central America (Howell and Webb 1995, page 213), and it is the bird that the Aztecs saw land on a cactus holding a snake in its talons at the site that became Tenochtitlan (present-day Mexico City) (Pearson 1917). Historically, Latin American cultures were fascinated with birds and their feathers, transporting and trading both live birds and their feathers large distances across Central America. There is evidence that this trade artificially altered the range of some bird species (Haemig 1978). Likely because of this fascination, the *Florentine Codex* includes many drawings of birds (Sahagún 1963, Figs. 42–187), and *quauhtli* seems to be a general term for eagle (Sahagún 1963: 40).

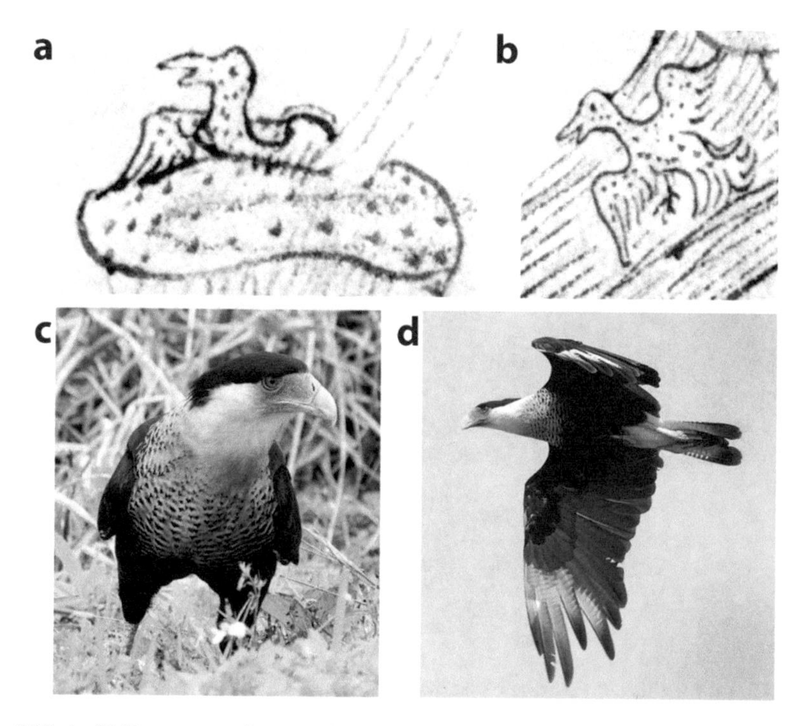

Fig. 6.10 (**a**, **b**) Two separate images of a bird of prey appear in the *Voynich Codex* in folio 86r. Both drawings emphasize a robust beak of a raptor and a spotted pattern on the breast similar to the markings of the crested caracara (*Caracara plancus audubonii*). (**c**) The crested caracara is also an important bird in Aztec history. (Source: US Fish and Wildlife Service Southwest Region). (**d**) The crested caracara. (Source: www.everimages.com)

Mammals

Armadillo (Phylum: Chordata; Class: Mammalia; Order: Cingulata) The image on the left side of folio 80v (Fig. 6.11a) shows an armored animal with large ears that is clearly an armadillo. Other historical illustrations of armadillos exist, including one in *De Simplicibus Medicamentis* (by Nicolás Monardes, 1574; Fig. 6.11e), a text based on medical discoveries in the New World. The nine-banded armadillo (*Dasypus novemcinctus* Linnaeus, 1758; Fig. 6.11d), the species in Monardes' drawing, is covered by ossified dermal plates that form movable bands, which are covered by thick, leathery skin (McBee and Baker 1982). All armadillo species have large, thick-skinned ears, short legs, and long claws (McBee and Baker 1982) and all these features are included in the Voynich drawing. Although there are 21 total extant armadillo species, nearly all are found in South America (Feldhamer et al. 2007: 308). The nine-banded armadillo is the only species found in North America and it is widely distributed in Mexico (Leopold 1959). Its range extends from northern and eastern South America, through Central America and Mexico, and into the southern USA (McBee and Baker 1982). Some South American species

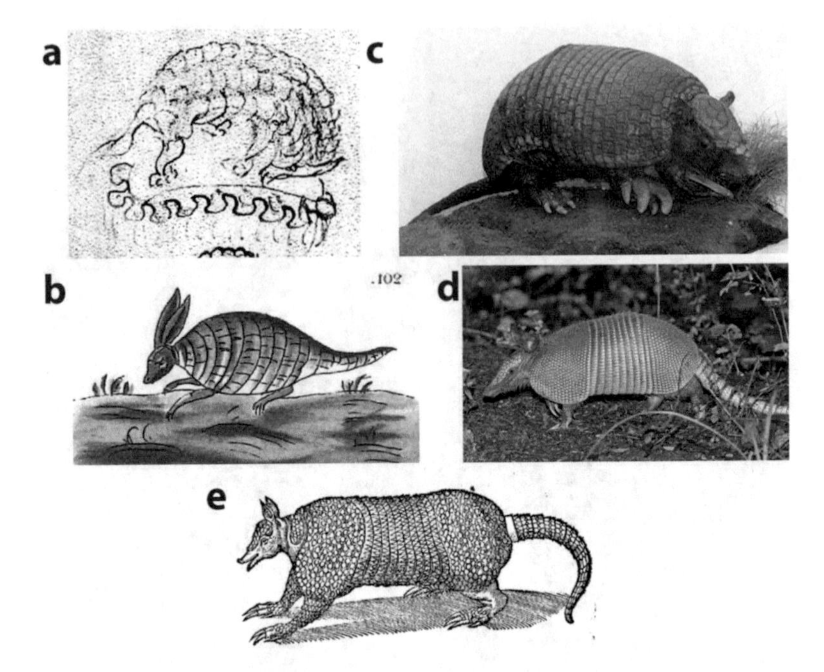

Fig. 6.11 (**a**) An armadillo illustration on the left side of folio 80v. (**b**) A similar drawing of an armadillo (*ayotochtli*) was included in the *Florentine Codex* by Sahagún. The *Voynich Codex* armadillo lacks the typical banded scute (bony plate) formation featured in the Sahagún illustration. The Voynich image may be one of the many armadillo species found in Central and South America, such as (**c**) the southern naked-tailed armadillo (*Cabassous unicinctus*), which appears to have more rounded scuta or may simply be a reflection of the artist's interpretation of the skin and may not include the banded formation, such as (**d**) the nine-banded armadillo (*Dasypus novemcinctus*). Both the Voynich and Sahagún armadillo drawings share similarities in the ears and the armor-like skin, which overall are most similar to the nine-banded armadillo. (Source: Tom Friedel). (**e**) *De Simplicibus Medicamentis*, by Nicolás Monardes in 1574, also included an illustration of a nine-banded armadillo

of armadillo, including the southern naked-tailed armadillo (*Cabassous unicinctus* Linnaeus, 1758; Fig. 6.11c), lack well-defined bands encircling the abdomen, as seen in the nine-banded armadillo, and instead are covered in overlapping scuta (or bony plates), similar to the drawing on folio 80v (Fig. 6.11a). The image in the *Voynich Codex* lacks the obvious bands of the nine-banded armadillo, but this may be a result of artistic license by the author because the ears, feet, and tail all resemble this animal and other contemporary drawings of it.

The only other animals covered in similar armor are pangolins (Manidae), or scaly anteaters. These mammals are found in Africa and Southeast Asia and are covered in a keratinized epidermis that resembles scales (Feldhamer et al. 2007: 310). Because pangolins lack external pinnae (ears) and the animal in the image has well-defined ears, it is unlikely that this image is a pangolin. Drawings of armadillos (*ayotochtli*) were included in the *Florentine Codex* (Sahagún 1963: 61, Fig. 201) and the *Pomar Codex* (López Piñero 1992), with all artists similarly emphasizing the ears. *Ayotochtli* was translated to "gourd-rabbit" by Sahagún (Fig. 6.11b) or "turtle-rabbit," as *ayotl* may refer to a water animal (Fernando Moreira, personal communication).

Cattle (Phylum: Chordata; Class: Mammalia; Order: Artiodactyla) Two images associated with the zodiac on folios 71v(1) and 71v(2) of the *Voynich Codex* (Fig. 6.12a, b), show cattle of Spanish descent. Tucker and Talbert (2013) identified the dark red bull as the Retinta breed (Fig. 6.12d) of cattle (*Bos taurus taurus* Linnaeus, 1758) and the pale red one, possibly a cow, as Andalusian Red (Fig. 6.12c). The Spanish initially brought cattle to the Caribbean islands including the Antilles islands and Puerto Rico as early as 1493 and 1499 respectively (Brand 1961; Rodero et al. 1992). The cattle were moved from the Antilles to Cuba, and then to Mexico and Central America around 1521 by Ruy López de Villalobos and Hernán Cortés (Brand 1961). These early cattle thrived in the New World because of a lack of disease and parasites and the breeds' tolerance for arid climates. Once in Mexico, cattle were generally released and allowed to roam free (Sponenberg and Olson 1992). They were initially introduced into south central Mexico and then moved north.

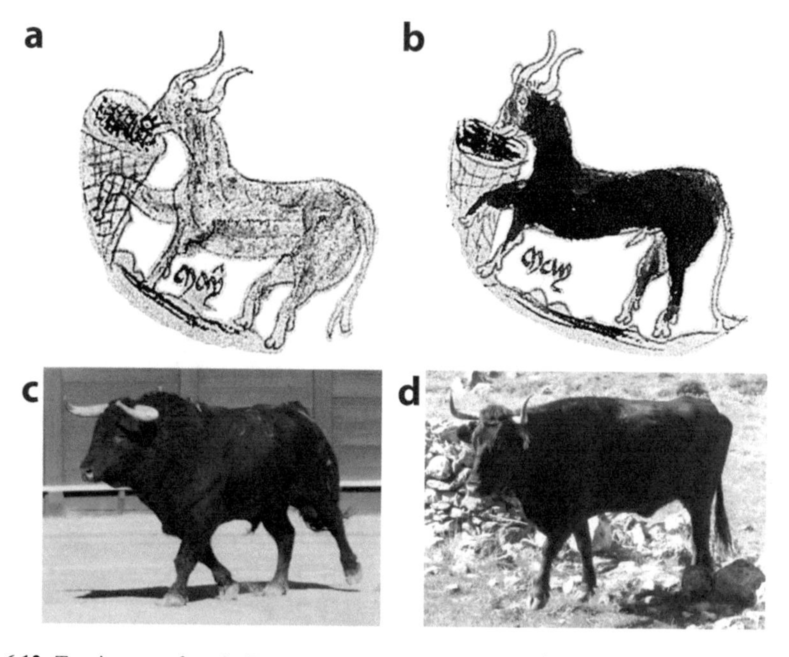

Fig. 6.12 Two images of cattle (*Bos taurus*) associated with the zodiac in folios 71v(1) and 71v(2) possibly show (**a**, **b**) a cow and bull, with upturned horns and red and dark coloration. The word *maŷ* or *may* was written under the cow and bull at a later date and indicates the month of May. (**c**) Andalusian Red was among several breeds of Spanish cattle imported to Mexico. (Source: Alexander Fiske Harrison). (**d**) Retinta breeds were transported from Spain by early explorers. Cattle were first introduced to the Caribbean islands and were subsequently transported into Mexico

Within 10 years of their introduction, ranches extended from Perote (in the present-day Mexican state of Veracruz) and Tepeaca (in the present-day Mexican state of Puebla) to the Valley of Toluca, near present-day Mexico City (Brand 1961). Christopher Columbus, who was responsible for introducing cattle into the Antilles in 1493, likely brought Retinta and Gallego breeds, in addition to the ancient Andalusian Red breed, which were used as fighting bulls. The Gallego cattle would have survived quite well because of their tolerance for drought and rough terrain (Lydekker 1908: 9). The Spanish cattle likely included a mix of red, tan, and black colors, and possessed upward curving horns similar to the images in the *Voynich Codex*. They were the ancestors of modern Texas longhorns, which were also commonly solid red and tan. Spotted longhorns were a result of artificial selection through selective breeding (Sponenberg and Olson 1992).

Coatimundi (Phylum: Chordata; Class: Mammalia; Order: Carnivora) The light colored, long-nosed, and long-tailed animal rendered in the image on the lower right of folio 79v (Fig. 6.13a) is a coatimundi. Two species of coatimundi occur in the New World: one (*Nasua nasua* Linnaeus, 1766) is found in northern South America (Gompper and Decker 1998), whereas the other (*Nasua narica* Linnaeus, 1766)

Fig. 6.13 (**a**) Another roughly sketched *Voynich Codex* animal from the pool scene, on the lower right of folio 79v, shares several notable morphological characteristics of a coatimundi (*Nasua* spp.) with (**b**) a contemporary drawing from the Casa del Deán mural (Morrill 2014: 184). The illustrations emphasize the animal's unique, morphological traits, including: (**c**) its long tail; (**d**) elongated and flexible nose, called a rhinarium; and tan coloration, which is especially common in juveniles

ranges from northern South America through Central America and Mexico into the southern USA (Gompper 1995). These mammals range in color from a pale tan (especially as juveniles) to reddish or dark brown and black (Gompper 1995). They have very short ears, an elongated snout that forms a proboscis (Fig. 6.13b, c), capable of a wide variety of movements (Gompper 1995), and a long tail (Fig. 6.13d). These mammals tend to interact with humans, and social groups are known to raid corn crops (Leopold 1959). The only other mammals with similar long and flexible rhinarium are the Russian desman (*Desmana moschata* Güldenstädt, 1777), tapirs (*Tapirus* spp. Brünnich, 1772), and pigs (*Sus* spp. Linnaeus, 1758), none of which look like the animal in folio 79v.

Jaguarundi (Phylum: Chordata; Class: Mammalia; Order: Carnivora) The dark brown or black cat-like animal with a noticeably long tail on folio 73r (Fig. 6.14a) appears to be a jaguarundi (*Puma yagouaroundi,* Geoffroy Saint-Hilaire, 1803; Fig. 6.14b), in agreement with Tucker and Talbert (2013), who identified this as a black Gulf Coast jaguarundi (*P. yagouaroundi cacomitli*). These big cats (or felids) have long bodies, narrow, small skulls, small ears, and short limbs, relative to body size (Oliveira 1998). The tail is especially notable at approximately two-thirds the length of the head and body, which is a high tail-to-body length ratio for felids. Color phases range from red, gray, and brown to black, and the color is generally uniform (Oliveira 1998). Jaguarundis occur from southern Texas through the lowlands and coasts of Mexico into Peru. They are found in a wide variety of habitats (Oliveira 1998) and respond well to disturbance (Reid 1997: 273). At present, they are the most commonly observed felid in Central America; thus, interactions with earlier human settlements were highly likely. Mountain lions or cougars, *Puma concolor* Linnaeus, 1771, also occur throughout the region, but are much lighter in color and have shorter tails.

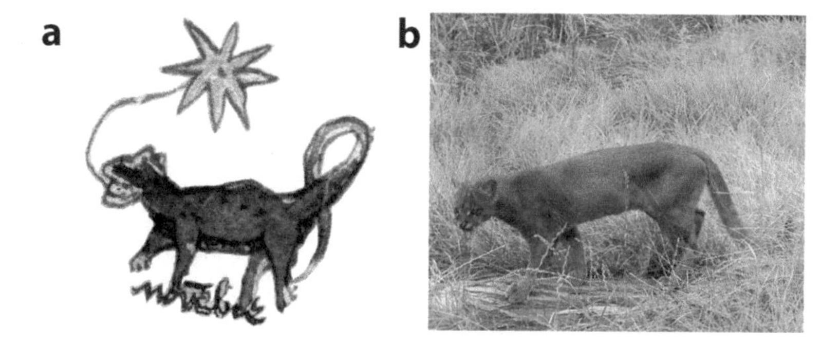

Fig. 6.14 (**a**) Folio 73r, another image associated with the zodiac in the *Voynich Codex*, shows a black cat with an exceptionally long tail, which likely is a jaguarundi (*Puma yagouaroundi*). These felids also possess relatively small skulls and shorter legs relative to their bodies. The word *novēbre* ("ē" indicates a missing "m"), representing November, has been added to the image at a later time. (**b**) A jaguarundi. (Source: Wikimedia Commons)

Ocelot (Phylum: Chordata; Class: Mammalia; Order: Carnivora) The image in zodiac folio 72r(1) depicts a spotted felid (Fig. 6.15a). Two spotted cats occur in Mexico: the ocelot (*Leopardus pardalis* Linnaeus 1758) and the jaguar (*Panthera onca* Linnaeus 1758). Tucker and Talbert (2013) identified this as the ocelot. The ocelot is a smaller species with spotted pelage and some striping patterns in the coat, especially on the face (Fig. 6.15b), which appears in the drawing and is not a characteristic of the larger jaguar (Seymour 1989). This species once ranged from the southern USA south into northern Argentina and Peru (Murray and Gardner 1997) and occurs in a variety of habitats. The *Florentine Codex* described ocelots as "princely," and the species was clearly well respected and held in high regard for its hunting prowess (Sahagún 1963: 1–2, Fig. 1). The ocelot is also among animals in lists of 20 "day names," part of ritual calendars used by Mesoamerican people, including the Aztecs and Mayas (Kelley 1960).

Paca (Phylum: Chordata; Class: Mammalia; Order: Rodentia) An animal sketched in the image on the lower right of folio 79v (Fig. 6.16a) appears to have rows of spots extending longitudinally and a long nose. Based on these characteristics,

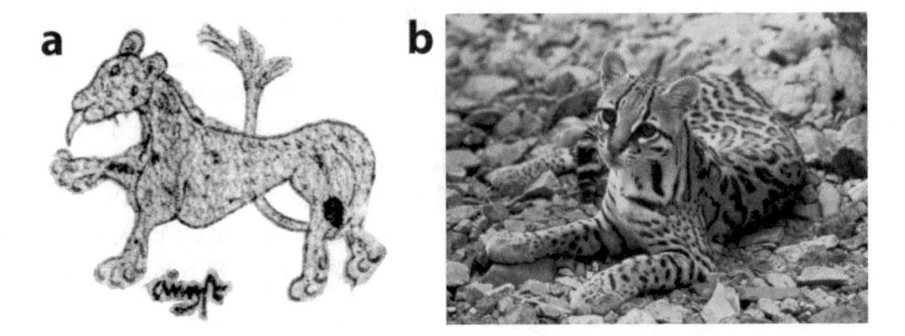

Fig. 6.15 (**a**) Another illustration of a felid, likely an ocelot (*Leopardus pardalis*), is associated with the Voynich zodiac. This drawing includes spots covering the animal's body with some striping on the head. Ocelots were highly regarded by cultures in Mexico and Central America. The position of the tail between the legs in the drawing is similar to medieval drawings of lions in the constellation Leo. The word *augst* under the ocelot indicates the Occitan word *agost* for August. (**b**) An ocelot. (Source: Wikimedia Commons)

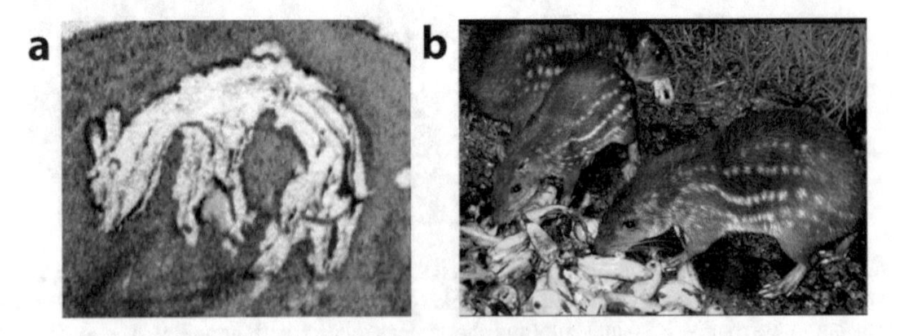

Fig. 6.16 (**a**) A smaller, rodent-like creature was included in the Voynich pool scene, folio 79v. This animal appears to have spotted lines running along its flanks and a reddish color, suggesting (**b**) a paca (*Agouti paca*). (Source: Wikimedia Commons)

it is most likely a paca (*Agouti paca* Linnaeus, 1766; Fig. 6.16b). These large rodents are generally brown, dark brown, or gray with white to light yellow spots extending along the flanks and arranged in rows (Pérez 1992). The square head extends to a long nose and the animal lacks a tail (details that are also reflected in the sketch). These animals live in tropical forests (Leopold 1959: 388) from Mexico to northern Argentina, and are hunted for meat (Pérez 1992).

Sheep (Phylum: Chordata; Class: Mammalia; Order: Artiodactyla) Two images associated with the zodiac, folios 70v and 71r (Fig. 6.17a, b), and another later in folio 116v (Fig. 6.17c) appear to be sheep. All the images show an ungulate with horns that curl back and a short tail. These images may show desert bighorn sheep (*Ovis canadensis nelsoni* Shaw, 1804; Fig. 6.17d), which are native from central Baja California to Colorado and eastern Texas (Leopold 1959: 524; Shakleton 1985), Mexican desert bighorn sheep (*O. canadensis mexicana*), which are native to western Sonora (a northwestern border state in Mexico), or domestic sheep (*O. aries* Linnaeus, 1758), introduced by the Spanish. Tucker and Talbert (2013) identified these zoomorphs as possibly the Mexican desert bighorn sheep, based in part on the cleft hooves and their sexually dimorphic nature (Mooring et al. 2003).

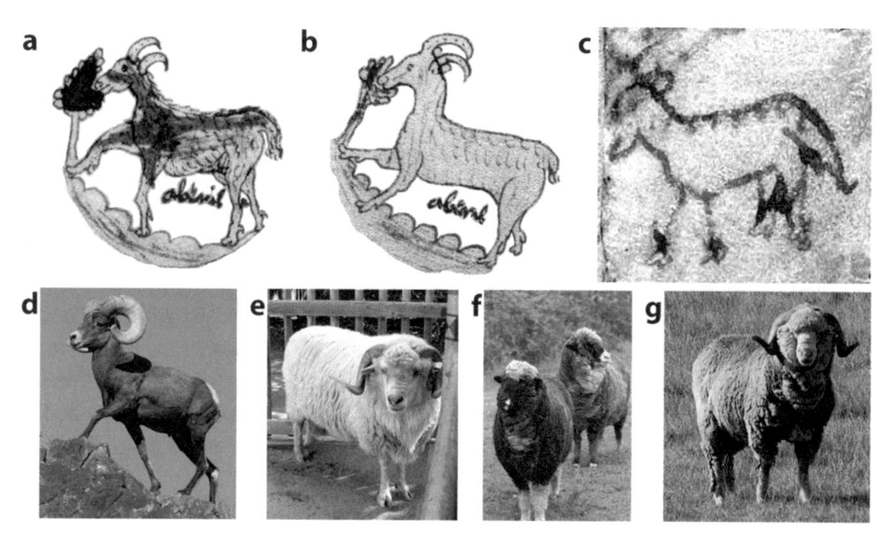

Fig. 6.17 Three images of sheep were included in the *Voynich Codex*, including: (**a**, **b**) two, possibly a ram and a ewe, associated with the zodiac, folios 70v and 71r and (**c**) one later on folio 116v, on the last page of the codex. The word *abéril* under the sheep represents the month of April. All images show animals with backward-curving horns and a short tail. (**d**) These may depict the native bighorn sheep (*Ovis canadensis*) or domestic sheep (*Ovis aries*), which were introduced by the middle of the sixteenth century from Spain, including the Spanish (**e**) Churro and (**f**) Merino breeds. It is difficult to determine if these represent native bighorn sheep or the introduced breeds, and it is unlikely that humans would have successfully translocated bighorn sheep from the very northern mountains of Mexico and southern USA. Therefore, it is more plausible that these represent domestic Spanish sheep. (Source: Wikimedia Commons: Peter Shanks). (**g**) A Merino sheep. (Source: Wikimedia Commons: Phillip Capper)

Coloration of desert bighorn sheep ranges from a reddish brown to a dark brown with lighter legs (Shakleton 1985). They have large horns that curl up and back and a short tail, similar to the image in folio 70v. The range of bighorn sheep does not extend very far into Mexico, making it unlikely that many people in this region would have interacted with this species. Because modern translocations are difficult and success rates are relatively low (Singer et al. 2001), it is unlikely that earlier cultures were moving these animals farther south into Mexico. However, domestic sheep were introduced into Mexico by the Spanish, were established by the mid-1500s (Watson 1984; Rodero et al. 1992; Melville 1997), and were primarily the Spanish Churro (Fig. 6.17e) and Merino breeds (Watson 1984; Rodero et al. 1992; Fig. 6.17f, g), which may have been the more likely inspiration for these Voynich drawings. Both breeds are adapted to semi-arid environments. The coloration of the Spanish Churro varies from gray to brown to black, whereas Merino breeds are white to light tan. The only other two possible wild sheep species present in the world that would resemble these images are the argali (*O. ammon*, Linnaeus, 1758), found in Central Asia and southern Siberia (Fedosenko and Blank 2005), and the mouflon (*O. orientalis* Linnaeus, 1758), found in the mountains of Europe and Asia (Clutton-Brock 1999: 70).

Conclusion

The identification of animals indigenous to the New World in the *Voynich Codex* supports and affirms botanical evidence (Tucker and Talbert 2013; Tucker and Janick 2016) that it is a sixteenth century, colonial New Spain manuscript created in central Mexico by someone familiar with both Spanish and Aztec cultures. There is no other community of animals in the world similar to the one portrayed in the *Voynich Codex*. The range of animals illustrates that the natural fauna of Mexico, and that of Spain, influenced the artist. The Spanish culture is indicated by the inclusion of the zodiac, Spanish breeds of cattle, and perhaps imported sheep, whereas the Aztec culture is expressed by the transformation of some of the Western zodiac signs by indigenous Mexican animals, in addition to the inclusion of gender separation, a completely new interpretation of the zodiac.

Literature Cited

Brand, D.D. 1961. The early history of the range cattle industry in northern Mexico. *Agricultural History* 35: 132–139.

Clutton-Brock, J. 1999. *Natural history of domesticated mammals*. Cambridge, UK: Press Syndicate of the University of Cambridge.

De Oliveira, T.G. 1998. *Herpailurus yagouaroundi*. *Mammalian Species* 578: 1–6.

De Sahagún, B. 1963. *Florentine Codex. General history of the things of New Spain. Book 11—Earthly things*. Trans. C.E. Dibble, and A.J.O. Anderson. Salt Lake City: University of Utah Press.

Echelle, A.A., and L. Grande. 2014. Lepisosteidae: Gars. In *Freshwater fishes of North America*, ed. M.L. Warren Jr. and B.M. Burr, 243–278. Baltimore: Johns Hopkins University Press.
Fedosenko, A.K., and D.A. Blank. 2005. *Ovis ammon*. *Mammalian Species* 773: 1–15.
Feldhamer, G.A., L.C. Drickhamer, S.H. Vessey, J.F. Merritt, and C. Krajewski. 2007. *Mammalogy: Adaptation, diversity, ecology*. 3rd ed. Baltimore: Johns Hopkins University Press.
Flores-Villela, O., and L. Canseco-Márquez. 2004. Nuevas especies y cambios taxonómicos para la herpetofauna de México. *Acta Zoologica, Mexico* 20: 115–144.
Fragoso, C., S.W. James, and S. Borges. 1994. Native earthworms of the north neotropical region: Current status and controversies. In *Earthworm ecology and biogeography in North America*, ed. P.F. Hendrix. Boca Raton: CRC Press.
García de León, F.J., L. Gonzáles-García, J.M. Herrera-Castillo, K. Winemiller, and A. Banda-Valdéz. 2001. Ecology of the alligator gar, *Atractosteus spatula*, in the Vicente Guerrero Reservoir, Tamulipas, México. *Southwestern Naturalist* 46: 151–157.
Gompper, M.E. 1995. *Nasua narica*. *Mammalian Species* 487: 1–10.
Gompper, M.E., and D. Decker. 1998. *Nasua nasua*. *Mammalian Species* 580: 1–9.
Gutiérrez-Yurrita, P.J. 2004. The use of crayfish fauna in México: Past, present…and future? *Freshwater Crayfish* 14: 30–36.
Haemig, P.D. 1978. Aztec emperor Auitzotl and the great-tailed grackle. *Biotropica* 10: 11–17.
Howell, S.N.G., and S. Webb. 1995. *A guide to the birds of Mexico and Northern Central America*. Oxford, UK: Oxford University Press.
Keane, A.H. 1908. *The world's peoples: A popular account of their bodily and mental characters, beliefs, traditions, political and social institutions*. New York: G.P. Putnam's Sons.
Kelley, D.H. 1960. Calendar animals and deities. Southwest. *Journal of Anthropology* 16: 317–337.
Kennedy, G., and R. Churchill. 2006. *The Vonich manuscript: the mysterious code that has defied interpretations for centuries*. Rochester, Vermont: Inner Traditions. (First published in 2004 by Orion press, London).
Lee, J.C. 2000. *A field guide to the amphibians and reptiles of the Maya world: The lowlands of Mexico, northern Guatemala, and Belize*. Ithaca/New York: Cornell University Press.
Leopold, A.S. 1959. *Wildlife of Mexico: The game birds and mammals*. Berkeley: University of California Press.
López Piñero, J.M. 1992. The Pomar codex (ca 1590): Plants and animals of the old word and from the Hernandez expedition to America. *Nuncius/Istituto e Museo di Storia della Scienza* 7: 35–52.
Lydekker, R. 1908. *A guide to the domesticated animals (other than horses) exhibited in the central and north halls of the British museum (natural history)*. London: British Museum.
McBee, K., and R.J. Baker. 1982. *Dasypus novemcinctus*. *Mammalian Species* 162: 1–9.
McClane, A.J. 1978. *McClane's field guide to freshwater fishes of North America*. New York: Henry Holt and Company.
Melville, E.G.K. 1997. *A plague of sheep: Environmental consequences of the conquest of Mexico*. Cambridge, UK: Cambridge University Press.
Mooring, M.S., T.A. Fitzpatrick, J.E. Benjamin, I.C. Fraser, T.T. Nishihira, D.D. Reisig, and E.M. Rominger. 2003. Sexual segregation in desert bighorn sheep (*Ovis canadensis mexicana*). *Behaviour* 140: 183–207.
Morrill, P.C. 2014. *The casa del Deán: New world imagery in a sixteenth century Mexican mural cycle*. Austin: University of Texas Press.
Murray, J.L., and G.L. Gardner. 1997. *Leopardus pardalis*. *Mammalian Species* 548: 1–10.
Nussbaum, R.A., and M. Wilkinson. 1989. On the classification and phylogeny of caecilians (Amphibia: Gymnophiona), a critical review. *Herpetology Monographs* 3: 1–42.
Pearson, T.G. 1917. Caracara, the Mexican eagle. *Art World* 3: 264.
Pérez, E.M. 1992. *Agouti paca*. *Mammalian Species* 404: 1–7.
Pianka, E.R., and Parker, W.S. 1975. Ecology of horned lizards: A review with special reference to *Phrynosoma platyrhinos*. *Copeia* 1975:141–162.
Reid, F.A. 1997. *A field guide to the mammals of Central America and Southeast Mexico*. 1st ed. Oxford, UK: Oxford University Press.
Rodero, A., J.V. Delgado, and E. Rodero. 1992. Primitive Andalusian livestock and their implication in the discovery of America. *Archivos de Zootecnia* 41: 383–400.

Savage, J.M., and M.H. Wake. 2001. Reevaluation of the status of taxa of Central American caecilians (Amphibia: Gymnophiona) with comments on their origins and evolution. *Copeia* 2001: 52–64.

Scarnecchia, D.L. 1992. A reappraisal of gars and bowfins in fishery management. *Fisheries* 17: 6–12.

Seymour, K.L. 1989. *Panthera onco*. *Mammalian Species* 340: 1–9.

Shakleton, D.M. 1985. *Ovis canadensis*. *Mammalian Species* 230: 1–9.

Sherbrooke, W.C. 2003. *Introduction to horned lizards of North America*. California Natural History Guide No. 64. Berkeley: University of California Press.

Singer, F.J., C.M. Papouchis, and K.K. Symonds. 2001. Translocation as a tool for restoring populations of bighorn sheep. *Restoration Ecology* 8: 6–13.

Sponenberg, D.P., and T.A. Olson. 1992. Colonial Spanish cattle in the USA: History and present status. *Archivos de Zootechnia* 41: 401–414.

Summers, A.P., and J.C. O'Reilly. 1997. A comparative study of locomotion in the caecilians *Dermophis mexicanus* and *Typhlonectes natans* (Amphibia: Gymnophiona). *Zoological Journal of the Linnaean Society* 121: 65–76.

Tucker, A.O., and J. Janick. 2016. Identification of phytomorphs in the Voynich codex. *Horticultural Reviews* 44: 1–64.

Tucker, A.O., and R.H. Talbert. 2013. A preliminary analysis of the botany, zoology, and mineralogy of the Voynich manuscript. *HerbalGram* 100: 70–75.

Wake, W.H. 2003. Tailless caecilians (Caeciliidae). In *Grzimek's animal life encyclopedia*, ed. Hutchins M., Duellman, W.E., Schlager N., vol. 6, 435–441. Amphibians. 2nd ed. Farmington Hills (MI): Gale Group.

Watson, W.A. 1984. The import and export of sheep and goats. *British Veterinary Journal* 140: 1–21.

Wyllie, C. 2010. The mural paintings of El Zapotal, Veracruz, Mexico. *Ancient Mesoamerica* 21: 209–227.

Young, K.V., E.D. Brodie Jr., and E.D. Brodie III. 2004. How the horned lizard got its horns. *Science* 304: 65.

Chapter 7
Nymphs and Ritual Bathing

Jules Janick and Arthur O. Tucker

Balneological Section

The Balneological (bathing) section (folios 75r, 75v, 76v–84v) is the most bizarre part of the *Voynich Codex* and is the source of continuous speculation. It includes 202 shapely nude women referred to as nymphs, mostly cavorting in pools with green water connected by tubes suggestive of plumbing or vascular systems (Fig. 7.1). Nymphs (some clothed) are also found in the zodiacs and are discussed in the following chapter. All the nymphs appear to be the same blonde woman. The images in the Balneological section are surrounded by text. These illustrations have been examined to determine if they might reveal the artist's intent, and to determine or confirm whether they are consistent with Mesoamerican, Spanish, and Aztec influences dating to the sixteenth century. Selected images of individual nymphs are shown in Fig. 7.2a–m.

Nymph Figures

The nymph illustrations are formulaic and show very similar-looking, curvaceous, petite women with blonde hair, which is usually cascading (Fig. 7.2e, k, and l), but sometimes short (Fig. 7.2g, j, and m). The women appear to be European. Blonde hair was considered beautiful in medieval Spain (Claudio da Soller 2005). In Tetela de Ocampo, Mexico, there was a sixteenth century painting of the blonde, martyred Saint Ursula. In addition, Queen Isabella of Castile was blondish and her long, wavy hair resembled the hairstyle of some of the nymphs. The drawings are presented in three-quarter or frontal view, some sexually explicit (Fig. 7.2c, i, and l). None are shown in back view, suggesting that portrayal of the buttocks was objectionable, perhaps offensive to the Aztec culture. Some nymphs have head coverings of various styles,

J. Janick, A. O. Tucker, *Unraveling the Voynich Codex*, Fascinating Life Sciences, https://doi.org/10.1007/978-3-319-77294-3_7

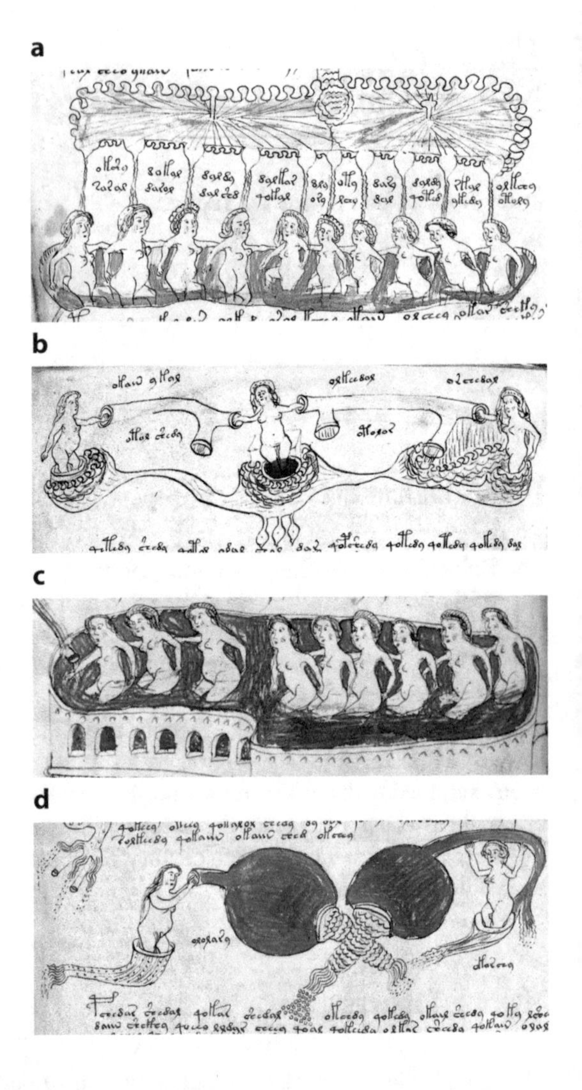

Fig. 7.1 Bathing nymphs and plumbing examples of the Balneological section of the *Voynich Codex*: (**a**) folio 75v (note cross); (**b**) folio 77v; (**c**) folio 78v; (**d**) folio 83v. The green color of the pools is suggestive of the presence of algae, such as *Chlorella*

including a crown with a cross (Fig. 7.2a) and a medieval hat (Fig. 7.3). Some hold a crucifix (Fig. 7.2b), or possibly a metal ring or collar (Fig. 7.2c–f), or a torpedo-like object (Fig. 7.2g–i). Other nymphs carry objects such as a spinning top (Fig. 7.2j), fruits (Fig. 7.2k), a seedling (Fig. 7.2l), or a stick or knife (Fig. 7.2m).

Crucifix

There is an undeniable Roman Catholic connection to the *Voynich Codex*, but it is not emphasized. Three obvious crosses in the Balneological section include a cross in a crown worn by a nymph (Fig. 7.2a), a nymph holding a cross (Fig. 7.2b), and a crude isolated cross in folio 75v (Fig. 7.1a). Putative crosses in the *Voynich Codex* are also

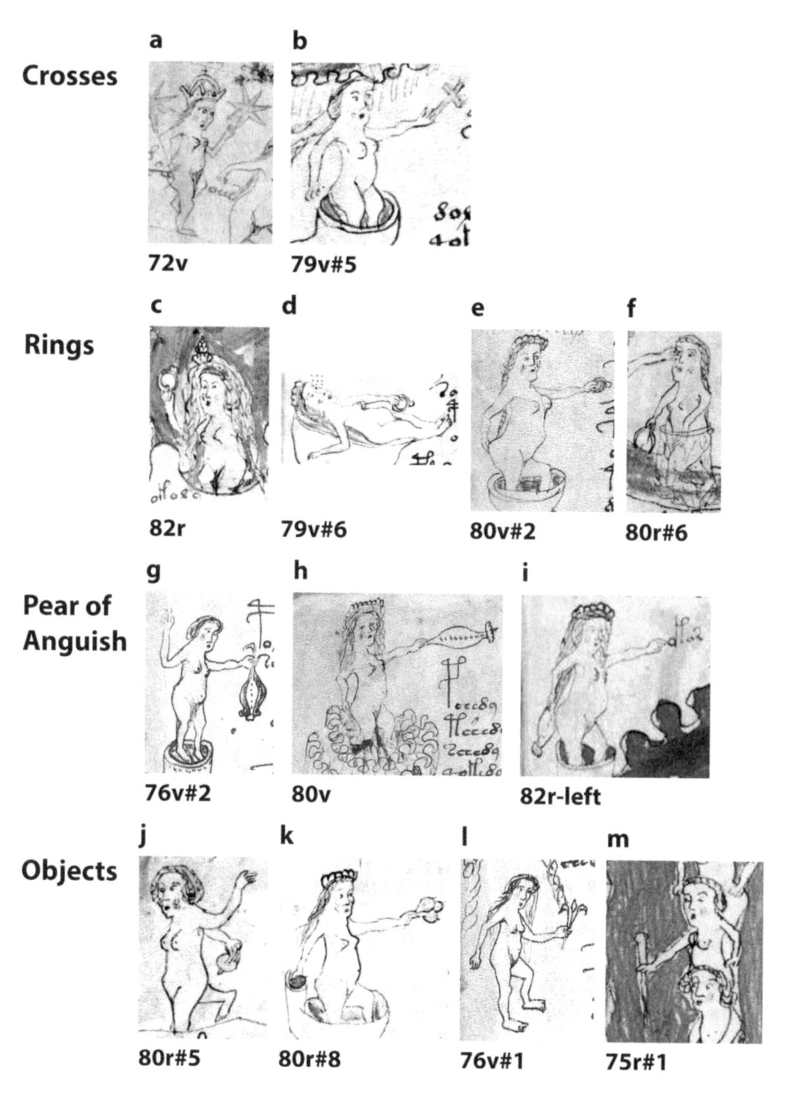

Fig. 7.2 **(a–m)** Selected nymphs of the Balneological section with folio designations from the *Voynich Codex*

found on the tops of *qubba* (domes) in circle A of folio 86v (see Chap. 10). Crosses can also be observed on some ciboria-like "maiolica" or apothecary jars (see Chap. 2).

Metallic Rings

Four nymphs each hold what appears to be a metallic ring (Fig. 7.2c–f) in which three (Fig. 7.2c–e) clearly show a knob or nub. The knob is unclear in the fourth ring (Fig. 7.2f) in which part is submerged in water. The identification of the three rings

Fig. 7.3 Nymph with a medieval hat

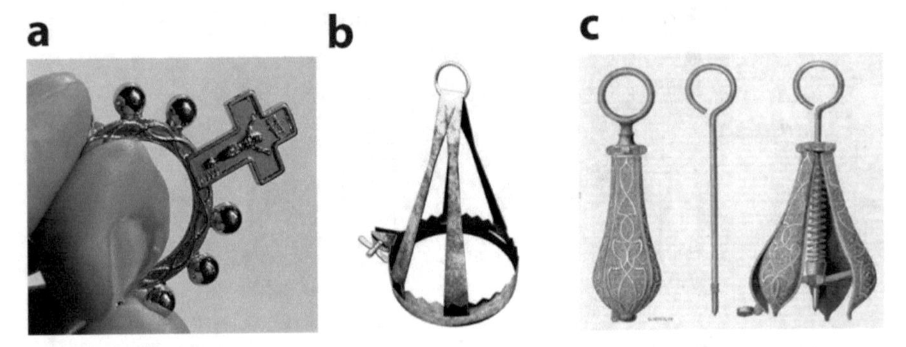

Fig. 7.4 Spanish religious artifacts and Inquisition torture devices: (**a**) ring rosary; (**b**) iron collar; (**c**) pear of anguish

with a knob is problematic. Lincoln Taiz and Taiz (2011) has suggested that the ring held by the nymph in Fig. 7.2d might be a rosary, and in fact this prone nymph was drawn just beneath the nymph holding a crucifix (Fig. 7.2b). Although most rosaries are beads on a string, there were solid ring rosaries in medieval Spain and perhaps Mexico, but these were small (Fig. 7.4a). A darker suggestion made by Arthur O. Tucker is that the ring is an iron collar (Fig. 7.4b), an instrument of subjugation or torture. Indeed, in 1596, iron collars were used to strangle two sisters in New Spain, Isabel de Carvajal de Andrade and Leonor de Carvajal, accused of being crypto-Jews in Mexico (Anonymous 1996).

Oblong, Torpedo-Shaped Objects

Three nymphs hold complex, oblong, torpedo-shaped objects (Fig. 7.2g–i). The most detailed (Fig. 7.2g) has a red outer sheath with a lip or ridge on one end and two flexible, string-like objects on the opposite end (held by the nymph). The sheath encloses an inner portion that extends below the lip with a line of six dots from top to bottom. We first surmised that this object resembled an ear of maize with the dots representing kernels, but we discarded this interpretation as unrealistic. The second image (Fig. 7.2h) is similar, with an inner line of nine dots; however, there is no indication of a red sheath, but the lip is shown and the inner portion also extends beyond the lip. The third image (Fig. 7.2i) is less clear, but shows the oblong device facing down with the lip or ridge evident. We surmise that these three illustrations might represent the same object and furthermore that its identification might be an important clue to understanding the meaning of the codex. The possibility of the rings (Fig. 7.2c–f) being iron collars as instruments of torture led us to conjecture that the torpedo-like objects might be "pears of anguish" (Fig. 7.4c). This torture device had a lip-like opening at one end and a screw to open its sheath at the opposite end. This horrible device, very variable in structure, was inserted into the mouth, anus, or vagina to extract confessions or inflict pain. Under this interpretation the dots could indicate the screw. Note that all the objects have a lip, and the screw extends from two of them. If the rings described above are iron collars, and if the torpedo-like objects are the pears of anguish, the presence of these objects of torture suggests that they might represent a cryptic protest statement from an artist of indigenous ancestry.

Plumbing or Vascular Systems

The perplexing plumbing in the Balneological section (Fig. 7.1) has been interpreted as having human gynecological allusions, such as fallopian tubes or ovaries (Strong 1945; Strong and McCawley 1947; Kennedy and Churchill 2006: 181), or plant vascular systems (Taiz and Taiz 2011). The precise meaning will only be clarified if the *Voynich Codex* can be translated. It is evident that the tubes and pipes must have symbolic meanings, but the precise interpretation is maddeningly obtuse. Granziera (2005) suggests that the Aztec cult of water and fertility might have been based on the concept that the Earth is like a vessel containing water, or like a womb filled with amniotic fluid. Important ritual celebrations were situated near rivers and springs, inside caves, or on mountain tops. We suggest that the vascular systems in the Balneological section might be consistent with the Aztec cult of water and fertility.

Ritual Bathing Allusions

The bathing nymphs portrayed in the *Voynich Codex* suggest symbolic examples of ritual bathing. The Aztecs, in contrast with the Spanish, were devotees of daily bathing (Sahagún 1981: 130–138). There are many images of bathing in Aztec culture (Fig. 7.7b, and Chap. 16, Fig. 16.5). Archeological remains of bathing pools can be found in Texcotzingo, the ancient palace of Nezahualcoyotl near Texcoco, in

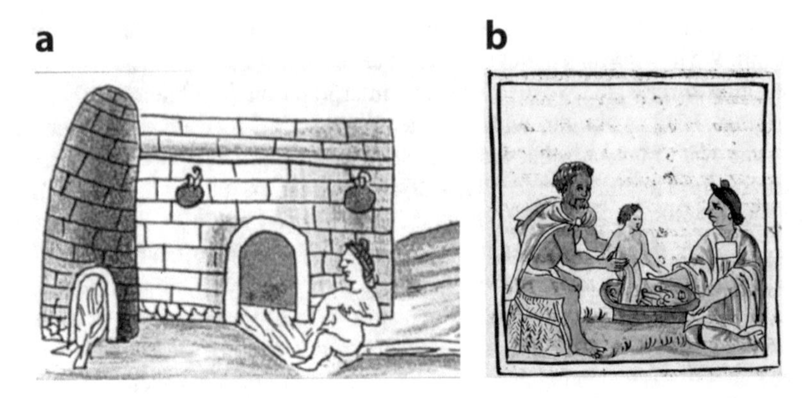

Fig. 7.5 Aztec bathing: (**a**) Aztec steam bath; (**b**) infant bathing. (Source: De Sahagún 1981)

Fig. 7.6 Archeological remains of Nezahualcoyotl bathing pools: (**a**) Nezahualcoyotl location near Texcoco; (**b**) Chapultepec Park, Mexico City

addition to the ancient Nezahualcoyotl baths in Chapultepec Park, Mexico City (Fig. 7.6). There are allusions to ritual bathing in the Tepantitla mural at Teotihuacan (dated between 250 and 560 CE), such as a scene under the Great Goddess, where many male figures (most wearing only loincloths) swim in a river associated with a mountain (Fig. 7.7a). Outside the mountain are four crouching male figures with speech emanating from their mouths, each with a hand between his legs grasping the hand of the figure behind him in a ritual "daisy chain" (Fig. 7.7b), which has a parallel to the connected bathing nymphs in the *Voynich Codex*. Next to the river, various plants (Fig. 7.7c) are being tasted or examined. Clearly, there are similarities between this scene and the *Voynich Codex*. Milbrath (2013) has identified ritual bathing in the *Borgia Codex* (Fig. 7.7d). Nudity and partial nudity are occasionally found in Aztec art (Fig. 7.8) and there is a scene that includes nude bathing in the Casa del Deán murals (see Chap. 16, Fig. 16.7c). Evidently, the Balneological section has images that are consistent with Aztec culture.

There are many scenes of nude bathing in Western art. Two famous examples are paintings by Lucas Cranach the Elder: *The Golden Age*, 1530 (Fig. 7.9a), and *The Fountain of Youth*, 1546 (Fig. 7.9b). In *The Golden Age*, nude men and women

Fig. 7.7 Tepantitla mural, Teohuacan (ca. 250–560 CE): (**a**) bathing; (**b**) chain of crouching men; and (**c**) tending herbs. (**d**) Ritual bathing in the *Borgia Codex*

Fig. 7.8 Aztec deity showing partial nudity

a

b

Fig. 7.9 Nude bathing scenes painted by Lucas Cranach the Elder: (**a**) *The Golden Age*, 1530; (**b**) *The Fountain of Youth*, 1546

cavort in a stream, while other nude couples hold hands and dance in a ring around a tree. The artist uses many artifacts to conceal the genitals, but both frontal and rear views are shown. In *The Fountain of Youth*, old women enter the pool and leave as young maidens.

Conclusion

The green water and bathing scenes in the *Voynich Codex* represent a fusion of Western iconography and Aztec customs. Any water feature on the outside in Mexico would be expected to be green because of air-borne algae, e.g., *Chlorella vulgaris* and *Chlorococcum humicola.* We consider that the bathing images are consistent with a Western-trained artist. The curvy nymphs indicate that the artist was familiar with and probably trained in Western art. We know of sixteenth century institutions for training indigenous Indians, such as La Escuela de San José de los Naturales in Mexico City, founded by Fray Pedro de Gante, and El Colegio de Santa

Cruz de Tlatelolco (also in Mexico City), which offered instruction utilizing Belgian and Spanish paintings. The lack of any rear view of the Voynich nymphs suggests Aztec sensibilities with regard to the human form. This section conveys much about the personality of the artist, which is discussed more fully in Chap. 14.

Literature Cited

Anonymous. 1996. Rare documents shed light on grisly Mexican Inquisition. 1 April 2017.

De Sahagún, B. 1981. *Florentine Codex. General history of the things of New Spain. Book 2—The ceremonies*. Trans. C.E. Dibble and A.J.O. Anderson. Salt Lake City: Univ. Utah Press.

De Sollar, C. 2005. The beautiful woman in medieval Iberia: Rhetoric, cosmetics, and evolution. PhD thesis, University of Missouri-Columbia.

Granziera, P. 2005. Huaxtepec: The sacred garden of an Aztec Emperor. *Landscape Research* 30 (1): 81–107.

Kennedy, G., and R. Churchill. 2006. *The Voynich manuscript: The mysterious code that has defied interpretation for centuries*. Rochester: Inner Tradition.

Milbrath, S. 2013. *Heaven and earth in ancient Mexico: Astronomy and seasonal cycles in the Codex Borgia*. Austin: University of Texas Press.

Strong, L.C. 1945. Anthony Askham, the author of the Voynich manuscript. *Science* 191: 608–609.

Strong, L.C., and E.L. McCawley. 1947. A verification of a hitherto unknown prescription of the 16th century. *Bulletin of the History of Medicine* 21: 898–904.

Taiz, L., and S.L. Taiz. 2011. The biological section of the Voynich manuscript: A textbook of physiology. *Chronica Horticulturae* 51 (2): 19–23.

Chapter 8
The Zodiac

Jules Janick

Astrology

The zodiac (Greek *zoidiakos* or circle of animals) involves 12 divisions based on the monthly appearance and disappearance of certain constellations (star patterns) in the night sky of Babylon in the fifth century BCE. Each division was assigned 30° of celestial longitude, based on the 360° of the ecliptic (Fig. 8.1). The 12 divisions (or constellations, or zodiac signs) were given mainly human or animal names (Fig. 8.2). The assumption of a relationship between astronomical phenomena, particularly the movement of the sun, moon, and planets, and the human condition led to astrology, a divination system based on the horoscope, the position of the planets at birth for each person. Astrology entered the scholarly tradition and connected with other studies, including astronomy, alchemy, medicine, and botany (especially for herbalist Nicholas Culpeper, 1616–1654). It reached a peak in the sixteenth century when European monarchs had their own personal astrologers, but was still treated with some skepticism by the Catholic Church. Astrology was completely discredited in the Age of Enlightenment, but is still casually followed by many, with horoscopes published daily in many local newspapers. The zodiac and astrology were a part of Spanish culture in the fifteenth and sixteenth centuries. In this chapter, traditional zodiac signs are compared with signs found in the *Voynich Codex* and the *Codex Mexicanus*.

The Zodiac in the Voynich Codex

There are 12 pages of zodiac images (folios 70r-73v; Figs. 8.3, 8.4, 8.5, 8.6, 8.7, 8.8, 8.9, 8.10, 8.11, and 8.12) in the Astrological section. Only 10 of the 12 zodiac signs are shown. Capricorn and Aquarius are missing, presumably because of a lost folio (74r, 74v). Eight signs (Pisces, Gemini, Cancer, Leo, Virgo, Libra, Scorpio, and Sagittarius) have their own circle. Aries and Taurus both have two circles (one for

J. Janick, A. O. Tucker, *Unraveling the Voynich Codex*, Fascinating Life Sciences, https://doi.org/10.1007/978-3-319-77294-3_8

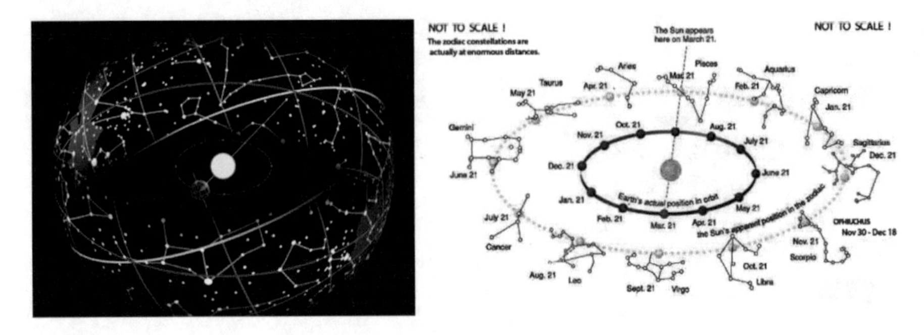

Fig. 8.1 The celestial geometry of the constellations

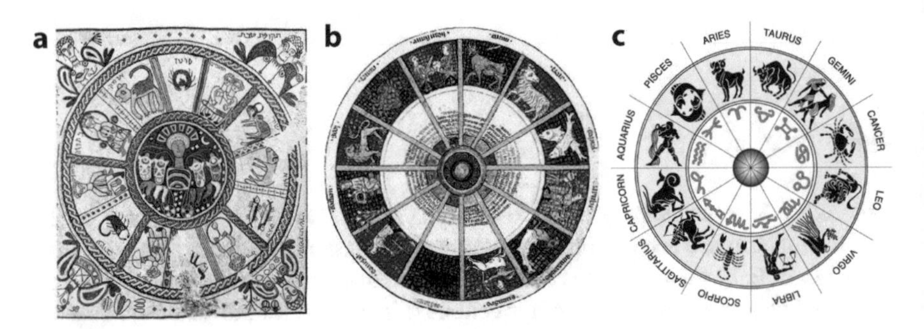

Fig. 8.2 Ancient, medieval, and modern zodiacs. (**a**) Beit Alfa, Israel, sixth century, (**b**) medieval, and (**c**) modern

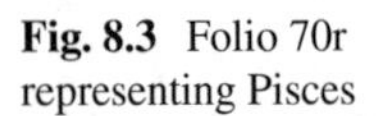

Fig. 8.3 Folio 70r representing Pisces

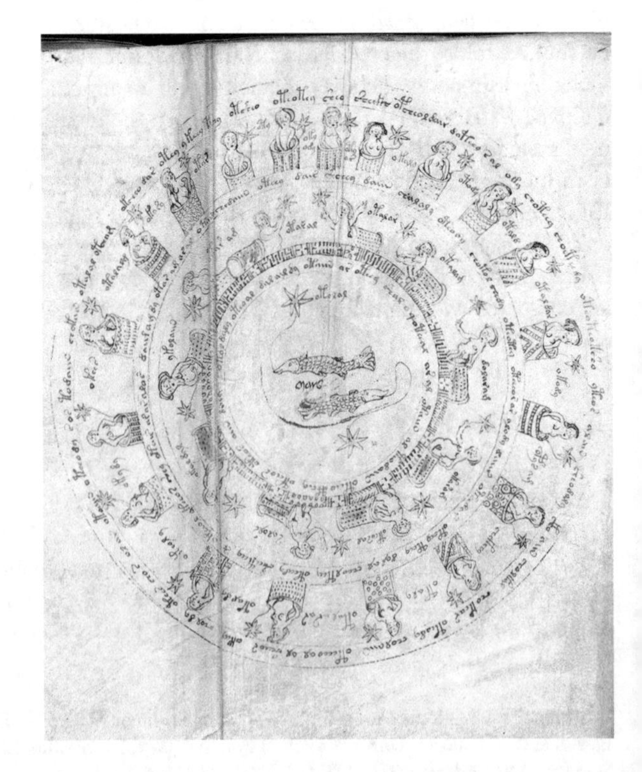

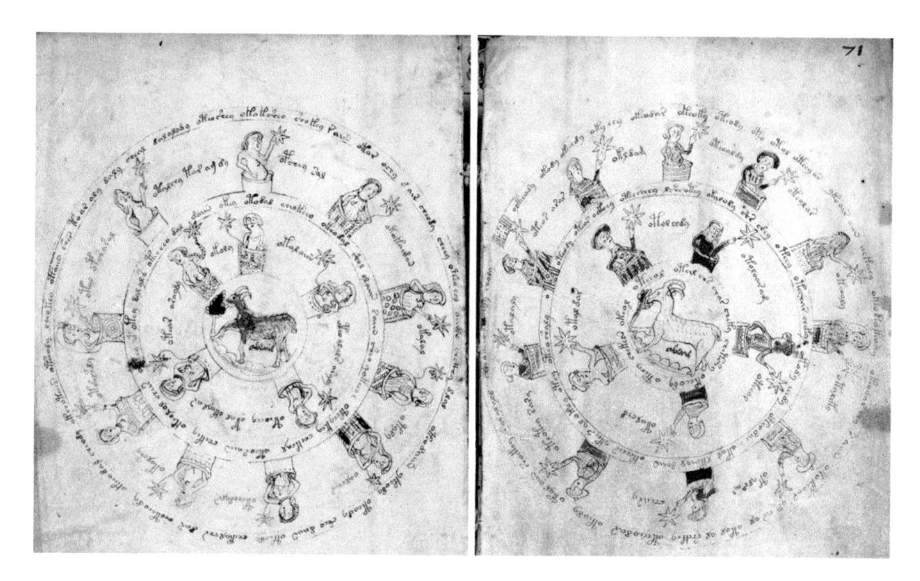

Fig. 8.4 Folios 70v (left) and 71r (right) representing Aries. Note that 15 nymphs in 71r are gowned

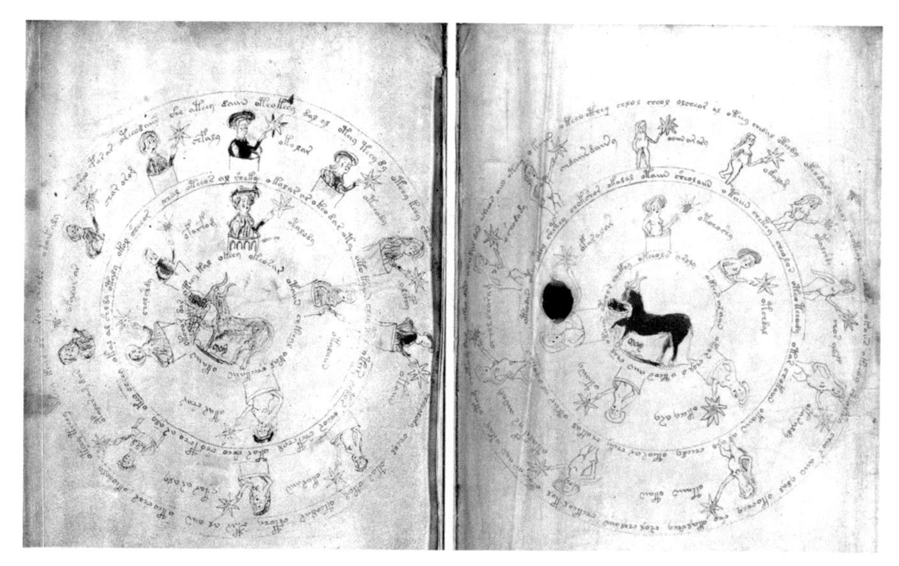

Fig. 8.5 Folio 71v(1) (left) and folio 71v(2) (right) representing Taurus. Note that the 15 nymphs in 71v(1) are gowned

each gender, male and female). The main symbol for most of the signs consists of a person, animal, or object, but for three signs (Pisces, Gemini, and Cancer) the symbols are paired. The symbol of each zodiac sign is drawn in the center of a circle, surrounded by two concentric rings filled with nymphs, with some zodiac signs showing nymphs in an outermost ring as well. In the concentric rings, the nymphs hold stars connected to strings, similar to balloons. Each nymph is labeled in Voynichese (see Appendix B). There also are Voynichese words within the rings.

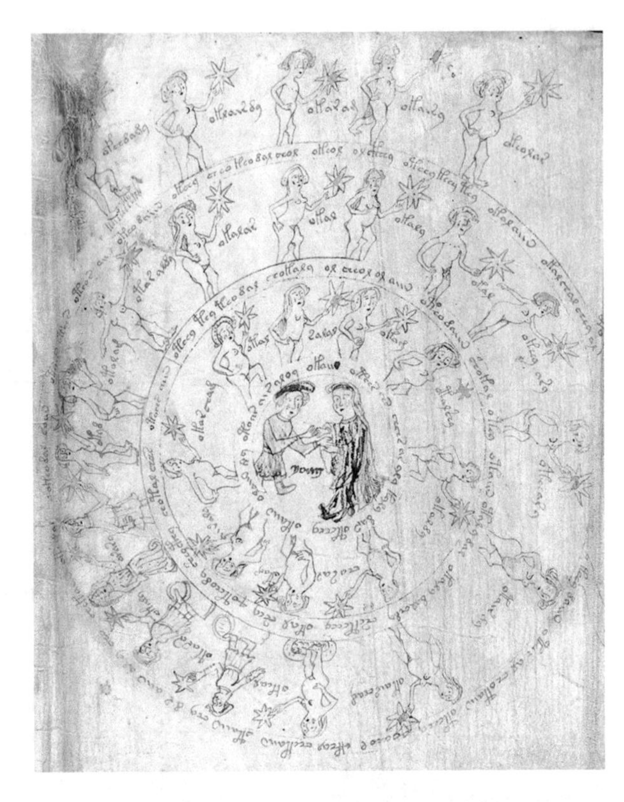

Fig. 8.6 Folio 71v(3) representing Gemini, but the "twins" are a clothed male and female

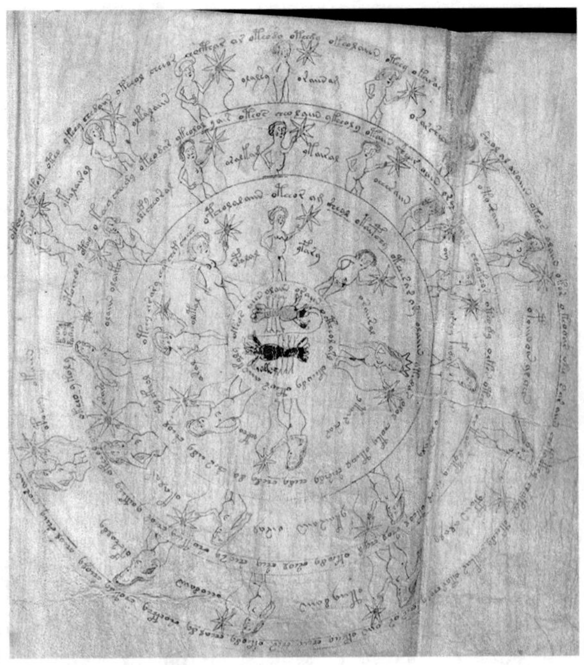

Fig. 8.7 Folio 71v(4) representing Cancer

Fig. 8.8 Folio 72r(1) representing Leo

Fig. 8.9 Folio 72r(2) representing Virgo

Fig. 8.10 Folio 72r(2) representing Libra

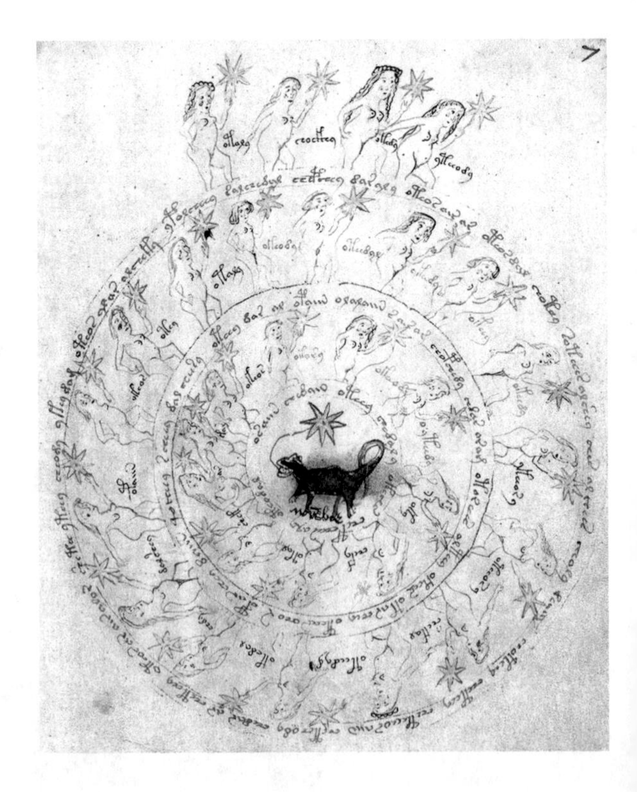

Fig. 8.11 Folio 73r representing Scorpio

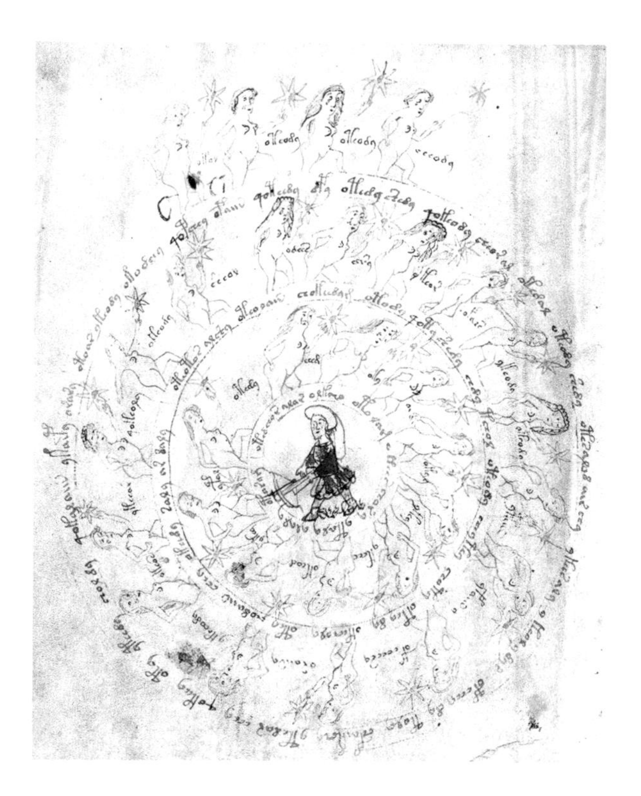

Fig. 8.12 Folio 73v representing Sagittarius

Of the 299 nymphs in the zodiac, 269 are nude, and 30 are gowned (15 in one of the Aries circles and 15 in one of the Taurus circles). In the middle or bottom of the sign, there are cursive words in a possible Occitan dialect (a Romance language spoken in southern France; Monaco; Italy's Occitan Valley; and Catalonia, Spain) that are clearly months, which are a later addition to the text (Palmer 2004–2012; Pelling 2006). We assume that these months were added by a librarian when the *Voynich Codex* was in the custody of either Rudolf II or possibly the Jesuits. The months are close to the appropriate dates for each sign (Fig. 8.1). The intriguing things about the Voynich zodiac are the use of New World animals in a number of signs, which is consistent with the codex being from New Spain (see Chap. 6), and the use of gender complementarity found in some of the signs, which is consistent with Aztec sensibility (or cosmic duality). The number of nymphs in each sign appears to represent days or degrees of celestial longitude, indicating a general knowledge of the Aztec calendar and Western astrology and astronomy.

Folio 70r (Fig. 8.3)

The first zodiac circle is labeled with the word *mars* (March) and represents Pisces (the fish). In the center, there are two elongated fish that closely resemble alligator gar (*Atractosteus spatula*), a species whose range extends from the Mississippi River Basin into Mexico (see Chap. 7). They perhaps represent a male and female.

Females are typically larger than males. Each fish has a line in its mouth terminating in a star. One is labeled (*ātlācâocâ*), which could be translated from Nahuatl as "still fished" (*ātlācâ* + spear thrower or fisherman, *oc* = still). In two outer, concentric rings, there are 29 nude nymphs, each arising from a round container.

Folios 70v and 71r (Fig. 8.4)
Two circles represent Aries (the ram) and are labeled with the word *abéril* (April). The darkly tinted, larger sheep (Fig. 8.4a), mid-circle in folio 70v, is probably a ram, whereas the other (mid-circle in folio 71r) is a ewe (Fig. 8.4b). Both are nibbling shrubs. They are most likely domestic sheep (*Ovis aries*, Linnaeus 1758) introduced by the Spanish, but could be Mexican desert bighorn sheep (*O. canadensis mexicana*), indigenous to western Sonora, a northwestern border state in Mexico (Tucker and Talbert 2013). There are 15 nymphs in each of the two concentric rings surrounding the symbols, nude in the ewe circle and clothed in the ram circle.

Folios 71v(1) and 71v(2) (Fig. 8.5)
There are two circles, each labeled *maŷ* or *may* (May), that represent Taurus (the bull). Each illustrates a different breed of cattle (*Bos taurus taurus,* Linnaeus, 1758). Folio 71v(1) shows a light red, horned cow (Fig. 8.5a) that resembles the Andalusian Red breed introduced by the Spanish into Mexico. Folio 71v(2) is a bull (Fig. 8.5b), clearly indicated by the penis, similar to the dark red Retinta breed, ancestor to the Texas longhorn (Tucker and Talbert 2013). There are 15 nymphs in concentric rings surrounding each symbol. The nymphs in the cow circle are clothed and arise from containers, whereas the nymphs in the bull circle are standing nudes.

Folio 71v(3) (Fig. 8.6)
This folio is labeled *yoni* (June) and represents Gemini (the twin brothers). However, in the *Voynich Codex*, the Gemini symbol is a couple holding hands, with the male dressed in a hat and tunic and the female in a blue gown. The replacement of the traditional male twins with a heterosexual couple is found in some medieval zodiacs, but also suggests Aztec sensibility to gender equivalence. Surrounding the symbol are 30 standing nude nymphs in three concentric rings.

Folio 71v(4) (Fig. 8.7)
The sign for Cancer (the crab) labeled *iollet* (July) is illustrated by a pair of Mexican crayfish (*Cambarellus montezumae*, Saussure, 1857), perhaps male and female. Although the classical symbol is a crab, some medieval images show a crayfish. Surrounding the symbol are 30 standing nude nymphs in three concentric rings.

Folio 72r(1) (Fig. 8.8)
This folio is labeled *augst* (August) for Leo (the lion), but shows an ocelot (*Leopardus pardalis*), indigenous to South America. Note that the tail, positioned between the legs, resembles the tail position in the medieval image of Leo (Table 8.1), indicating that the artist was aware of the classic depiction of this zodiac symbol, the lion. There are 30 nude nymphs in concentric rings surrounding the symbol.

Table 8.1 Names and zodiac signs in medieval manuscripts, the *Voynich Codex*, and the *Codex Mexicanus*

Constellation (dates)	Traditional medieval sign[a]	*Voynich Codex*				*Codex Mexicanus*
		Voynich folio and image	Sign	Month names[b] **Voynich** *Occitan* English	No. of nymphs	
Pisces (Feb. 19–March 20)	Fish 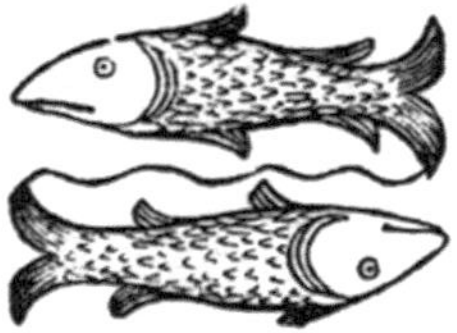	70r 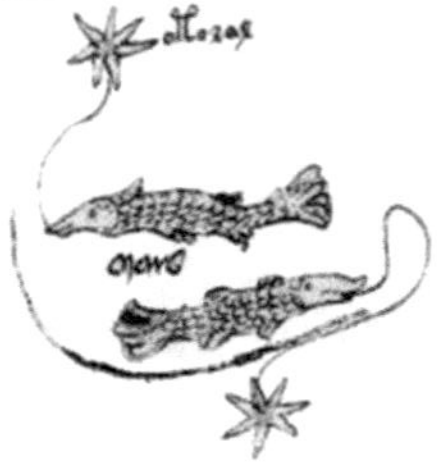	Alligator gar	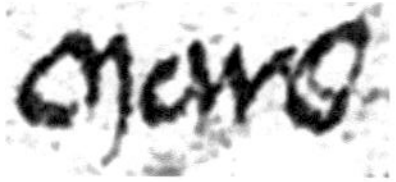 **mars** *març* March		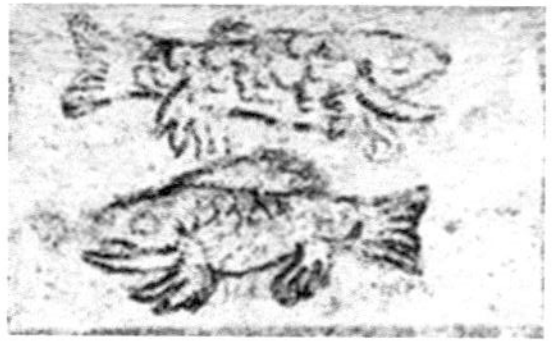
Aries (March 21–April 19)	Ram 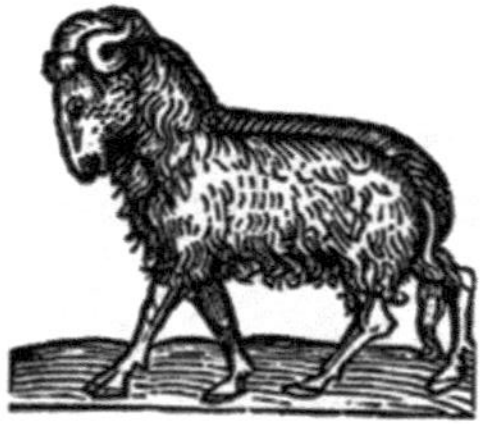	70v 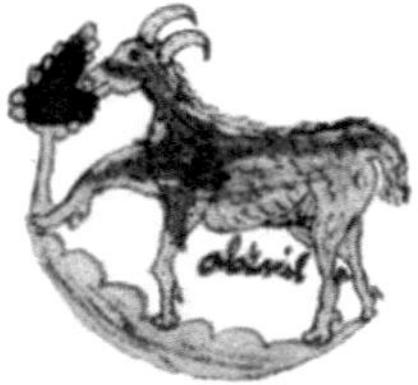	Sheep (ewe)	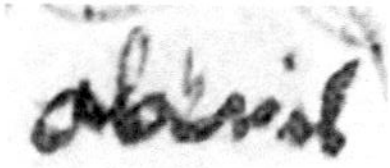 **abéril** *abril* April	15	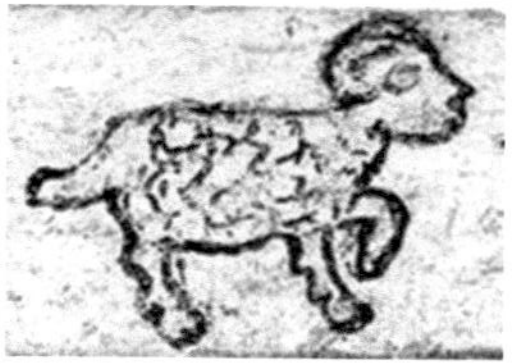
	195	71r 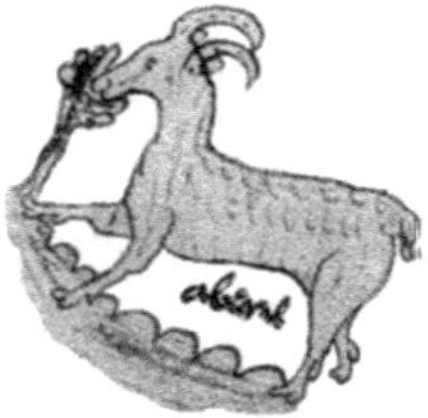	Sheep (ram)	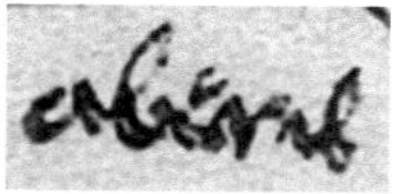 **abéril** *abril* April	15	

(continued)

Table 8.1 (continued)

Constellation (dates)	Traditional medieval sign[a]	*Voynich Codex*				*Codex Mexicanus*
		Voynich folio and image	Sign	Month names[b] **Voynich** *Occitan* English	No. of nymphs	
Taurus (April 20–May 20)	Bull 	71v(1) 	Cow	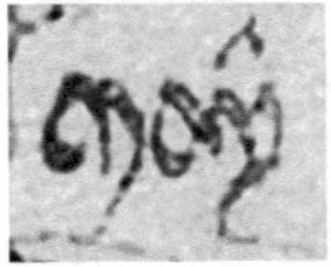 **maŷ** *mai* May	15	
		71v(2) 	Bull	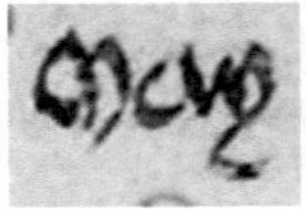 **may** *mai* May	15	
Gemini (May 21–June 20)	Twins 	71v(3) 	Man and woman (rather than male twins)	 **yony** *junh* June	30	

Cancer (June 21–July 22)	Crab[c] 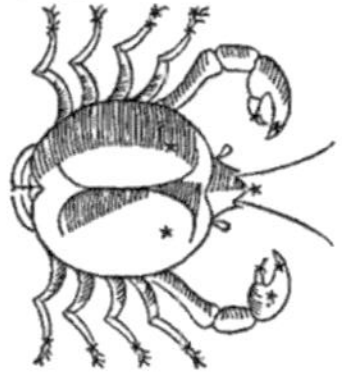					
	Crayfish 	71v(4) 	Mexican crayfish (male & female?)	 **iollet** *julhet* July	30	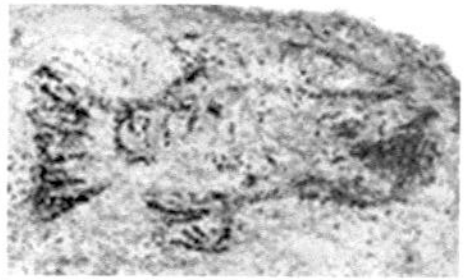
Leo (July 23–Aug. 22)	Lion 	72r(1) 	Ocelot	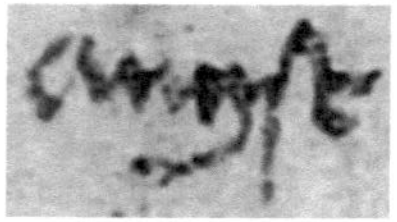 **augst** *agost* August	29–30	

(continued)

Table 8.1 (continued)

Constellation (dates)	Traditional medieval sign[a]	*Voynich Codex*				*Codex Mexicanus*
		Voynich folio and image	Sign	Month names[b] **Voynich** *Occitan* English	No. of nymphs	
Virgo (Aug. 23–Sept. 21)	Virgin 	72r(2) 	Gowned woman	 **septē[m]bre** *septembre* September	29	
Libra (Sept. 22–Oct. 23)	Scales 	72v 	Scales	 **octē[m]bre** *octobre* October	30	
Scorpio (Oct. 24–Nov. 21)	Scorpion 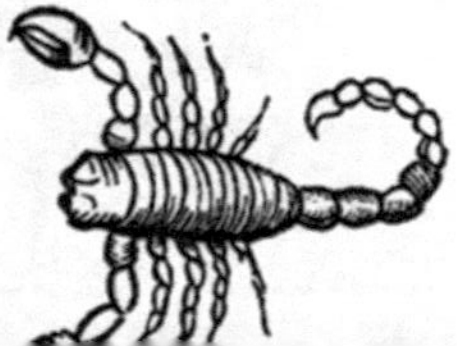					

	Cat-like creature[d]	73r	Jaguarundi	**novē[m]bre** *novembre* November	30	
Sagittarius (Nov. 22–Dec. 21)	Archer	73v	Crossbow archer	**decē[m]bre** *decembre* December	30	
Capricorn (Dec. 22–Jan. 19)	Goat	74r missing				

(continued)

Table 8.1 (continued)

Constellation (dates)	Traditional medieval sign[a]	*Voynich Codex*				*Codex Mexicanus*
		Voynich folio and image	Sign	Month names[b] **Voynich** *Occitan* English	No. of nymphs	
Aquarius (Jan. 20–Feb. 18)	Water bearer 	74v missing				

[a]Signs (except for crab and cat-like creature) are 1491 engravings from *Liber Astronomiae* by Guido Bonatti, ca. 1277
[b]Script identified as Catalan names by Sean B. Palmer (2004–2012) and as Occitan by Nicholas Pelling (2006)
[c]Traditional sign is of a crab, but many medieval signs are crayfish. Source: *Poetica Astronomicus*, 1482
[d]Some signs in northern Europe show a cat-like creature. Source: Psalterium of 1250–1270, Cambridge Ms. B.11.4

Folio 72r(2) (Fig. 8.9)

Virgo (the virgin or maiden) is associated with *septē[m]bre* (September; note that "ē" is pronounced "em"), symbolized by a woman in a long, blue, voluminous gown wearing a blue hat, holding a star, and standing on turf with a flower. Surrounding the symbol are 29 nude nymphs within two concentric rings.

Folio 72v (Fig. 8.10)

The sign Libra (the scales) is labeled *octē[m]bre* (October) and represented by the traditional scales of justice. There are 30 standing nymphs in two concentric rings surrounding the symbol.

Folio 73r (Fig. 8.11)

The traditional zodiac sign labeled *novē*[m]*bre* (November) is Scorpio (the scorpion). In northern Europe, where scorpions are rare, the sign is sometimes an animal with a cat-like face and long tail (Table 8.1). In the *Voynich Codex*, the sign is represented by a black cat-like creature that resembles a jaguarundi (*Puma yagouaroundi*, Geoffroy Saint-Hilaire, 1803), with a string in its mouth connected to a star. The jaguarundi's long, curved tail suggests the curved tail of a scorpion. There are 30 standing nude nymphs within the rings around the symbol.

Folio 73v (Fig. 8.12)

The sign probably labeled *decē*[m]*bre* (December) representing Sagittarius (the archer) shows a male in a blue tunic and cream hat with a plume, holding a crossbow, rather than the traditional centaur with a bow (Fig. 8.12). There are 30 standing nude nymphs, 10 in the inner concentric ring, 16 in the outer ring, and four on the outer rim.

The sequence of the zodiac in the *Voynich Codex* begins with Pisces and is different from the Western Gregorian calendar, which begins 1 January. The beginning of the year in Babylon, based on the lunar calendar, was usually in March, and the Aztec calendar usually began around 23 February. Thus, the beginning of the zodiac in March may represent Aztec sensibilities rather than a misbinding of the missing folio page. Otherwise, the month-by-month sequence is correct.

The Astrological section includes 10 of the 12 zodiac signs. We assume that the two missing zodiac signs would have been in the missing folio (74r, 74v), with Capricorn (the goat, associated with January) on the *recto* side and Aquarius (the water carrier, associated with February) on the *verso* side. If an estimated 30 or 31 nymphs each (for the two signs, Capricorn and Aquarius) were added to the 299 nymphs already shown (in the other 10 signs), there would be about 359 or 360 in total, which suggests that the nymphs might have been symbols for days or degrees of celestial longitude. The Aztec annual calendar was 360 regular days (18 months of 20 days), plus five "unlucky" days (Smith 1996: 256).

Of particular significance are the facts that the animals in the *Voynich Codex* associated with Pisces, Cancer, Leo, and Scorpio are animals indigenous to Mesoamerica or, in the case of Aries and Taurus, could be Spanish imports. Thus, it is clear that the Voynich zodiac represents a fusion of Western and Mesoamerican symbols. Five of the signs (Pisces, Aries, Taurus, Gemini, and Cancer) appear to be associated with male and female forms. This attempt at gender inclusion, referred to as cosmic duality by Rogers (2007), is a feature of the Aztec culture.

Words Associated with the Zodiac in the Voynich Codex

The 12 zodiac circles have nymphs holding stars labeled with Voynichese names, in addition to Voynichese words embedded around the rings. There were a total of 1136 words: 289 associated with nymphs holding stars (nymph names) and 848 associated only with rings (ring words). All the words were decoded based on the syllabary of Tucker and Talbert (2013) and then alphabetized. The nymph names are listed in Appendix B.

In both the nymph names list and the ring words list, many names or words were found more than once. Of the 289 nymph names ascertained, only 259 were unique (89.6%). For example, [Voynich glyphs] was found five times and [Voynich glyphs] was found three times. Both were decoded as *ātlaāchi*. In Nahuatl, *atl* = water and *achi* = a little bit (Karttunen 1983). Of the 848 ring words, 607 were unique (71.6%). Thus, the frequency of duplications in the ring words (28.24%) was almost triple that in the nymph names (10.1%).

We also compared the unique nymph names and the unique ring words with the 48 most common words in the *Voynich Codex*, based on the list of Kevin Knight (2009), as shown in Table 8.2. Of the 259 unique nymph names, seven (2.70%) were in the Knight list: [Voynich glyphs] (*ātlaai*), [Voynich glyphs] (*ātlachi*), [Voynich glyphs] (*ātlocâ*), [Voynich glyphs] (*ātlocâ*), [Voynich glyphs] (*ātloll*), [Voynich glyphs] (*mchi*), and [Voynich glyphs] (*oe*). Of the 607 unique ring words, 32 (5.77%) were in the Knight list. Thus, the most frequent Voynich words were twice as frequent in the ring words as in the nymph names. This indicates that the ring words were probably words associated with prose, where duplicate words would be expected and where more common words would be found, in contrast to nymph names, which must be proper nouns. These results would be analogous to the common word frequency in Shakespeare prose versus a list of characters in a play. We conclude that the ring words were text and the nymph names were proper nouns.

Zodiac Mexicanus

A zodiac (Fig. 8.13) is found in the *Codex Mexicanus* (1571), an early colonial, Mexican, pictorial manuscript, housed in the Bibliothéque Nationale de France in Paris. From internal evidence, Manuel Aguilar-Moreno (2007: 271) dates it to 1570–1590, but the last entry is 1583. This zodiac shows associations with the four Aristotelian elements (Brotherston 1998): air (breath emanating from a head), water (an ice crystal producing hail and rain), fire (a crown of flames), and Earth (a plot of ground pierced by the triangular point of a digging stick; Fig. 8.14). Aquarius, Gemini, and Libra are associated with air; Pisces, Cancer, and Scorpio with water; Aries, Leo, and Sagittarius with fire; and Taurus, Virgo, and Capricorn with Earth. As in the *Voynich Codex*, one traditional sign was modified: a fish was exchanged for the traditional crab in Cancer (Table 8.1 and Fig. 8.13). The transformation of Cancer into a fish is appropriate for its association with water. Some of the signs

Most frequent 48 words in Knight		Frequency of 11,689 (%)	Presence in Voynich zodiac		Most frequent 48 words in Knight		Frequency of 11,689 (%)	Presence in Voynich zodiac	
Voynichese	Decoded		Nymph names	Ring words	Voynichese	Decoded		Nymph names	Ring words
[Voynichese]	aca	4.59	0	3×	[Voynichese]	mae	1.87	0	1×
[Voynichese]	ae	3.11	0	0	[Voynichese]	mai	2.94	0	5×
[Voynichese]	atlaai	1.51	2×	21×	[Voynichese]	mchi	1.28	0	2×
[Voynichese]	atlaai	1.20	0	0	[Voynichese]	mcui	1.20	0	1×
[Voynichese]	atlachi	1.01	1×	3×	[Voynichese]	mi	1.33	0	2×
[Voynichese]	atlachi	1.33	3×	3×	[Voynichese]	n	2.08	0	0
[Voynichese]	atloca	1.18	4×	3×	[Voynichese]	noll	1.23	0	5×
[Voynichese]	atloca	1.22	0	0	[Voynichese]	oca	2.22	0	12×
[Voynichese]	atloe	1.10	0	7×	[Voynichese]	oe	2.99	1×	2×
[Voynichese]	atloe	1.21	0	6×	[Voynichese]	oll	4.01	0	19×
[Voynichese]	atloll	1.81	0	4×	[Voynichese]	quaca	1.29	0	0
[Voynichese]	atloll	1.32	1×	10×	[Voynichese]	quatlaachi	2.61	0	2×
[Voynichese]	atol	1.23	0	0	[Voynichese]	quatlaai	2.63	0	1×
[Voynichese]	camachi	1.02	0	0	[Voynichese]	quatlachi	2.33	0	0
[Voynichese]	chi	2.31	0	5×	[Voynichese]	quatli	1.26	0	1×
[Voynichese]	choca	2.16	0	3×	[Voynichese]	quatloca	1.63	0	0
[Voynichese]	choe	2.72	0	4×	[Voynichese]	quatloe	1.30	0	0
[Voynichese]	chol	1.81	0	1×	[Voynichese]	quatlol	2.39	0	0
[Voynichese]	choll	7.38	0[a]	7×	[Voynichese]	quatloll	2.24	0	0
[Voynichese]	i	1.29	0	0	[Voynichese]	tsa	1.11	0	0
[Voynichese]	maaca	1.47	0	3×	[Voynichese]	tsaai	1.23	0	0
[Voynichese]	maai	1.49	0	2×	[Voynichese]	tsaca	1.59	0	2×
[Voynichese]	maca	3.39	0	2×	[Voynichese]	tsachi	3.64	0	6×
[Voynichese]	machi	4.29	0	2×	[Voynichese]	tsai	2.42	0	0

[a]choll mahua = [Voynichese]

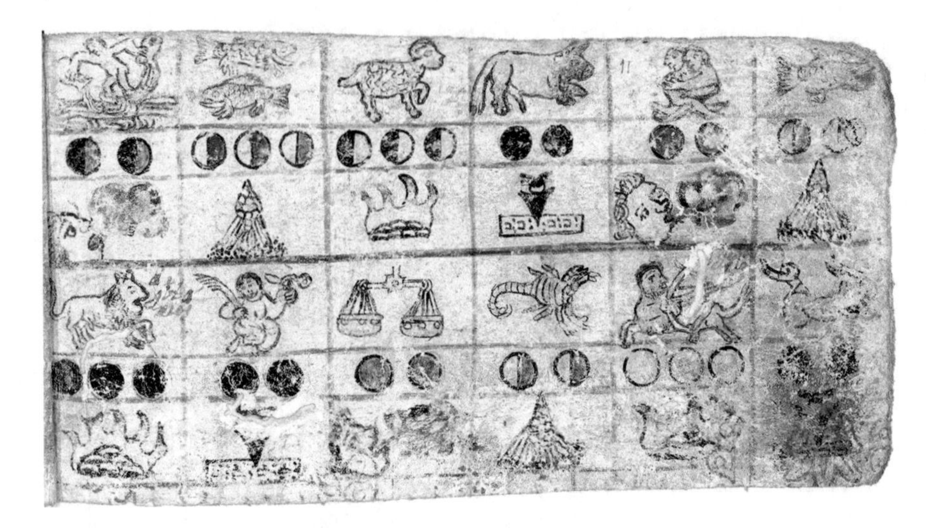

Fig. 8.13 The zodiac in the *Codex Mexicanus*

Fig. 8.14 An Aztec digging stick

show changes from the traditional symbols, displaying Aztec sensibilities. Aquarius shows two water jars, reflecting Aztec duality. Gemini (the twin brothers) is shown as a copulating couple sitting astride face-to-face, which has also been found in pre-Hispanic manuscripts, according to Gordon Brotherston (1998). As expected, there are allusions to medieval European zodiacs. Virgo is represented as a maiden with a frond and a flower similar to a fifteenth century medieval image (Fig. 8.15). Thus, the zodiac in the *Codex Mexicanus*, as in the *Voynich Codex*, shows a fusion of European and Aztec influences.

Fig. 8.15 Medieval maiden representing Virgo

Conclusion

The zodiac in the Astrological section provides evidence that the *Voynich Codex* is a Mesoamerican work based on the substitution of indigenous animals and the use of gender to separate some of the traditional zodiac signs. The number of nymphs located in concentric rings surrounding the zodiac symbols probably represents days. The zodiac in the *Codex Mexicanus* is not directly related to the *Voynich Codex*, but demonstrates a similar fusion of Western and Aztec iconography that was brought about by the clash of cultures in New Spain.

Literature Cited

Aguilar-Moreno, M. 2007. *Handbook to life in the Aztec world.* New York: Oxford University Press p. 271.

Brotherston, G. 1998. European scholasticism analyzed in Aztec terms: The case of the Codex Mexicanus. Boletim do CPA 5/6 Campinas, Jan/Dec.169–181.

Codex Mexicanus. 1571–1590. World digital library. https://www.wdl.org/en/item/1528t4/view/1/1.

Karttunen, F. 1983. *An analytical dictionary of Nahuatl.* Norman: University of Oklahoma Press.

Knight, K. 2009. *The Voynich manuscript.* MIT. http://www.isi.edu/natural-language/people/voynich.pdf. 1 May.

Montemurro, M.A., and D.H. Zanette. 2013. Keywords and co-occurrence patterns in the Voynich manuscript: An information-theoretic analysis. *PLoS One* 8 (6): e66344. https://doi.org/10.1371/journal.pone0066344.

Palmer, S.B. 2004–2012. Voynich manuscript: Months. http://inamidst.com/voynich/months.

Pelling, N. 2006. *The curse of the Voynich: The secret history of the world's most mysterious manuscript*. Surbiton: Compelling Press.

Rogers, R.C. 2007. The resilience of Aztec women: A case study of modern Aztec myths. *Journal of Humanities & Social Sciences* 1 (2): 1–18 footnote 4.

Smith, M.E. 1996. *The Aztecs*. Oxford, UK: Blackwell.

Tucker, A.O., and R.H. Talbert. 2013. A preliminary analysis of the botany, zoology, and mineralogy of the Voynich manuscript. *HerbalGram* 100: 70–85.

Chapter 9
Astronomical Images

Jules Janick

Aztec Astronomy

Astronomy was vital to the Aztecs' religious beliefs, rituals, and complex calendars. They designed temples for astronomical measurement and built them specifically to observe the heavens from fixed points. Aztec nobility and priests carried out celestial observations and calculations (Fig. 9.1). The rising and setting of the sun, stars, planets, and other heavenly bodies were tracked daily and by seasons, with the sun being of prime interest. The Aztecs placed great importance on Venus, and also identified several constellations (Table 9.1). The celestial bodies of significance were the sun, moon, planets, stars, and constellations (Fig. 9.2).

Sun

The central figure in Aztec cosmology was the sun (*tonatiuh*). The Aztecs believed that four previous suns had been destroyed, and the present one was also destined for destruction (Caso 1959).

Moon

The Nahuatl moon (*metzli*) was considered by the Aztecs to be both a coward and a copy of the sun. The moon was associated with rain, vegetation, menstruation, fertility, and reproduction (Aguilar-Moreno 2007: 309).

J. Janick, A. O. Tucker, *Unraveling the Voynich Codex*, Fascinating Life Sciences, https://doi.org/10.1007/978-3-319-77294-3_9

Fig. 9.1 The Aztec astronomer. (Source: *Codex Mendoza*, Berdan and Anawalt 1997)

Table 9.1 Aztec constellations

Nahuatl name	Constellations, planets, and stars from the *Florentine Codex*	Western name	Western constellations
mamahuaztli		Cancer	
tianquiztli		The Pleiades in Taurus	
citlatlachtli		(Celestial ball court) Gemini	
citlalxonecuilli		Aries?	
citlalcolotl		Scorpio	

(continued)

Table 9.1 (continued)

Nahuatl name	Constellations, planets, and stars from the *Florentine Codex*	Western name	Western constellations
colotlixayac (citlalcolotl?)		Orion?	
ozomatli yohualiqui mamalhuaztli		Orion's belt	
colotlixayac (citlalcolotl)			
xitlalzonecuilli (xonecuilli)		Little Dipper	
texcatlipoca		Big Dipper	

Source: Aguilar-Moreno (2007)

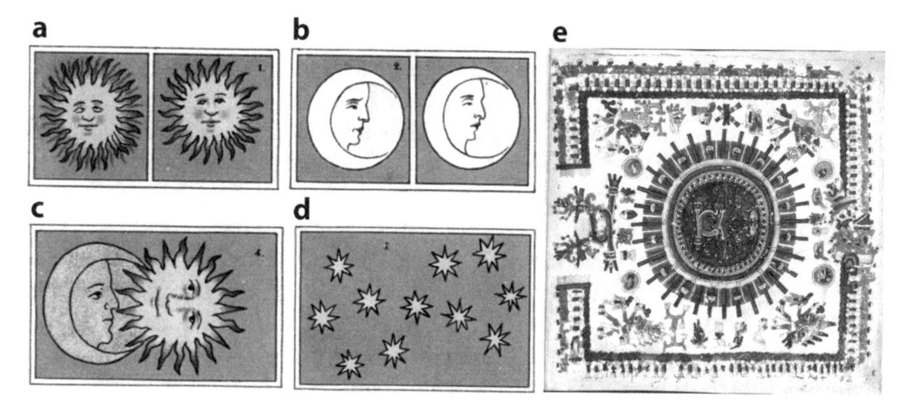

Fig. 9.2 Astronomical symbols in the *Florentine Codex*. (**a**) Sun, (**b**) moon, (**c**) eclipse of the moon, and (**d**) stars. (Source: Sahagún 1953). (**e**) Venus in the *Codex Borgia*. (Source: Milbrath 2013)

Planets

The Aztecs recognized Mercury, Saturn, Jupiter, Mars, and Venus. Venus was an intrinsic part of the Aztec calendar and was tied in many ways to the sun. It was recognized as both an evening star and as a morning star. The Nahuatl names for Venus were *citlalpol,* or sometimes *totonametl*, depending on the context.

Stars

Many stars were named, but in most cases these historic records were lost. The star cluster known to us as the Pleiades (*tianquiztli* or gathering place) in the constellation Taurus was vitally important. Its appearance at the zenith at midnight marked a 52-year cycle, celebrated by the Aztecs during the New Fire ceremony. A number of constellations were also recognized (Table 9.1).

Cosmological Illustrations in the Voynich Codex

The Cosmological section in the *Voynich Codex* includes 16 magic circles that feature combinations of the Earth, sun, moon, planets, and stars (Fig. 9.3). Human faces or figures are embedded in four folios: folio 57v (not considered a cosmological circle; see Chap. 5, Fig. 5.3); folio 67v(1) (15 small heads); folio 85v(1) (perhaps representing four ages of humans); and folio 86v(2) (four hidden figures holding objects perhaps representing seasons). The cosmological images are difficult to explain without a translation of the text.

Earth

The T and O (T-O) map is made up of a circle divided once along the diameter and then once again at a right angle from the diameter (Fig. 9.4). It was also referred to as a Beatine or Beatus map because it was attributed to the Spanish monk Beatus of Liébana, a *comarca* (region) of Cantabria. The T-O map represents the physical world first described by the seventh century scholar Saint Isidore of Seville (Spain) in his *Etymologiae*. In many medieval maps, the symbol relates to three continents: Asia (large subsection), Europe, and Africa. The inclusion of a T-O map describing the four winds in the *Florentine Codex* by Bernardino de Sahagún (see Fig. 9.4c), which is restricted to Aztec images, indicates that it was clearly a common feature known by the Spanish clergy in New Spain. Thus, the presence of a T-O map in Sahagún is consistent with the hypothesis that the *Voynich Codex* might be Mesoamerican.

A T-O map is found in four locations in the *Voynich Codex* (Fig. 9.5).

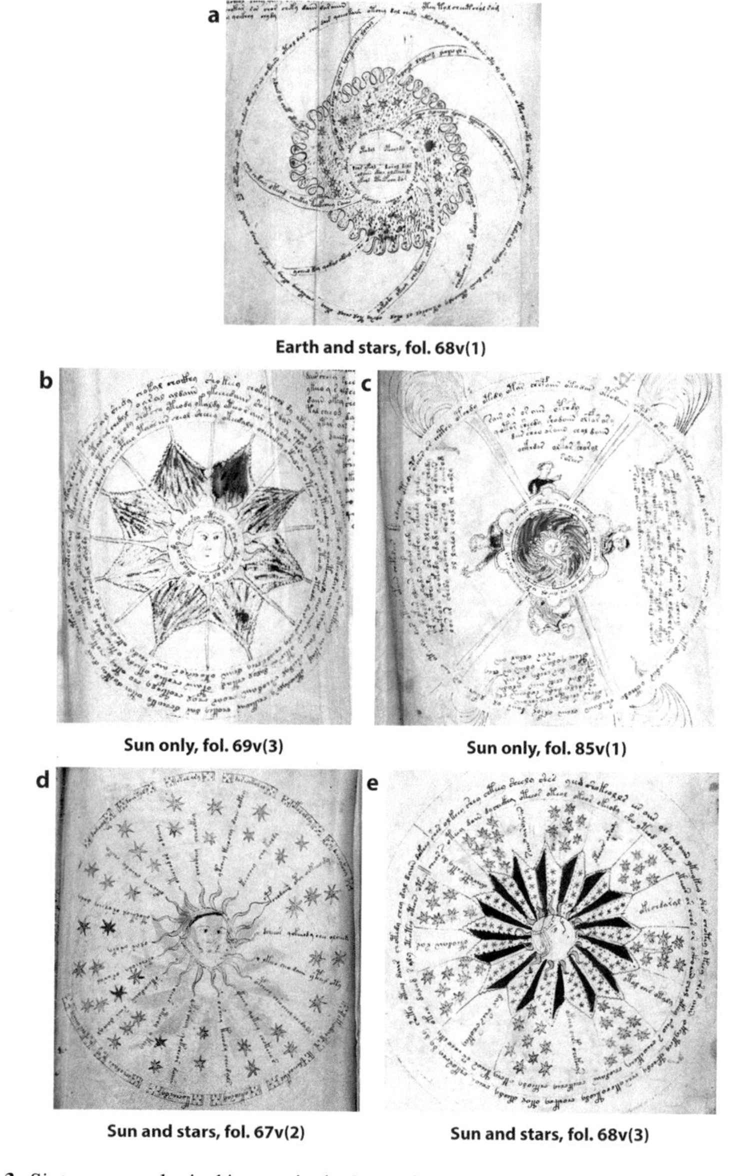

Fig. 9.3 Sixteen cosmological images in the *Voynich Codex*. (**a**) Earth and stars; (**b**, **c**) sun only; (**d**, **e**) sun and stars; (**f**) moon only, note four embedded figures highlighted in *red*; (**g**, **h**) sun and moon with named stars; (**i**) moon with stars; (**j**) moon with named stars; (**k**) planet, perhaps Venus; (**l**, **m**) planets; (**n**) planet with moon phases; (**o**) planet with stars; (**p**) sun, moon, Earth, planet, and stars

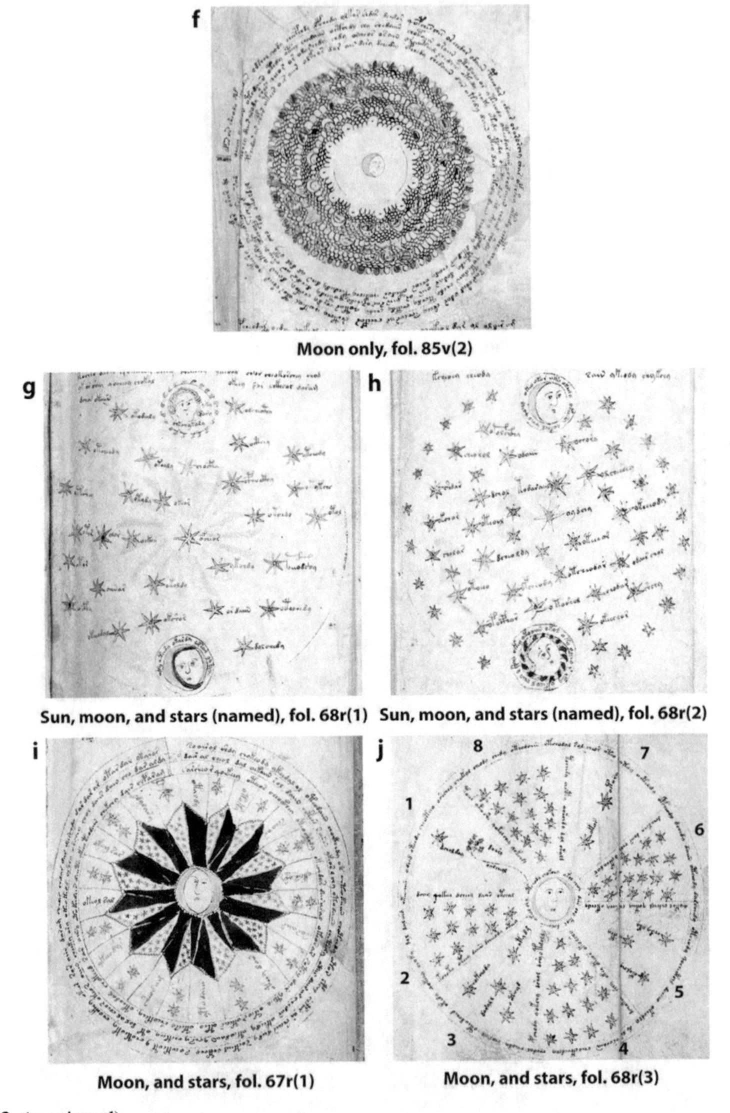

Moon only, fol. 85v(2)

Sun, moon, and stars (named), fol. 68r(1)

Sun, moon, and stars (named), fol. 68r(2)

Moon, and stars, fol. 67r(1)

Moon, and stars, fol. 68r(3)

Fig. 9.3 (continued)

Folio 67v(1) (Fig. 9.5a) A T-O map is shown in color in the lower left along with the rising and setting sun and a central, elongated, starfish-shaped form that we assume is Venus. The four sets of orbs in each corner contain three or four heads in a chain in four configurations, which are unexplainable, but (according to Aztec tradition) may represent the previous four world epochs. It shows similarities to the Aztec calendar stone (see Chap. 16, Fig. 16.9).

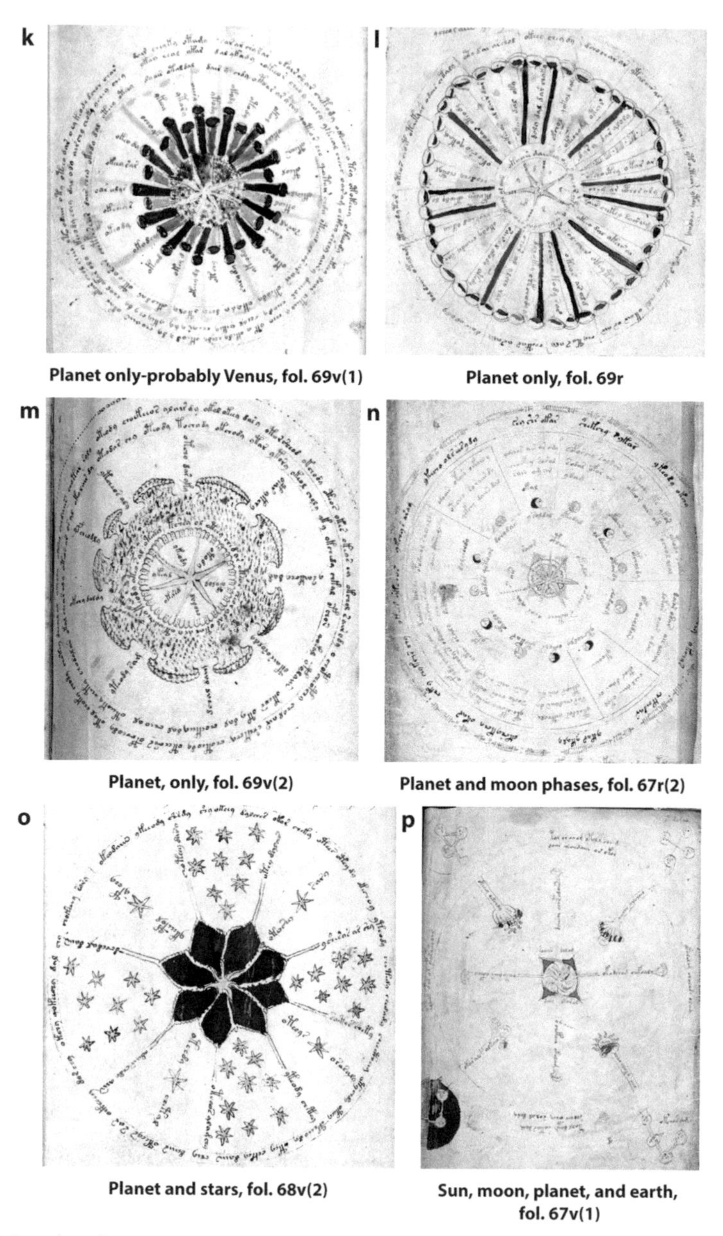

Planet only-probably Venus, fol. 69v(1)

Planet only, fol. 69r

Planet, only, fol. 69v(2)

Planet and moon phases, fol. 67r(2)

Planet and stars, fol. 68v(2)

Sun, moon, planet, and earth, fol. 67v(1)

Fig. 9.3 (continued)

Folio 68v(1) (Figs. 9.5b and 9.6a) A T-O map is covered by a mantle of stars, with eight spirals: four emanating from the circular map and four from the mantle. The mantle of stars is very similar to that found in circle A of folio 86v, which represents Puebla de los Angeles, also called the Celestial City of Jerusalem (see Chap. 10).

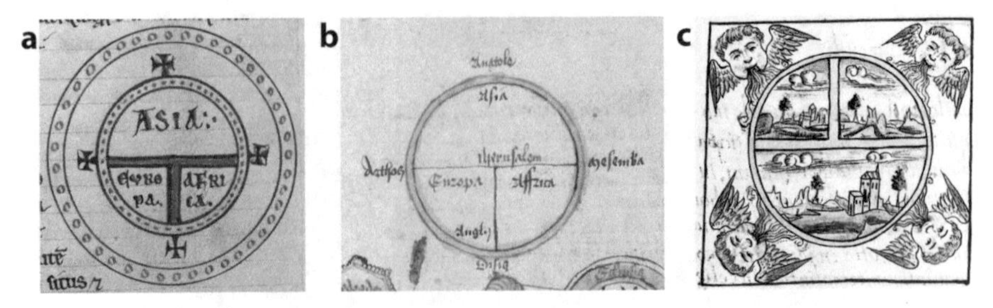

Fig. 9.4 T-O maps. (**a**, **b**) Classic medieval images, and (**c**) image of four winds in the *Florentine Codex*

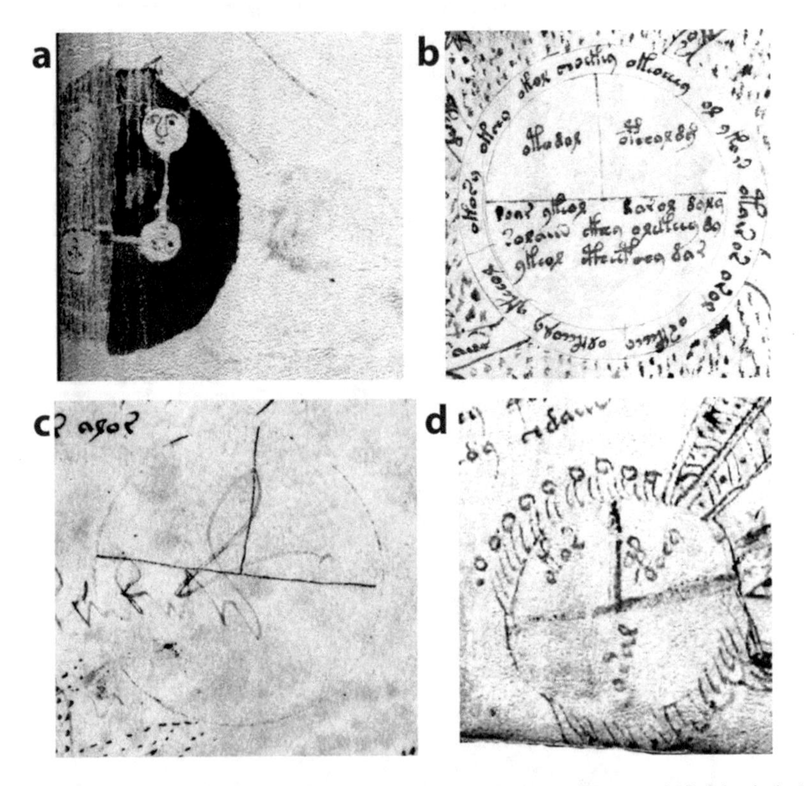

Fig. 9.5 T-O maps in *Voynich Codex* folios. (**a**) 67v(1), (**b**) 68v(2), (**c**) 86r(3), and (**d**) 86v circle 1

Folio 86r(2) (Fig. 9.5c) A T-O map is very lightly sketched in the middle of the page, which also shows strange, spewing, fruiting bodies that some have considered to be phallic-like.

Folio 86v (Fig. 9.5d) A T-O map is shown in circle 1 of the kabbalah Tree of Life, which probably represents the sephiroth Malkuth, meaning Kingdom or Earth (see Chap. 10).

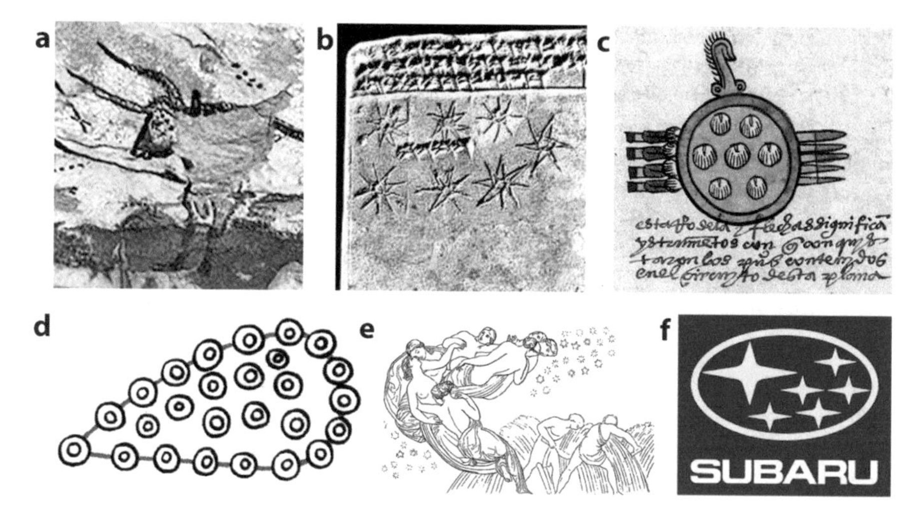

Fig. 9.6 Illustrations of the Pleiades. (**a**) Neolithic cave painting, (**b**) Sumerian tablet, (**c**) Aztec shield, (**d**) the Pleiades from the *Florentine Codex*, (**e**) Seven Sisters, and (**f**) Subaru logo

Sun Imagery

The Sun, represented by a face with rays, is found in seven cosmological folios. The sun is centric in four folios: 67v(2), 68v(3), 69v(3), and 85v(1), demonstrating the key role of the sun in Aztec traditions. The heliocentric position of the sun was incompatible with Christian or Islamic cosmology until Galileo.

Moon Imagery

Images of the moon representing the night sky are included in seven cosmological folios. The moons are shaped either as heads (without rays) or as crescents.

Planet Imagery

We believe that the elongated, starfish-shaped symbols in the Cosmological section most likely represent Venus, but other planets well known in both Aztec and Spanish astronomy may be involved. We cannot assign names to images of planets with starfish symbols in the center, but Nicolas Pelling (2006) believes that folio 69r is Mercury based on the 46 spokes, which may represent 46 goal years, the number of Earth years that coincides with planet years. We are unaware whether Aztec astronomers would know the goal years of different planets. Symbols of Venus can be found in the pre-Columbian *Codex Borgia* (Fig. 9.2e). We were struck by the similarity between the symbols of Venus in the *Codex Borgia* and *Voynich Codex* folio 69v(1) (Fig. 9.3k). The symbol of Venus in Book 7 in the *Florentine Codex* (1540–1590) is a typographical error, according to Susan Milbrath (personal correspondence). This image has been suggested to be the constellation Gemini (Aguilar-Moreno 2007).

Star Imagery

Stars are the most prevalent astronomical symbol and are labeled in folios 68r(1) (24 stars), 68r(2) (29 stars), and 68r(3) (10 stars + the Pleiades). The names have been deciphered from the Voynichese names and may be Arabic. This is discussed in detail in Chap. 13.

The Pleiades The star cluster known as the Pleiades is a striking feature in the night sky and was recognized by very ancient cultures as far back as the Neolithic period (Fig. 9.6). Galileo, the first astronomer to view the Pleiades through a telescope, included a sketch showing 36 stars in this cluster. In the New World, the Pleiades was recognized by the Aztec, Sioux, and Cherokee cultures. The Aztecs closely followed its appearance and disappearance in the night sky as part of their cosmology (Malmstrom 1997). The star cluster is most obvious to the naked eye in the night sky during December and January in both the Northern and Southern Hemispheres.

The Pleiades illustration in Voynich folio 68r(3) (Fig. 9.3j) consists of eight segments, four with stars only and four with labeled stars. The four segments with named stars are surmised to represent the overhead sky (perhaps Puebla or its vicinity) in four seasons (winter, spring, summer, and fall): perhaps 15 December (segment 1), 15 March (segment 3), 15 June (segment 5), and 15 September (segment 7). Segment 1 of folio 68r(3) shows three items:

1. A seven-star cluster assumed to be the Pleiades labeled 8oa?9 (*chāoei*).
2. A curved line under the seven stars, suggesting movement underscored with the name oa{cccoq (*āocâmaācâ*).
3. A large star labeled oa{cccoq (*chmācâhoi*; see Chap. 13).

The Pleiades is a common site in winter and is invisible in summer, as it is only present in the sky during the day. In the mid-fall and mid-spring, the star cluster is near the horizon and is often invisible because of haze on the horizon. If the seven small stars in segment 1 of folio 68r(3) represent the Pleiades, the large, bright star to its left may be the very bright star Aldebaran (in Taurus) or possibly Betlegeuse (in the right shoulder of the constellation Orion), which is below and slightly west of the Pleiades in the sky. If we assume that segment 1 represents the overhead night sky near Mexico City in winter, then the named stars in the four segments are likely to be the following:

Segment 1 shows the Pleiades and Aldebaran, the brightest star in the constellation Taurus, during winter.

Segment 3 shows four bright stars (Regulus, Denebola, Algieba, and Zosma – the four brightest stars in the constellation Leo), which are overhead in spring. The Pleiades is on the western horizon and barely visible.

Segment 5 shows the three brightest stars – Sargas, Antares, and Acrab – in the constellation Scorpio, which is overhead in summer. The Pleiades is only visible in the sky during the day and therefore invisible in the night sky.

Segment 7 shows two stars, Sadalsuud and Sadelmelik, the two brightest stars in the constellation Aquarius, which is overhead in fall. The Pleiades is on the eastern horizon and barely visible.

The stars in these four segments have Voynichese name labels. The decoded names for stars in segments 3, 5, and 7 did not correspond to any of the stars' actual names (see Chap. 13). We assume that the unnamed stars in segments 2, 4, 6, and 8 merely represent the starry sky between seasons.

The Pleiades and the Calendar The Aztecs had two calendars, a 365-day calendar called *xiuhpohualli* (solar calendar) and a 260-day calendar called *tonalpohualli* (sacred calendar). The 365-day calendar had 18 months of 20 days with five "unlucky" days (*nemontani*). The year began with the Pleiades directly overhead in the winter night sky. Every 52 years, the Pleiades appeared directly overhead at midnight, based on conjunction of the star cluster with the sun and the Earth. On the other side of the Earth, the sun was at the 12 o'clock noon position. The 260-day sacred calendar was divided into 13 named periods (numbers or weeks with 20 sequentially repeating days, each with its own day sign; Fig. 9.7). Thus, any day could be distinguished by its unique combination of week number and day sign.

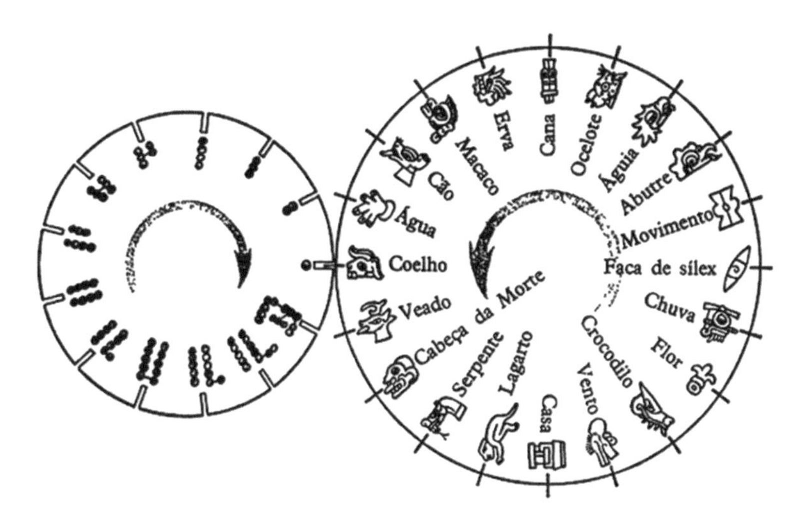

Fig. 9.7 The 260-day calendar based on 13 weeks (*tricena*) and 20 day signs, producing 260 days with a unique combination of day signs and week numbers

Fig. 9.8 The New Fire ceremony. (**a**) Fire drill placed on the chest of a sacrificial victim. (Source: Sahagún 1981); and (**b**) fire distributed to the temple. (Source: *Codex Borbonicus*)

The two calendars (365- and 260-day) coincided every 52 years. This 52-year cycle was critical because the Aztecs (and the Mayans) believed that it could potentially mark the end of the world and required a human sacrifice to keep the world turning and prevent the cessation of the heavens. On this date, the New Fire, or Binding of the Years, ceremony occurred. The once-in-a-lifetime event involved the heart sacrifice of a warrior with burning wood bundles representing the past 52 years placed on his chest cavity (Fig. 9.8a), while the populace cut and bled themselves. The fire was subsequently brought to the Great Temple at Tenochtitlan (Fig. 9.8b) to be shared by the entire community, who rekindled all fires that had previously been extinguished.

The Number Four in the Voynich Codex

In Aztec religion and cosmology, the number four holds special significance with sacred, ritualistic connotations. In Book 10 of the *Florentine Codex*, the number four appears in virtually every chapter designating some ritual (e.g., 4 days of fasting associated with the sun, 4 days of penitence, and 4 days of fire associated with the moon). At King Moctezuma I's inauguration ceremony (in 1440), he was advised to arise at midnight to observe the stars that precisely marked the four cardinal points of the sky: *yohualitqui mamalhuaztli* (Orion's Belt), *citlaltlachtli* (celestial ball court – Gemini and other stars to the right), *colotl ixayac* (the Scorpion), and *tianquiztli* (the Pleiades) (Aguilar-Moreno 2007: 308). Specifically important to the Aztecs were the four cardinal directions. Graphically, the Aztecs placed east on top, north to the left, west on the bottom, and south to the right (Aguilar-Moreno 2007: 303).

The number four comes up many times in the *Voynich Codex*, especially in the Cosmological section.

1. Folio 57v (not strictly in Cosmological section; see Chap. 5, Fig. 5.3): there are four human figures with arms outstretched, and four lines of text radiating from a central shape.
2. Folio 6v (Fig. 9.3p): there are four slim shafts originating from the center star shape. In addition, there are four "puffs of vapor" (D'Imperio 1978) coming in from the diagram's four corners, and at the head of each rests either a sun or a moon. Outside of the diagram, there are four instances of circular, perhaps lunar figures connected by stems and arranged in various configurations.
3. Folio 68v(1) (Fig. 9.3a): from the center circle, four curved passages of text emanate. Four more curved passages of text emerge from the surrounding starry swirl.
4. Folio 68v(2) (Fig. 9.3d): the diagram is divided into eight segments, four of which include a star and text, and the other four contain only stars.
5. Folio 68r(3) (Fig. 9.3j): there are eight segments of the diagram, four of which contain constellations with accompanying text, and the other four include merely stars.
6. Folio 85v(1) (Fig. 9.3c): this is the diagram that relies most heavily on the use of the number four. The diagram is divided into quadrants by four stems, at the end of which spout what appears to be wind, indicating the four winds. Inside the circle, there are four human figures. Mary E. D'Imperio (1978) suggests that these four figures might represent the four ages of man, as one bent figure seems to be holding a linked chain that resembles a cane.
7. Folio 85v(2) (Fig. 9.3f): embedded in the diagram are four human figures, each holding various items.

Conclusion

The Aztecs clearly had a formidable knowledge of astronomy, including information on the sun, moon, planets, and stars, including the Pleiades. All these celestial bodies are found in the *Voynich Codex*. Many of the illustrations in the Codex, such as T-O maps and images of the sun, moon, and stars, are found in Sahagún and the ceiling murals of Tecamachalco (see Chap. 15, Fig. 15.4). The Aztecs held the number four in high ritualistic esteem, and so did the Voynich illustrator. From the evidence above, we conclude that the Cosmological section of the *Voynich Codex* is compatible with a Mesoamerican origin.

Literature Cited

Aguilar-Moreno, M. 2007. *Handbook to life in the Aztec world*. New York: Oxford University Press.

Berdan, F.F., and P.R. Anawalt. 1997. *The essential Codex Mendoza*. Berkeley: University of California Press.

Caso, A. 1959. *Aztecs: People of the sun*. Oklahoma City: University of Oklahoma Press.

D'Imperio, M.E. 1978. *The Voynich manuscript: An elegant enigma*. Fort George, Meade: National Security Agency/Central Security Service.

de Sahagún, B. 1953. Florentine Codex. General history of the things of New Spain. Book 7–The sun, moon, and stars, and the binding of the years. Trans. A.J.O. Anderson and C.E. Dibble. Salt Lake City: University of Utah Press.

———. 1981. *Florentine Codex. General history of the things of New Spain. Book 2—The ceremonies*. Trans. C.E. Dibble and A.J.O. Anderson. Salt Lake City: University of Utah Press.

Malmstrom, V.H. 1997. *Cycles of the sun, mysteries of the moon: The calendar in Mesoamerican civilization*. Austin: University of Texas Press.

Milbrath, S. 2013. *Heaven and earth in ancient Mexico: Astronomy and seasonal cycles in the Codex Borgia*. Austin: University of Texas Press.

Pelling, N. 2006. *The Curse of the Voynich: The secret history of the world's most mysterious manuscript*. Surbiton: Compelling Press.

Chapter 10
Kabbalah Map of Motolinía's Angelopolis

Jules Janick and Arthur O. Tucker

Folio 86v, the Rosette

Folio 86v (sometimes referred to as the Rosette or Fertilization/Seed page) could be a key to decipherment and understanding of the *Voynich Codex* because it has many figures accompanied by text. The large diagram is on a sheet of vellum equivalent to six pages pasted together (Fig. 10.1a). It includes eight medium-sized circles (connected to each other by pathways) surrounding a large central circle, which is directly connected to four of them. There are also three small circles, including one on the lower left that is completely disconnected from the others. In addition, there are two sun images, one in the upper left and the other in the lower right corner. If the diagram was rotated 135° to the right so that the suns were in an east–west position and the small (disconnected) circle was in the north position, the reoriented diagram (Fig. 10.1b) would bear a striking resemblance to the *sephiroth* of the Tree of Life (Fig. 10.1c), a critical Jewish mystical symbol that enters the world of kabbalah (tradition or practice) in late medieval times (Encyclopedia Judaica 1971). The direction of the written text associated with each circle suggests a cyclical arrangement, as shown in Fig. 10.1d. To simplify the discussion, the circles in folio 86v have been labeled with numbers or letters, as suggested by Thomas Ryba, a theologian at Purdue University in Indiana. Note that there are a total of 12 circles, 11 of which are connected: a small circle at the bottom (circle 1); eight medium-sized circles (circles 2–9) in a ring around a large central circle (circle A); and a small circle that resembles a volcanic caldera (circle B) connected to the bottom of circle 6. A small disconnected circle (circle C) hovers above circle 6.

J. Janick, A. O. Tucker, *Unraveling the Voynich Codex*, Fascinating Life Sciences, https://doi.org/10.1007/978-3-319-77294-3_10

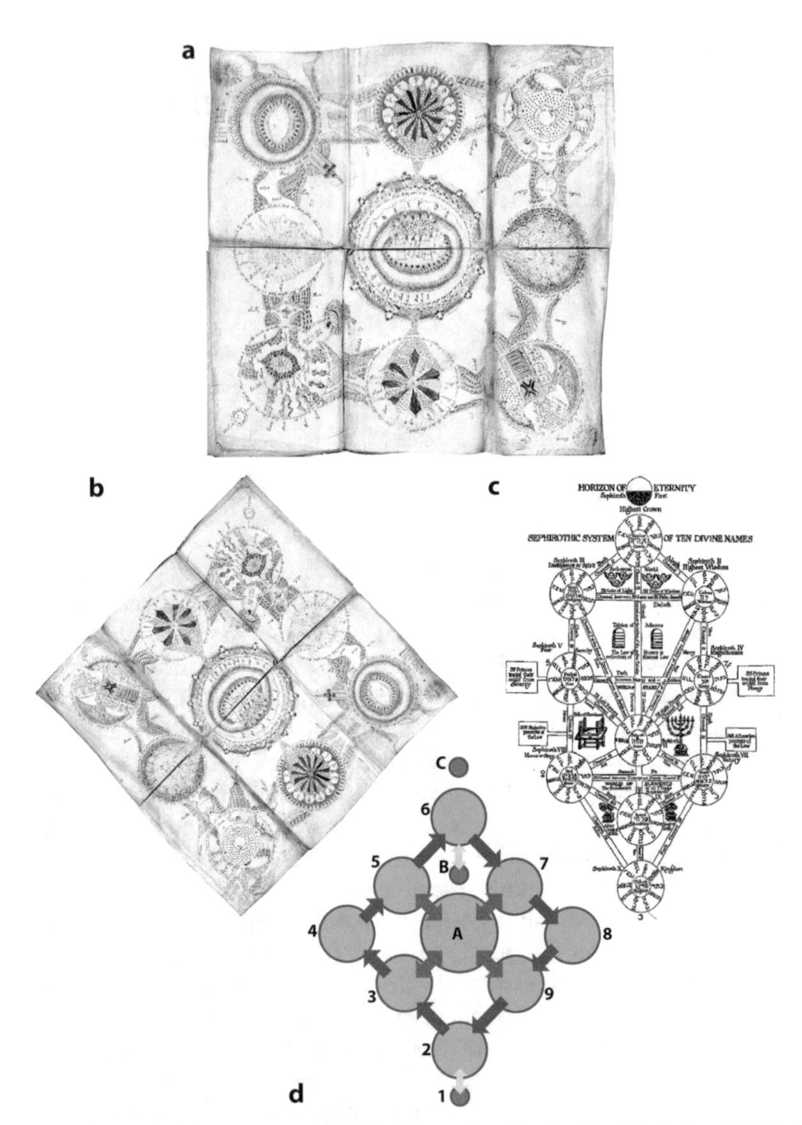

Fig. 10.1 The Voynich Rosette page. (**a**) Folio 86v as found in the *Voynich Codex*; (**b**) folio 86v reoriented with the suns in an east–west position and the small disconnected circle at north; (**c**) Sephirothic Tree of Life of the kabbalah, published in the seventeenth century by the Jesuit priest Athanasius Kircher. (Source: *Oedipus Aegyptiacus*, 1652–1654). (**d**) Schematic diagram of folio 86v with the labeling of the circles and the direction of the text

The Kabbalah Sephirothic Tree of Life

Kabbalah is an ancient Jewish mystical discipline codified in Spain in the thirteenth century text, *Zohar* (the *Book of Splendor*), written by Moses de Leon (1250–1306), a Spanish rabbi. The Sephirothic Tree, a visually descriptive symbol of kabbalah,

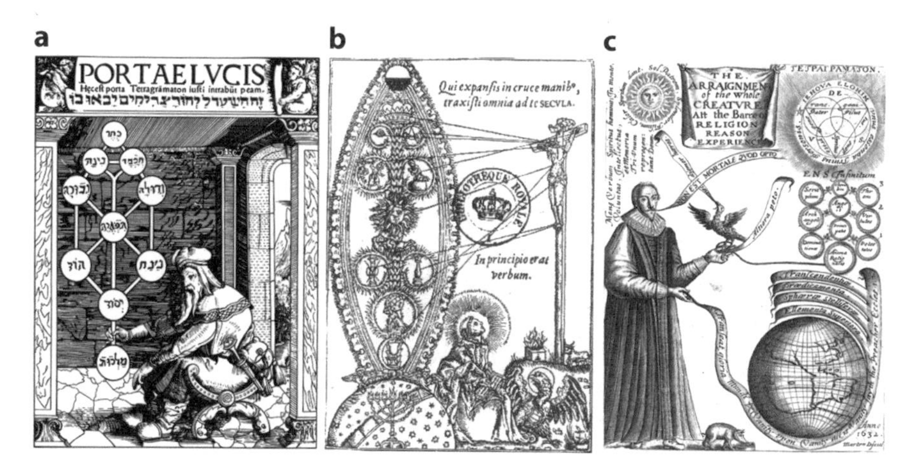

Fig. 10.2 Christian and philosophic kabbalah. (**a**) Diagram of the Tree of Life in Portae Lucis (Portal of Light), a Latin work of 1516 by the Jewish convert Paolo Ricci; (**b**) Christian appropriation of kabbalah; (**c**) portrait of Alexander Henderson by Martin Droeshout (1632), a kabbalistic image with a very large Ain Soph sephira with the names Elohim and Jehovah (God), in addition to the Father, Son, and Holy Spirit, in Latin

consists of ten interrelated *sephiroth* that are considered to be divine emanations (or attributes) of God. The Tree of Life usually is represented as ten *sephiroth*, illustrated as circles and arranged in three columns (as shown in Fig. 10.1) that represent a spiritual map. The *sephiroth* are traditionally numbered and named in Hebrew, as follows: 1 = *Kether* (Crown), 2 = *Chokhmah* (Wisdom), 3 = *Binah* (Intelligence), 4 = *Chesed* (Love), 5 = *Geburah* (Fear), 6 = *Tiphareth* (Beauty), 7 = *Netzach* (Triumph), 8 = *Hod* (Glory), 9 = *Yesod* (Foundation), and 10 = *Malkuth* (Kingdom or Earth). The diagram often shows an 11th sephira circle, called *Ain Soph* (or *Ein Sof*, "without end" or infinite, understood to be the Godhead), hovering, disconnected above the rest. The ten *sephiroth* are connected by mystical pathways and their relationships have been a source of spiritual and occult interest up to the present day.

In the fifteenth to seventeenth centuries, kabbalah was incorporated into and harmonized with Christian theosophy (Fig. 10.2), probably by Jewish *conversos* (converts), and referred to as *cabala* (in Spanish) (Blau 1944; Dedopulos 2012). The Franciscan friar Francesco Giorgio of Venice (1460–1541) authored two large volumes on kabbalah, *De Harmonia Mundi* (1525) and *Problemata* (1536), which were read extensively in the sixteenth century. Before the expulsion of Jews from Spain in 1492, the Iberian Peninsula was home to most major kabbalists. In the early 1500s, Franciscan friars, some perhaps of *converso* background, introduced kabbalah into New Spain. During the sixteenth century, the Spanish Inquisition in Spain banned all forms of kabbalah along with other occult sciences.

The Inquisition in New Spain existed as early as 1522, but King Philip II of Spain signed royal degrees on 16 August 1564 and 25 January 1569, which formally created the Tribunal of the Inquisition in New Spain. During the first period (1522–1536), inquisitions were conducted by local officials. With the arrival of the first bishops (led by Fray Juan de Zumárraga) in New Spain, the second period (1536–1569) witnessed a rise in inquisitions in which early bishops and archbishops served as inquisitors. The third period (1571–1820) saw the establishment of the official Tribunal of the Inquisition (Perry and Cruz 1991; Villa-Flores 2006; Don 2010; Chuchiak 2012).

Folio 86v as the Sephirothic Tree of Life

There is a good fit between the Sephirothic Tree and the Voynich folio 86v diagram (Table 10.1). Thus, the highest circle (circle C) shown in Fig. 10.1d, which has no direct connection to the other circles in the diagram, would represent *Ain Sof* (God). Evidence that supports this comes from a right angle-like symbol (ℰ; see Fig. 10.13b) in circle C, which has mystified interpreters of this diagram. As the Hebrew alphabet may be written with ball-and-stick elements, this symbol may be interpreted to represent the letter *dalet* (ד) in Hebrew ("D" in the Latin alphabet). It has been associated with the number four and has long been a symbol for the tetragrammaton, YHWH, the four letters in the name of God, pronounced Yahweh in Hebrew and Jehovah in English.

Similarly, the small bottom circle (circle 1) in folio 86v (Fig. 10.3a) would represent *Malkuth* (Kingdom or Earth). This is supported by its T-O symbol (⊖), which consists of a circle with three subsections based on a diameter and a single radius at a right angle to the diameter. It is referred to as a Beatine or Beatus map, attributed to the Spanish monk Beatus of Liébana, a *comarca* (region) of Cantabria. The T-O map represents the physical world first described by the seventh century scholar (Saint) Isidore of Seville, Spain, in his *Etymologiae* (see Chap. 14) (Barney et al. 2006).

Table 10.1 Numbers and names of Sephirothic Tree of Life circles and folio 86v circles of the *Voynich Codex*

Sephira number	Sephira name	Voynich circle	Suggested Aztec city	Associated volcano
–	*Ain Soph*	C		
1	*Kether*	6	Tecamachalco (Tepeaca?)	Pico de Orizaba (Citlaltepetl)
2	*Chokhmah*	7		
3	*Binah*	5		
4	*Chesed*	8		
5	*Geburah*	4	Tlaxcala	La Malinche (Matlalcuéyetl)
6	*Tiphareth*	A	Puebla de los Angeles	
7	*Netzach*	9		
8	*Hod*	3	Vera Cruz-Cempoala	
9	*Yesod*	2	Huejotzingo	Popocatepetl
10	*Malkuth*	1		
		B		Pico de Orizaba (Citlaltepetl)

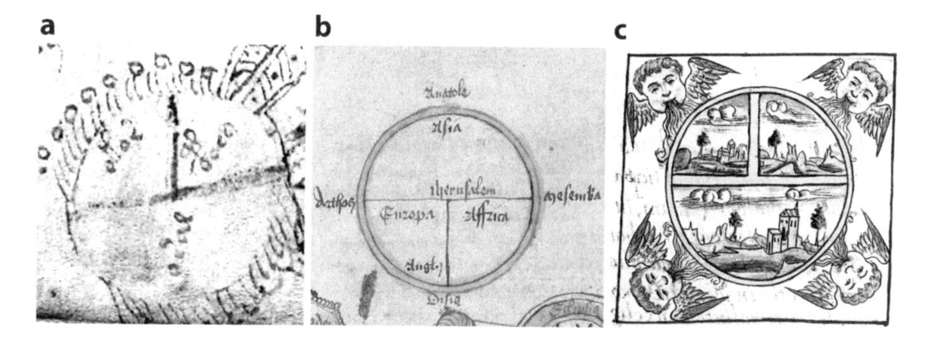

Fig. 10.3 T-O maps. (**a**) Circle 1 of folio 86v; (**b**) medieval T-O map showing Asia, Europe, and Africa. Source: HM 64, Digital Scriptorium, Huntington Catalog Database, Huntington Library, San Marino, California. (**c**) A T-O map showing the four winds from the *Florentine Codex* of Sahagún, Book 7. (Source: Sahagún 1953)

In many medieval maps, the symbol relates to three continents: Asia (the large subsection), Europe, and Africa (Fig. 10.3b), and sometimes to the Holy Trinity. A T-O map (describing the four winds) is also found in the *Florentine Codex* of Bernardino de Sahagún (Fig. 10.3c). The meaning of Voynichese words in circle 1 are: (*ānocâ*), with Anoca being a small town in the Mexican state of Jalisco; (*ātlāe*), which may be associated with water; and (pai), which is unknown (but *pa* means "far away" in Nahuatl).

The large central circle A, considered to be *Tiphareth* or Beauty, is only directly connected to circles 3, 5, 7, and 9. Furthermore, the small circle B resembles a volcanic caldera and we believe that it is not a *sephira*, but a location marker. (There are also volcanoes in circles 2 and 4). In conclusion, without circle B, the folio 86v diagram consists of ten connected circles and one disconnected circle, which makes it identical in form to the Tree of Life shown in Fig. 10.1c. We conclude that the Sephirothic Tree of kabbalah influenced the Voynich artist's folio 86v diagram, but there is no evidence that this was its main purpose.

The Folio 86v: Circles as Cities

There is another explanation for folio 86v that is not mutually exclusive with a kabbalah interpretation. Folio 86v may be considered a map superimposed on a Sephirothic Tree that represents clusters of cities, connected by a network of roads or pathways, and volcanic landmarks (Fig. 10.4). We also interpret folio 86v as a map of the cities surrounding Puebla de los Angeles (circle A), including Huejotzingo (circle 2), Tlaxcala (circle 4), Tecamachalco-Tepeaca (circle 6), and Vera Cruz-Cempoala (circle 8). The circles are described in more detail below.

Evidence that folio 86v represents a series of cities comes from the history of Nueva España (New Spain) in the sixteenth century. The Franciscan missionary

Fig. 10.4 Cities and volcanoes in New Spain combined with the circles of folio 86v

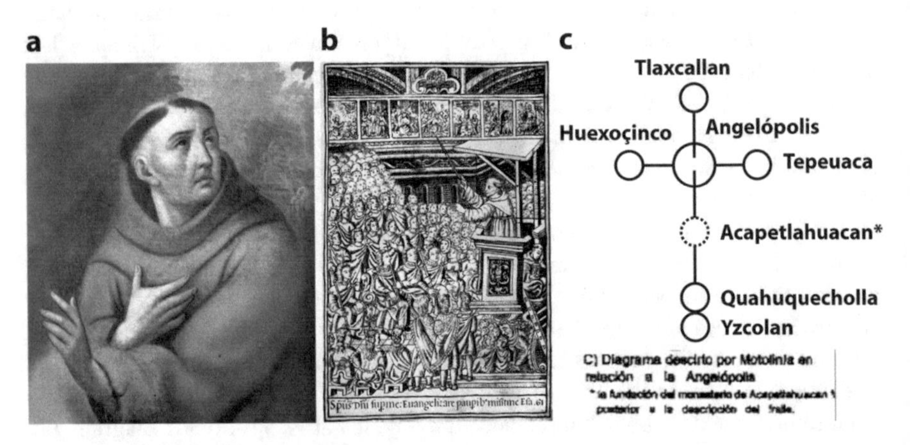

Fig. 10.5 (**a**) Franciscan friar, Toribio de Benavente (1482–1568) known as Motolinía. (**b**) The friar's method of teaching indigenous students using Spanish, Italian, and Flemish pictures. (**c**) Angelopolis (Puebla de los Angeles) configured into a cross, based on Motolinía. (Source: Sambhu 2013)

Toribio de Benavente (1482–1568), known as Motolinía, from the Nahuatl meaning "he is poor" (Fig. 10.5a, b), was one of the Twelve Apostles who arrived shortly after Hernán Cortés and attempted to help to convert the indigenous people. He entered the Valley of Mexico (Anahuac) in 1524 and, by 1530, resided at the convent at Huejotzingo near Tlaxcala and Texcoco. That same year (1530), the Franciscans attempted to establish a New Jerusalem or City of the Angels (Puebla de los Angeles or Angelopolis) to convert a pagan city into the capital of a new Israel (Lara 2004, 2008). The location is described, as follows:

> *The site which the city of Los Angeles occupies is very good and the region is the best in all New Spain. Five leagues to the north lies the city of Tlaxcallan and five leagues to the west is Huexotzinco. To the east, five leagues, is Tepellacac (Tepeyacac)* [Tepeaca]. *To the south*

lies the hot region. Here, seven leagues away, are Itzocan [Izucar de Matamoros] *and Cuauhquechollan* [*Quahuquecholla/Huaguechula*]*; Chololl an* [Cholula] *and Totomiahuacan are two leagues and Calpa is five leagues distant. All these towns are large. To the east, forty leagues away, is the harbor of Vera Cruz, and the distance to Mexico is twenty leagues. The highway from the harbor of Vera Cruz to Mexico passed through the City of Los Angeles.* (Motolinía 1951: 199)

It is intriguing that in one recent publication (Sambhu 2013), some of these cities, as described by Motolinía, have been configured as a cross (Fig. 10.5c), a concept not dissimilar to the kabbalah diagram.

Circle 2. (Fig. 10.6)

The most detailed circle (Fig. 10.6) has many elements, which suggests that it represents Huejotzingo, a small city in the central Mexican state of Puebla (Fig. 10.7a). There are alternate spellings on maps and documents of the sixteenth century, such as Huexucinco, Huexutcinco, and Huexotzinco, and, found in a map from 1691, the spelling Guazocingo (Fig. 10.7b). Prominent in circle 2 is a fortified structure with crenelated battlements that resembles the fortified monastery of San Miguel Arcángel at Huejotzingo, which was built as early as 1530 (Motolinía 1951).

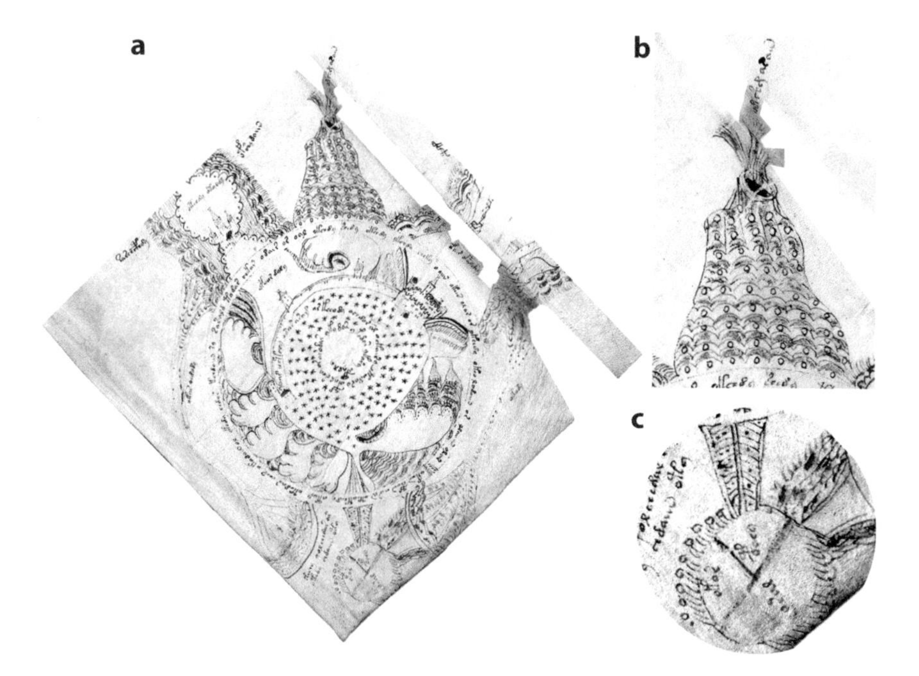

Fig. 10.6 In folio 86v. (**a**) Circle 2 (representing Huejotzingo, at the top) is connected to small circle 1; (**b**) an enlargement of an erupting volcano; (**c**) an enlargement of circle 1 shows a T-O map

Fig. 10.7 Huejotzingo. (**a**) Contemporary map of central Mexico; (**b**) map of 1691 with Guazocingo, an alternative spelling of Huejotzingo. (Source: copper-plate engraving from Jan Karel Donatus Van Beecq 1638–1722)

San Miguel Arcángel was re-constructed from 1544 to 1570 and, at the time, called the "Queen of the Missions" (Fig. 10.8) (Granados and MacGregor 1934). This *convento* (or fortified monastery) has a unique architecture that is a mix of medieval and Renaissance styles with plateresque and Moorish elements. Fray Juan de Alameda based his 1544 design heavily on Italian models:

> *Native encomienda laborers built the convento that Alameda designed based on his familiarity with European monastic compounds. The friar also relied on European architectural treatises including Leon Battista Alberti's De re aedifacatoria (1485), Vitruvius's De architectura (1486), Sebastino Serlio's Regole generali di architettura (1537, Spanish edition 1565), and Dieto de Sagredo's Medidas del romano (1526). The variety of structural and formal solutions displayed at Huejotzingo and New Spain's other monastic compounds suggest, however, that treatises and European models were not the architect's only sources of inspiration. Constructing monuments for this new Christian república de indios meant deploying an array of forms to create an ideal mission setting* (Donahue-Wallace 2008).

The merlons in the drawing of circle 2 (Fig. 10.8a) are M-shaped (the architectural term is "swallowtail") and this feature has been used by many (e.g., Pelling 2006) to infer that the fortress represents an Italian fifteenth century structure. However, the monastery in Huejotzingo was not completed until 1570, when presumably the triangular pierced merlons (an Islamic influence) were added. The merlons on the stepped buttresses (Fig. 16.8c; Chap. 16), which must predate the triangular ones because of construction techniques, are trapezoidal or cubic pyramidal merlons, but could be described as "swallow-tailed" (Garcia Granados and Mac

Fig. 10.8 Landmarks in circle 2. (**a**) Illustration of a fortified monastery in circle 2 of folio 86v with merlons (also called Ghibelline battlements or priest caps); (**b**) contemporary picture of the front view of San Miguel Arcángel in Huejotzingo showing an attached aqueduct; (**c**) contemporary views of the San Miguel Arcángel in Huejotzingo showing triangular pierced merlons along the walls, and swallowtail battlements on the stepped buttresses; also see Fig. 16.8c in Chap. 16; (**d**) view of Popocatepetl from the roof of the monastery. (Source: Granados and MacGregor 1934). (**e**) Three towers and a single tower with names in circle 2

Gregor 1934: 148; Angulo Iñiguez 1983: 137; Perry 1993: 96). We suggest that the drawing of the fortress might predate the finished date of 1570, and the buttress merlons were used by the Voynich artist as a model for the sketch of the fortress in circle 2. The Voynich artist also used swallowtail merlons in a sketch found in the passageway to circle 6 (Fig. 10.11).

An attached aqueduct was fed by runoff from the slopes of Popocatepetl (smoking mountain) via the Xopanac River. The smoking volcano, obviously Popocatepetl, can be seen in the rim of circle 2, with the following letters emanating from it: (*o ch o ts ha o n o l*), which may be expected to refer to the spewing lava. We interpret *ocho* to refer to the Spanish number eight. The other letters, *ts ha o n o l,* are somewhat similar to *ts'onc'oyal* = *embarra* in Spanish or to "muddy" or "dirty" in English (Larsen 1955: 68). Thus, the letters from the volcano could be interpreted as "eight muddy emanations." It should be noted that there were eight eruptions of Popocatepetl from 1518 to 1571 (in 1518, 1519–1523, 1528, 1530, 1539–1540, 1542, 1548, and 1571) (Volcano Discovery 2017).

Connected to the fortress is a long wall with towers, one of which contains the name (*āhuāāe chocâi*) (Fig. 10.8e). In the center of the inner circle are four towers with Moorish domes. In addition, there are 118 six-pointed stars, evenly distributed, and in their midst is a spiral of words. In the passageway to circle 9 there are a number of structures, and in the passageway to circle 3 there are three towers with the name [Voynich script] [Voynich script] (*ātlaachi ātlācâchi*) above them. *Atlacachi* refers to spear-throwers or fishermen in Nahuatl:

> *Settlers of the islands and shores were called atlacachi-chimeca, which probably means 'los chichimeca con atlatl'* [chichimeca with spear throwers] *or 'los del cordelo/linaje de perro hombres del agua'* [those of the cordage/water dog men lineage (Molina 1977: 8, 78, 95)], *and their tools consisted of the net, the atlatl* [spear-thrower], *and the harpoon* (Staedtler and Hernández 2003).

The entire circle 2 is ringed by words. Elements in the circle representing fire, air, water, and Earth suggest kabbalah and Aristotelian allusions of the universe. Fire and dust are alluded to by the smoking volcano, Popocatepetl. Five strange finger-like clouds may represent wind or air and alternating blue-and-yellow stripes suggest water.

The association of the fortified building as the monastery in Huejotzingo is more than a conjecture. First, very few of the monasteries in New Spain had crenelated battlements, which are found in the present structure at Huejotzingo (Perry 1993), and none of the others had swallowtail merlons. Second, a convincing case can be made that the fortified structure in circle 2 is the monastery San Miguel Arcángel in Huejotzingo, as its name in Voynichese is written above the structure. The Voynichese letters [Voynich script] above the fortress can be transliterated as *hu o x e ā tl o câ p i*. *Huexotl* is the Classical Nahuatl stem for "willow," whereas *capi* is a sixteenth century Spanish word meaning a "capital city." Thus, Huoxeatlocapi would literally be "willow capital," against the Classical Nahuatl derivation of Huejotzingo (or its variations: Huexucinco, Huexutcinco, or Huexotzinco), which literally translates as "where the willows grow." We suggest that this decipherment is a smoking gun that does three things:

1. It confirms the decipherment of Voynichese symbols based on plant names (see Chap. 5).
2. It verifies circle 2 as the city Huejotzingo.
3. It confirms that most of the words found in circle 2 are not Classical Nahuatl, but may be based on a similar extinct dialect or language.

Circle A. (Fig. 10.9)

We are convinced that the central largest circle (circle A) represents Puebla de los Angeles (or Angelopolis, or the New Jerusalem), now known as Puebla. The central circle is surrounded by eight others and is directly connected to circles 3, 5, 7, and 9. Circle A is composed of three rings: the outer ring contains 12 sets of six raised pipes evenly distributed (an unseen spike may be in the crease); a thin middle ring has cell-like structures; and an inner ring contains a three-dimensional image of a mantle covered with five- or six-pointed stars supported by six *qubba* (domed tombs), one of which one clearly represents a ciborium, a goblet-shaped vessel for holding Holy Communion wafers.

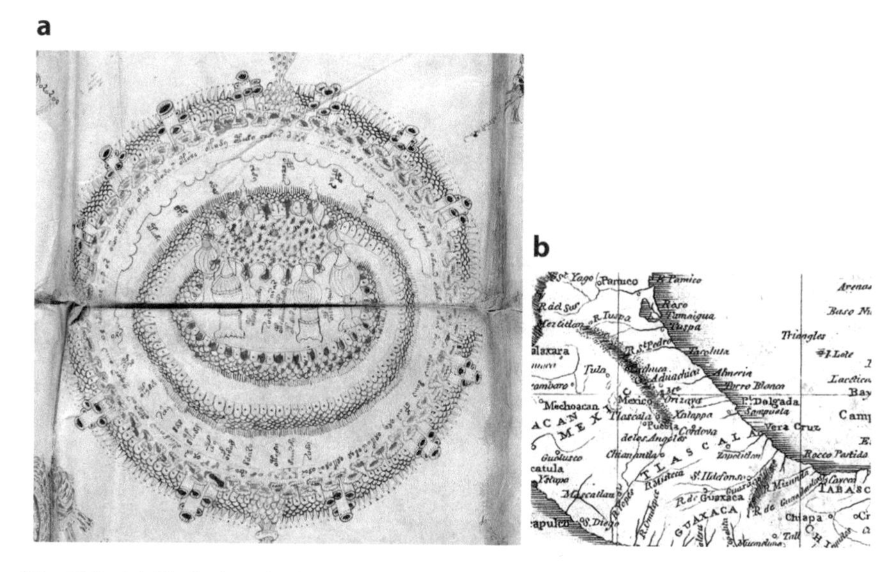

Fig. 10.9 (**a**) Circle A as Puebla de los Angeles; (**b**) 1733 map (Spanish dominions in North America) of central Mexico indicating the location of Puebla de los Angeles

This image is obviously theological in nature. The starry mantle may allude to one in the Old Testament that is said to have passed from Elijah, the prophet, to his successor, Elisha. It is also suggestive of the starry mantle of Our Lady of Guadalupe, shown in a painting enshrined in the Basilica of Mexico City (ca. 1531).

In circle A, five sets of words are written between the six *qubba* of the starry mantle: (*ātlchācho āahi*), (*nāts/tz?ān oeoe*), (*ātlacholl ātlachi*), (*itlachoeoi*), and (*āhumoch oll*). Some of these words have Nahuatl cognates. For example, with (*ātlchācho āahi*), *atl* = "water" and *chaco* = "sprinkled." The last term, *aahi,* is unknown, but may suggest "holy" or "blessed" water. In the Church of San Francisco at Puebla, there is a chapel with the "incorrupt" remains of Blessed Sebastian di Aparicio y del Pardo (1502–1600). Similarly, the words *altlacholli-atlachi,* may refer to a fisherman of Cholula. As discussed earlier, *atlacachi* (in Nahuatl) refers to spear-throwers or fishermen.

At one time, Puebla was envisioned as the New Jerusalem or Celestial City of Jerusalem. According to Motolinía, the city was founded in 1530 with the hope of it being "*settled by people who instead of all waiting to have Indians assigned to them, would devote themselves to tilling the field and cultivating the land in the Spanish way.*" A 1733 map of Mexico shows the location of Puebla de los Angeles (Fig. 10.9b). There are similar representations of Jerusalem. A Psalter fragment from 1200 (Fig. 10.10a) shows a map of Jerusalem with *qubbas* and crucifixes to mark each holy site (Harvey 2012). A map of Jerusalem with *qubbas* in the walls surrounding the city (Fig. 10.10b) was found in the ceiling of a church in Tecamachalco completed in 1564, based on an etching by Hans Holbein.

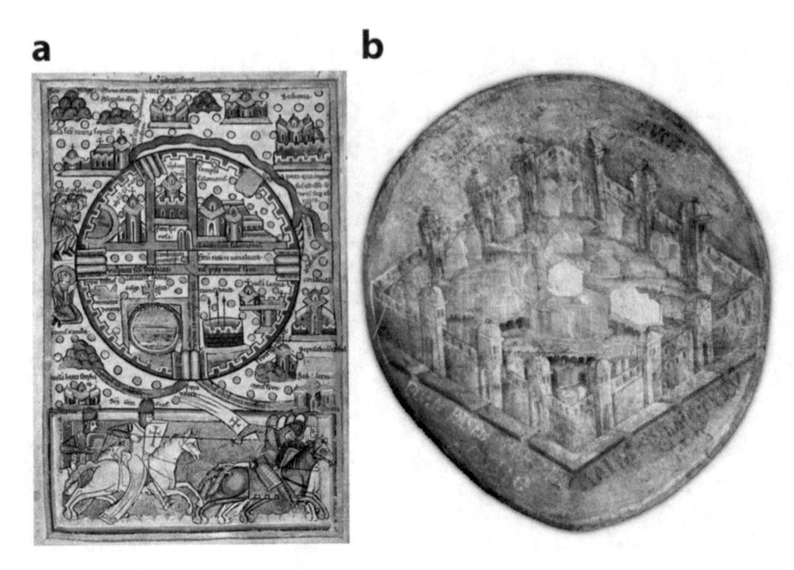

Fig. 10.10 (**a**) Abbreviated map of Jerusalem from 1200. (Source: *Koninklijke Bibliotheek* ©, National Library of the Netherlands). (**b**) Map of Jerusalem found in the ceiling of a church in Tecamachalco completed in 1564, based on an etching by Hans Holbein

This circular arrangement of six *qubbas* around a central holy area also bears an uncanny resemblance to the map of Cholula (Fig. 10.11) from the *Relación Geográfica* of 1581 (Mundy 2000). It shows a central Franciscan monastic complex (Ciudad S. Gabriel, Ancta yglessia, Cabila) with Cholula's largest pre-Hispanic pyramid, the *Tlachihualtepetl*, pictured upper right and labeled "*Tollan Cholulā*." Six *conventos* (fortified monasteries) are positioned around the complex, each with a crucifix and *qubba*-like mound to the rear (probably representing previous pagan temples). Clockwise from upper right on the Cholula map, the *conventos* are:

1. Sanct. Andres Cabezera – possibly Ex Convento de San Andrés, in Calpan, Puebla state, or Ex Convento San Diego, San Andrés, in Cholula, Puebla state.
2. Sanct. Pablo Cabezera – possibly Ex Convento de San Pablo de los Frailes, in Puebla de los Angeles, Puebla state.
3. Sanct. Maria Cabezera – possibly Ex Convento de Santa María, in Tepapayeca, Puebla state, or Santa Maria de la Concepción, in Atlihuetzía, Tlaxcala state.
4. Sanct. Juan Cabezera – possibly Ex Convento de San Juan Bautista, in Cuauhtinchán, Puebla state.
5. Sanctjago [Santiago] Cabezera – possibly Ex Convento de Santiago Apóstol, in Cuilapam, Oaxaca state.
6. Sanct. Miguel Tecpan Cabezera – possibly San Miguel Tecpan, in Jilotzingo, Mexico state.

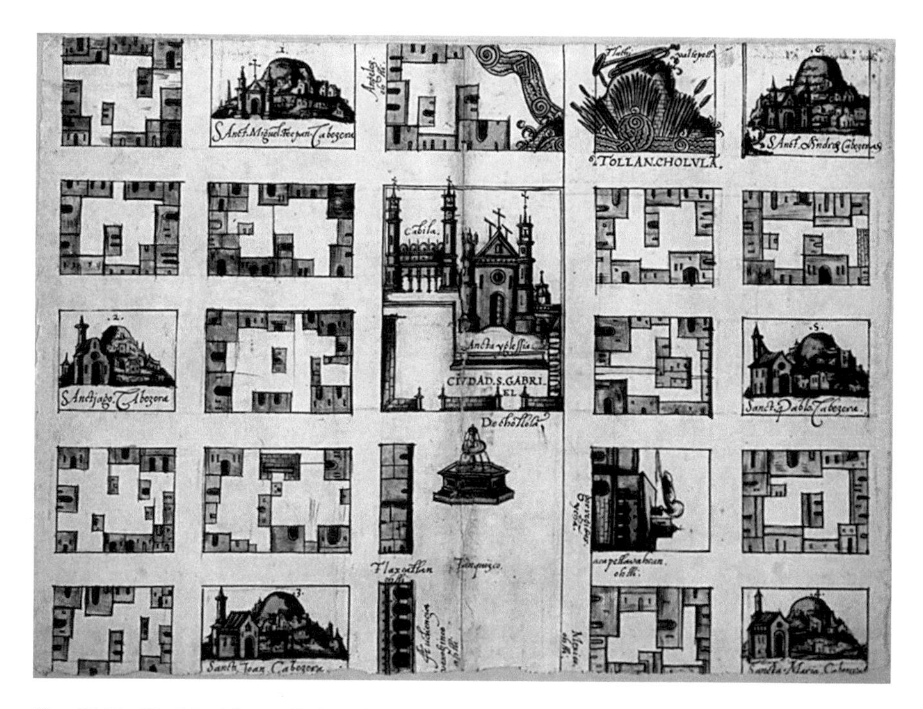

Fig. 10.11 Cholula (also called de Chollola) from the *Relación Geográfica* of 1581. (Source: Mundy 2000)

Circle 4. (Fig. 10.12a)

We suggest that circle 4 might represent Tlaxcala, the city entered by Cortés on his way to Tenochtitlan, the name of which in Nahuatl means "the place of maize tortillas." The population was 300,000 in the sixteenth century. There was a large fountain and square in the center of the colonial city, which may be represented by the blue cross with a circle at its center. Next to the blue cross is a large rectangular area with rows that seem to represent the rows of a prominent *tienda* (marketplace) behind the fountain and square. This matches the plan of central Tlaxcala in the sixteenth century (Fig. 10.12b). A stone fountain in the center of plaza was erected in 1548 by Fray Francisco de las Navas and Diego Ramírez, the second registered *corregidor* (chief magistrate). Water from the Zahuapan River came to this fountain by a conduit, and was used as a source of drinking water for miles around. *Tiendas* (shops) were located at the southwestern and southeastern sides of the plaza as early as 1549 (Gibson 1952). Circle 4 has six poles. There are two patches filled with stars. The volcano on the edge pointing to circle A would be La Malinche (also known as Matlalcuéyetl), the largest volcano in Tlaxcala.

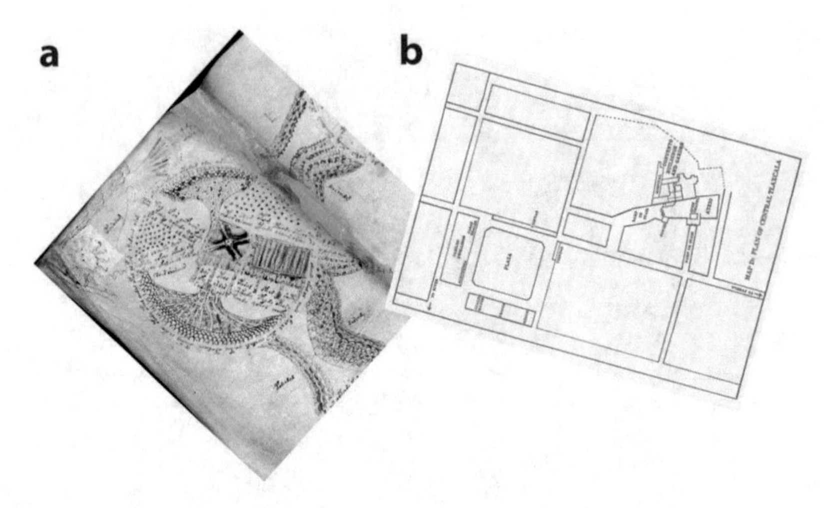

Fig. 10.12 (**a**) Circle 4 (Tlaxcala); (**b**) map of Tlaxcala. (Source: Gibson 1952)

Circle 6 and Circle B. (Fig. 10.13)

We propose that circle 6 (Fig. 10.13a) under the disconnected circle C represents Tecamachalco (in Nahuatl, the name means "jaws of stone") in the Mexican state of Puebla, rather than the nearby city of Tepeaca. The Franciscans arrived in Tecamachalco in 1541. The church, Asuncion de Nuestra Senora (Fig. 10.13c), was dedicated in 1551 and completed by 1557. During this time Andrés de Olmos, Francisco de las Navas, and Motolinía resided in this dusty outpost and taught the local indigenous people not only the rudiments of Christianity, but also the tenets of European art. At the same time, the diseases brought by the Spanish resulted in an unremitting loss of life for the indigenous population. The sixteenth century saw three major epidemics in Tecamachalco (in 1520, 1542, and 1577). The authorities estimated that 90% of the local native population had disappeared by 1580. For their part, the Franciscans fervently believed that the evangelization of the indigenous people of the New World was essential to precipitate the second coming of Christ. Not surprisingly, Motolinía drew parallels between the epidemics and famine that decimated the native people and the disastrous events described in the Book of Revelation, which announced the imminent return of Christ. The Apocalypse of St. John gave meaning to this tragedy of human suffering (Marr and Kirkacofe 2000; Acuna-Soto et al. 2002).

Asuncion de Nuestra Senora contains 27 paintings in Northern Renaissance style by the *tlacuilo* (indigenous artist) christened Juan Gerson, completed in the mid-sixteenth century. Gerson's paintings, on native *amatl* paper, appear to be based on woodcuts from Bibles published in Germany and France in the fifteenth and sixteenth centuries. Flemish and Italian paintings were used as models to train sons

Fig. 10.13 (**a**) Circle 6 (Tecamachalco) connected with circle B (below, with caldera). Above circle 6 is small disconnected circle C. (**b**) Close-up of circle C. (**c**) Asuncion de Nuestra Senora church with connected aqueduct, and (**d**) the church's interior ceiling with paintings by the *tlacuilo* (indigenous artist) Juan Gerson

of the Nahua nobility in the colleges set up by the friars (Fig. 10.5b). Gerson's paintings include many scenes from the Hebrew Bible, such as a panel called *Jerusalem and the City of God* (Carmelo Arredondo et al. 1964; Azpeitia 1972; Boone and Cummins 1998).

The central ring in circle 6 suggests Asuncion de Nuestra Senora's vaulted ceiling, which has four ribs with a Franciscan rosette in the center (Fig. 10.13d), representing the five wounds of Christ. A number of strange shapes with blue-and-white stripes in the circle are unexplainable. In the large passageway associated with circle 6, there is a long structure with swallowtail merlons. We suggest that this structure could very well have been the aqueduct connected to a church that now is in ruins. A number of aqueducts in colonial New Spain had merlons. A current photograph of Asuncion de Nuestra Senora (Fig. 10.13c) shows a connected aqueduct. We consider this further evidence that circle 6 represents Tecamachalco.

Circle B is connected to the bottom of circle 6, without any words. It probably represents the stratovolcano, Pico de Orizaba (*citlaltepetl*), the highest mountain in Mexico, 47 km from Tecamachalco and dormant since 1687.

Circle 8. (Fig. 10.14)

This circle (Fig. 10.14a), composed of two rings, is associated with the Vera Cruz area (presently spelled Veracruz; including Cempoala, also spelled Zempoala) on the East Coast of Mexico based on five elements:

1. The outer rings resemble the rings in Cempoala (Fig. 10.14c).
2. The six large pipes emanating from the outside of the circle could represent the six chimneys in the ruins of the Temple of the Chimneys at Zempoala (Fig. 10.14b), an ancient Mesoamerican archeological site north of Vera Cruz.
3. The structure in the passageway to circle 9 is interpreted to be an ancient lighthouse/monastery in Vera Cruz.
4. The large oval shape in the inner circle may represent the large fresh water lake, Laguna Catemaco, in Vera Cruz.
5. A Voynichese name, oHcz8a2a2, in the orange tinted area in the inner circle, can be deciphered as *āltmchonon* = "water seller," based on Nahuatl cognates: "atl" + "m" + "aca" = "water" + "someone" and *chonon* = "seller." The residents

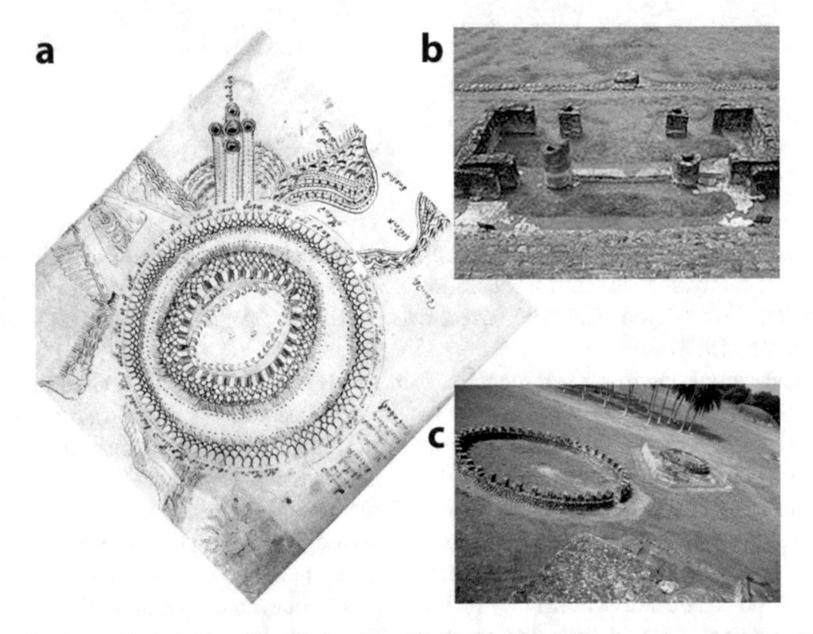

Fig. 10.14 (**a**) Circle 8 (Vera Cruz-Cempoala); (**b**) remains of five chimneys from the ruins of the Temple of the Chimneys, Cempoala; (**c**) rings at Cempoala

of various towns had nicknames (Furbee et al. 2010). Names associated with Vera Cruz include *chono trago* or *aguardiente* sellers (bootleggers), which would be redolent of water sellers. In addition, Cempoala derives from Cempoalli, which means "abundant waters" (Robelo 1902).

Vera Cruz was a bustling city when Cortés landed, but was severely decimated by smallpox. By 1577, only 30 households remained; these were relocated and the town abandoned. Only two inhabitants were recorded in 1600. The Temple at Zempoala was covered by forest and fell apart, only to be rediscovered in 1891 by Francisco del Paso y Troncoso, but the bulk of the work was done in the 1940s by García Payón. The temple was important in predicting the phases of the moon. The initial Franciscan *conventos* (fortified monasteries) in Vera Cruz were wooden, but were decimated by two fires in the seventeenth century. They also served as lighthouses at some point and were built on a promontory at the port above the gulf. The latest version of the Franciscan monastery was later nationalized and, at present, is a Holiday Inn. Approximately 35% of the fresh water in Mexico is in Vera Cruz and Laguna Catemaco, a freshwater lake there, is within sight of the Gulf of Mexico (Carrasco 1999).

Circles 3, 5, 7, 9. (Fig. 10.15)

These four circles alternate with the circles that represent cities. Circle 3 is very strange, with an outer circle consisting of small ball-like forms and associated with 16 protuberances or undulations. A circular object in the center is connected to two wing-like forms. The inner area is filled with blue marks. The 16 undulations resemble the seven legendary caves of Chicomoztoc (Fig. 10.16), the place of origin of the Aztecs (Rossell 2006; Carrasco and Sessions 2007). Pico de Orizaba in the Mexican state of Puebla has many small caves that were used for rituals before the conquest and circle 3 may refer to this area.

The three circles, 5, 7, and 9, resemble wheels with seven, eight, and 13 spokes that terminate in an area with eight, eight, and ten words respectively. The 26 words in these three circles are not found in the most frequent words of Voynich (Knight 2009; see Chap. 8, Table 8.2). In contrast, 13 of the 77 words around the rims of circles 5, 7, and 9, are among the most frequent words of Voynich, and 15 are among the most frequent optimal or thematic words of Montemurro and Zanette (2013). Thus, the words in circles 5, 7, and 9 must be unique nouns. As circles 2, 4, 6, and 8 are cities, we speculate that the words in circles 5, 7, and 9 might be related to villages or towns (Falling Rain Genomics 2015). The 26 words in these circles have been alphabetized and compared with cities in the area surrounding Puebla: one, *atla*, is a town in Puebla and some others seem close (see Appendix A). The words *altaachi* and *antlachi* refer to spear throwers or fishermen in Nahuatl and may be small lake villages (Staedtler and Hernández 2006). Thus, the eight circles surrounding circle A (the New Jerusalem or Celestial City of Jerusalem, founded by

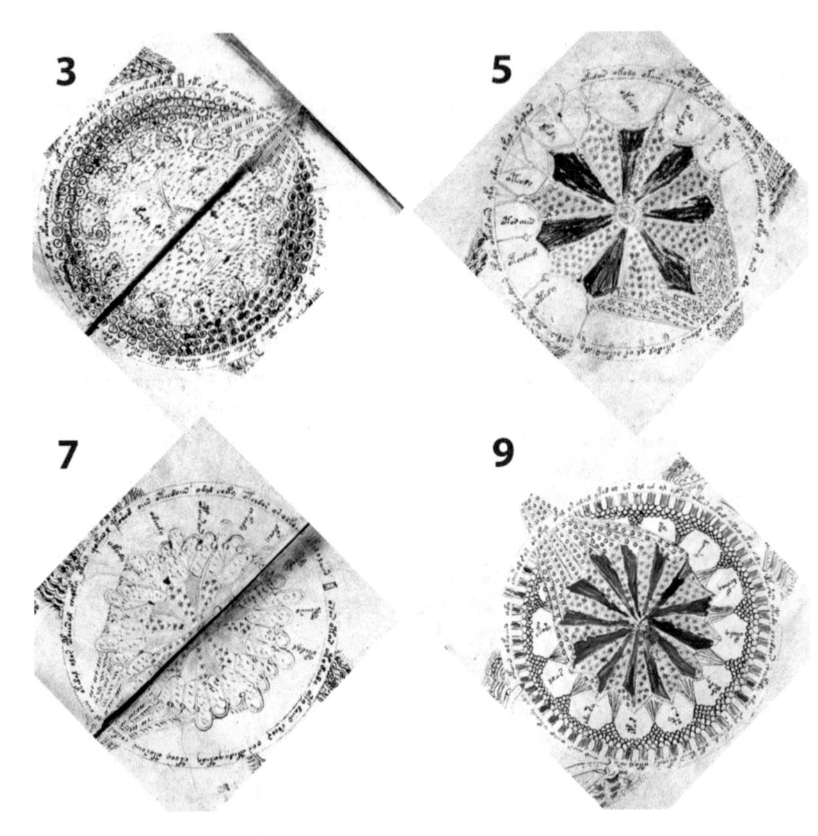

Fig. 10.15 Circles 3, 5, 7, and 9 in folio 86v

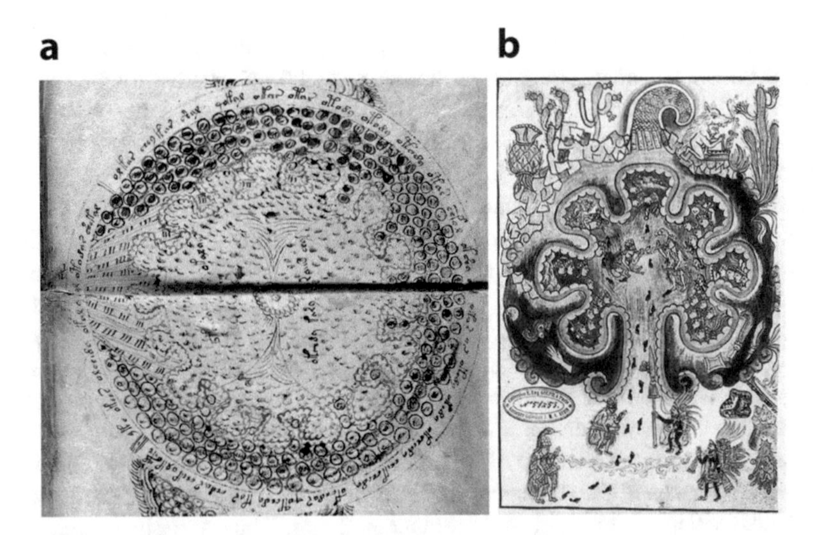

Fig. 10.16 (**a**) Circle 3 of folio 86v, showing undulations similar to (**b**) the seven legendary caves of Chicomoztoc. (Source: *Códice Historia tolteca chichimeca*, folio 16r)

Motolinía in 1530) make sense as a metaphoric map surrounding Puebla. It seems probable that many of the small villages or towns referred to could have disappeared or have had their names altered.

The similarities of the words in circles 5, 7, and 9 lead to another conjecture, i.e., the words may be incantations involving drought because, of the 26 words, 15 begin with the term or *ātl*, a Nahuatl cognate concerned with water, and four words begin with or *āhu*, which may refer to something dry, as *ahuoj* = "dry *arroyo"* (creek). Note that the word *atlachi* or *atlaachi* is found in all three circles. The repeated droughts in the sixteenth century (1545–1575), the worst in 500 years (Acuna-Soto et al. 2002), were among the causes of increased human sacrifice among the Aztecs and the words could very well have been associated with incantations for rain.

Conclusion

Folio 86v of the *Voynich Codex* is a complex figure that involves two concepts:

1. A kabbalistic Sephirothic Tree of Life.
2. A map associated with Puebla de los Angeles, the New Jerusalem established by the Franciscan friars, including Motolinía.

It includes four encircling cities, Huejotzingo, Tlaxcala, Tecamachalco or Tepeaca, and Vera Cruz-Cempoala (all mentioned by Motolinía) and three associated volcanoes of central Mexico. The diagram is evidence that the artist of the *Voynich Codex* was involved with Catholic mysticism linked to the Jewish kabbalah. The circle identified as Huejotzingo, which means "where the willow grows," was labeled with the deciphered name "willow" (Nahuatl) plus "capital" (sixteenth century Latin American Spanish), confirming the symbol decoding based on plant names (see Chaps. 5 and 11). We believe that folio 86v may be a key to the decipherment of the *Voynich Codex*, as many of the illustrations are associated with the text.

Literature Cited

Acuna-Soto, R., D.W. Stahle, M.K. Cleaveland, and M.D. Therell. 2002. Megadrought and megadeath in 16th century Mexico. *Emerging Infectious Diseases* 8: 360–362.

Angulo Iñiguez, D. 1983 (1932). *Arquitectura Mudéjar Sevillana de los siglos XIII, XIV y XV*. Sevilla: Servicio de Publicaciones del Ayuntamiento de Sevilla (originally published 1932 as "Discurso Inaugural" at the Universidad de Sevilla)

Azpeitia, R.C. 1972. *Juan Gerson, Pintor indigena del siglo XVI—simbolo del mestizage—Tecamachalco Puebla*. Mexico: Fondo Editorial de la Plastica Mexicana.

Barney, S.A., W.J. Lewis, J.A. Beach, and O. Berghof. 2006. *Etymologies of Isidore of Seville*. 1st ed. Cambridge, UK: Cambridge University Press.

Blau, J.L. 1944. *The Christian interpretation of the cabala in the renaissance*. New York: Columbia University Press.
Boone, E.H., and T. Cummins, eds. 1998. *Native traditions in the postconquest world*. Washington, DC: Dumbarton Oaks Research Library and Collection.
Camelo Arredondo, R., J. Gurría Lacroix, and C. Reyes Valerio. 1964. *Juan Gerson, Tlacuilo de Tecamachalco*. México: Departamento de Monomuntos Coloniales, Instituto Nacional de Anthropolgia e Historia.
Carrasco, D. 1999. *The Tenochca empire of ancient Mexico: The triple alliance of Tenochtitlan, Tetzcoco, and Tlacopan*. Norman: University of Oklahoma Press.
Carrasco, D., and S. Sessions, eds. 2007. Cave, city, and eagle's nest: An interpretive journey through the *Mapa de Cuauhtinchan No. 2*. Albuquerque: University of New Mexico Press.
Chuchiak, J.F. 2012. *The inquisition in new Spain, 1536–1820: A documentary history*. Baltimore: Johns Hopkins Press.
de Benevente Motolinía, T. 1951. *Motolinía's history of the Indians of New Spain*. Trans. F. B. Steck. Washington, DC: Academy of American Franciscan History.
de Sahagún, B. 1953. Florentine Codex. General history of the things of New Spain. *Book 7—The sun, moon, and stars, and the binding of the years*. Trans. A.J.O. Anderson and C.E. Dibble. Salt Lake City: University of Utah Press.
Dedopulos, K. 2012. *Kabbalah: An introduction to the esoteric heart of Jewish mysticism*. London: Carlton Books.
Don, P.L. 2010. *Bonfires of culture: Franciscans, indigenous leaders, and inquisition in early Mexico, 1524–1540*. Norman: University of Oklahoma Press.
Donahue-Wallace, K. 2008. *Art and architecture of Viceregal Latin America, 1521–1821*. Albuquerque: University of New Mexico Press.
Encyclopedia Judaica, 1971. Kabbalah. Keter, Israel.
Falling Rain Genomics. 2015. Directory of Cities, Towns, and Regions in Mexico. http://www.fallingrain.com/world/MX/. Accessed 3 Sept 2015.
Furbee, N.L., R. Jímenez Jímenez, T. López Mendez, M.B. Sántiz Pérez, H. Aguilar Méndez, and J. Méndez, 2010. Geographic Ideology: Nicknames for Chiapas towns: A Lexographic Bouquet for Bob Laughlin. American Anthropological Association, New Orleans, November 19. Abstract. http://sssat.missouri.edu/docs/B-Geographic-Ideology_Abstract-HandoutAAA19Nov2010.docx. Accessed 3 Sept 2015.
Garcia Granados, R., and L. MacGregor. 1934. *Huejotzingo, la ciudad y el convento Franciscano*. México: Talleres Gráficos de la Nación.
Gibson, C. 1952. *Tlaxcala in the sixteenth century*. Stanford: Stanford University Press.
Harvey, P.D.A. 2012. *Medieval maps of the holy land*. London: British Library.
Knight, K. 2009. The Voynich manuscript. MIT. http://www.isi.edu/natural-language/people/voynich.pdf. 1 May 2014.
Lara, J. 2004. *City, temple, stage: Eschatological architecture and liturgical theatrics in New Spain*. Notre Dame: University of Notre Dame Press.
———. 2008. *Christian texts for Aztecs: Art and liturgy in colonial Mexico*. Notre Dame: University of Notre Dame Press.
Larson, R. 1955. *Vocablularia huasteco del estado de San Luis Potosi*. Mexico: Instituto Linquitico de Veran.
Marr, J. S., and J. B. Kirkacofe. 2000. Was *hue cocoliztli* a hemorrhagic fever? *Medical History* 44: 341–362.
Montemurro, M.A., and D.H. Zanette. 2013. Keywords and co-occurrence patterns in the Voynich Manuscript: An information-theoretic analysis. *PLoS ONE* 8 (6): e66344. https://doi.org/10.1371/journal.pone0066344.
Mundy, B.E. 2000. *The mapping of new Spain: Indigenous cartography and the maps of the Relaciones Giográficas*. Chicago: University of Chicago Press.
Pelling, N. 2006. *The curse of the Voynich: The secret history of the World's most mysterious manuscript*. Surbiton, Surrey: Compelling Press.
Perry, R. 1993. *Mexico's fortress monasteries*. Santa Barbara: Espadana Press.

Perry, M.E., and A.J. Cruz. 1991. *Cultural encounters: The impact of the inquisition in Spain and the new world*. Berkeley: University of California Press.
Robelo, Cecilio. 1902. *Nombres Geographicas Mexicanos des Estado de Veracruz*. Cuernavaca (Cuauhnahuac): L.G. Miranda, impresor.
Rossell, C. 2006. Estilo y escritura en la Historia Tolteca Chichimeca. *Desacatos* 22: 65–92.
Sambhu [José Antonio Ramón Calderón]. 2013. El Corazón del Corazón de América. https://shivashambho.files.wordpress.com/2014/10/el-corazc3b3n-del-corazc3b3n-de-amc3a9rica.pdf. Accessed 3 Sept 2015.
Staedtler, M.C., and M.F. Hernández. 2006. Hydraulic elements at the Mexico-Texcoco lakes during the postclassic period. In *Lisa J. Lucero and Barbara W. Fash. Water management: Ideology, ritual, and power*, 155–170. Tucson: The University of Arizona Press.
Villa-Flores, J. 2006. *Dangerous speech. A social history of blasphemy in colonial Mexico*. Tucson: University of Arizona Press.
Volcano Discovery. 2017. Popocatepetl volcano news & eruption update. https://www.volcanodiscovery.com/popocatepetl/news.html. 22 May 2017.

Part III
Decipherment

Chapter 11
Cryptological Analyses, Decoding Symbols, and Decipherment

Jules Janick and Arthur O. Tucker

Historical

Cryptology or secret writing is an ancient technique to secure written communication between two parties and has become an important part of communication in warfare. It now is an essential part of our lives in internet communication, passwords, and encryption to secure privacy. Cryptology has a long history dating to the ancient Greeks and was popular in both the late medieval and Renaissance eras. Roger Bacon (1214–1292), who Wilfrid Voynich thought might be the author of the manuscript that now bears his name, was interested in secret writing and listed seven cipher methods involving substitutions of one letter for another. As a result of this conjecture, many of the early Voynich researchers surmised that the text can only be understood as a cipher, an encryption that requires either a simple substitution of one letter or a complex procedure that requires a set of rules to decode. However, the deciphering of the *Voynich Codex* is mysterious because the symbols have not been decoded and the underlying language is unknown. The most common language suggested has been some form of Latin (Newbold and Grubb 1928; Feeley 1943; Friedman 1962; Brumbaugh 1978; Rugg 2004a, b; Landsmann 2008; Kircher and Becker 2012; Hoschel 2012; Sullivan 2012). However, many other languages have been suggested, including: Arabic (Zandbergen 2004–2017); Chinese (Guy 1991; Stolfi 2006; Vonfelt 2014); Dutch (Kircher and Becker 2012; Maat 2012); medieval English (Strong 1945; Currier 1970–1976; Kircher and Becker 2012; Rivero 2012; Rugg 2004a, b); French (Friedman 1962), Gaelic (Friedman 1962; Kircher and Becker 2012); Germanic (Child 1976/2002, 2007; Gwynn 2006); Kircher and Becker 2012; Huth 2012); Greek (Williams 1999); Hebrew (Finn 2004; Burisch 2012); Italian (Friedman 1962; Kircher and Becker 2012; Packheiser 2013); Nabataean (Gibson 2003); Nahuatl (Comegys 2001; Kircher and Becker 2012; Tucker and Talbert 2013; Comegys 2013); Mixtec (Tucker and Talbert 2013); Sanskrit, (Zandbergen 2017); Slavic (Stojko 1978); Spanish (Rivero 2012; Tucker and Talbert 2013; Comegys 2013); Swedish (Kircher and Becker 2012); polyglot

J. Janick, A. O. Tucker, *Unraveling the Voynich Codex*, Fascinating Life Sciences, https://doi.org/10.1007/978-3-319-77294-3_11

(Levitov 1987); Pre-Welsh, Dutch, Scottish (Ackerson, cited in Dragoni 2012); and Taino (Tucker and Talbert 2013) (see Chap. 1, Table 1.1).

Is the Voynich a Hoax?

One conjecture was that the *Voynich Codex* was an elaborate hoax. Brumbaugh (1978) assumed that it was a forgery to fool Rudolf II, and Rugg (2004a, b) declared that it was gibberish. These theories have been tested by modern cryptographical analysis based on the frequency of letters or words. An extensive computer analysis by Casanova (1999) in a French doctoral thesis concluded that the *Voynich Codex* had the qualities of a synthetic language whose alphabet was subject to transformations. It was suggested that at least two languages, perhaps four, were involved. Montemurro and Zanette (2013) reported on an information analysis of the manuscript. They demonstrated that the language organization is complex, but that the distribution of words is compatible with real language sequences. For example, the distribution of letters follows Zipf's law, as in other languages, which is incompatible with sixteenth century cipher methods. Their conclusion was that the statistical features of the *Voynich Codex* support a genuine message and that the text is neither nonsensical nor a hoax. This is in agreement with a study by Amancio et al. (2013), based on a statistical approach to the words in the *Voynich Codex*, indicating that they are mostly compatible with natural languages and incompatible with random texts.

Previous Decipherment Attempts

The history of attempts at decipherment of the *Voynich Codex* has been reviewed by D'Imperio (1978), and by Kennedy and Churchill (2006). The following summary draws heavily on these two accounts.

William R. Newbold

In 1926, this distinguished philosophy professor at the University of Pennsylvania first analyzed the manuscript at the request of Wilfrid Voynich, who believed that Roger Bacon was the author. Newbold agreed with this assertion, and the assumption was made that the language was Latin. Newbold's work was published after his death, but the complex system proposed was almost impossible to follow. However, his explanation and system were demolished as "*baseless and should be definitely and absolutely rejected*" by his friend John M. Many, a professor of English at the University of Chicago, in a 1931 article published in *Speculum*.

Joseph Martin Feely

A lawyer named Joseph Martin Feely published a book, *Roger Bacon: Cipher,* in 1943, concerning guessing at words based on context. These were used to confirm Bacon's authorship. No one has accepted his system.

Leonell C. Strong

A cancer scientist at Yale University became interested in Rev. Dr. Hugh O'Neill's (1944) article, dating the manuscript to post-1493. He used a complex cipher system based on medieval English and came up with Anthony Ascham (fl. 1553), brother of Roger Ascham. His explanations have not made sense to cryptologists and have not been accepted.

Robert S. Brumbaugh

A professor of medieval philosophy at Yale made use of key sequences in folios 1r, 17r, 47v, 66r, and 76r, in addition to a second ring of 57v and a sentence in folio 116v. He assumed that the plaintext is an artificial language based on Latin, and that the work is that of a forger. His botany is highly suspect.

William and Elizebeth Friedman

William Frederick Friedman, a cryptographer famous for breaking the Japanese "Purple" cipher code and special assistant for the National Security Agency, and his wife, the former Elizebeth Smith, became the modern avatars of cryptological techniques. William Friedman became interested in the *Voynich Codex* and set up a number of study groups, but the research was never formally published, although Elizebeth Friedman authored an article in the Washington Post (1962) concerning Voynich. Her conclusions are worthy of repetition:

- The number of basically different symbols employed in the manuscript is quite small, perhaps 20 or even fewer. However, the variation and affixes may make multiple forms of a basic character, which may suggest counting them as different symbols.
- The symbols are grouped into "words" and the number of different "words" is quite limited.
- The words are generally short, averaging about four and a half symbols. Rarely are they more than seven or eight symbols in length.

- There is a very large number of repetitions of single "words" and of groupings of two, three, or more words.
- There is a very large number of "words" that differ from one another by only one or two symbols.
- Certain words occur in successive repetition, that is two, three or four times, as mentioned above.
- The text is homogeneous, with the same "words" appearing in all sections, whether botanical, astrological, biological, or astronomical.
- Words of just one or two symbols appear very rarely.
- Certain symbols appear most frequently as the initial symbols of "words," rarely as central ones. Others appear most frequently as the final symbols, rarely as central ones.
- Certain symbols appear so infrequently as to suggest that they might be extraneous to the text or are errors, made perhaps by the author himself or by some scribe who transcribed the original.

Elizebeth Friedman disputed O'Neill's results and mentioned 16 plants identified as European by the "great" Dutch botanist Holm. We disagree (see Chap. 4). She noted that some plants are composites of more than one plant.

William Friedman's conclusions were weirdly put in the form of an anagram as follows: *PUT NO TRUST IN ANAGRAMATIC ACROSTIC CYPHERS. FOR THEY ARE OF LITTLE REAL VALUE — A WASTE — AND MAY PROVE NOTHING — FINIS*. There were various incorrect solutions, but in 1970, the anagram was revealed: *THE VOYNICH MSSS WAS AN EARLY ATTEMPT TO CONSTRUCT AN ARTIFICIAL OR UNIVERSAL LANGUAGE OF THE* A PRIORI *TYPE. FRIEDMAN*. An a priori type of language refers to a forgotten synthetic language completely created from scratch. Kennedy and Churchill came up with a reasonable conclusion: "*The text has yet to yield up its secrets. Perhaps the answer to Voynich is in the enigmatic illustrations.*" We agree.

The lack of success of cryptology, in our judgment, has been hampered by the mistaken assumption that the *Voynich Codex* is a fifteenth century manuscript, despite the clear evidence that the plants and most of the animals are Mesoamerican. The symbols have been decoded by Tucker and Talbert (2013) based on an association of labeled plants using their names in Nahuatl, sixteenth century Latin American Spanish, Huastec, Mixtec, or Taino, as shown in Chap. 5. Although the decoding was successful for a number of words and cognates in Nahuatl and other languages, it failed to decipher the bulk of the text. We have come to the conclusion in confirmation with the cryptologist William Friedman that Voynichese is a mixed synthetic language somewhat similar to Esperanto (an a posteriori synthetic language), and was probably invented from various contemporary languages extant in sixteenth century New Spain, including Nahuatl, Taino, and other languages of central Mexico or a Nahuatl lingua franca (Dakin 1981), and also including some words of Spanish and perhaps Arabic (for astronomical star names).

Decoding Voynichese Symbols

Decipherment of the plaintext of the *Voynich Codex* requires the symbols to be decoded and the language identified. The simplest interpretation of Voynichese symbols is that they are a phonetic or syllabary of either a real language and/or an invented language.

One explanation for the use of symbols was that the language involved did not have a written form, and the author preferred not to transliterate it with a local alphabet, such as Spanish. There are at least two explanations for this decision. The most obvious is that the author wanted to keep the work secret, perhaps because of fear of the Inquisition in New Spain. Another might be that the author used indigenous translators not versed in Spanish (as did Francisco Hernández) to assist in botanical identification and invented new symbols, as did George Gist (Sequoyah), a three-quarter Cherokee silversmith who created a phonetic syllabary of 86 characters associated with Cherokee sounds in 1821 by modifying letters from Latin, Greek, and Cyrillic that he had encountered (see Chap. 5). Whatever the cryptological method, the language must be known for the plaintext to be understood. However, the language of the *Voynich Codex* is unknown.

Decoding the symbols relies on techniques used to decipher lost languages. The famous examples of this process are the decipherment of Sumerian cuneiform, Egyptian hieroglyphics, Linear B, and Mayan glyphs. These decipherments are based on similarities to related languages aided by the availability of bilingual texts (e.g., the Rosetta Stone), and/or knowledge of the culture to provide names of rulers or cities. However, decoding the alphabet does not necessarily decipher the language if the language is lost or unknown. Decipherment steps would be first to transliterate the symbols to create an alphabet or syllabary and then discover the language or languages involved. As we have not found a related text, the decipherment of the language is based on clues to the meaning of the illustrations.

There have been many attempts to decode the alphabet and most are shown in Table 11.1. All results are very different, indicating that only one (or none) is correct. Obviously most of the "guessed" words used to develop the alphabet decoding must have been in error. Unfortunately, many of the techniques employed to decipher the Voynichese symbols (such as the two study groups of Friedman, as cited by D'Imperio 1978) have not been published; thus, the reasoning cannot be explained or analyzed. Four examples of techniques used to decode the symbols are compared below.

Joseph Martin Feely (1943) has assumed that the two words at the top of folio 78r are gynecological terms and speculates that the words olffcz8g. 8go (translated as *femminino*) and olffczcg (translated as *femmin*) refer to ovaries (D'Imperio 1978). According to D'Imperio, Feely attempted to find various Latin words that might represent this figure, but his analysis has not been accepted.

Table 11.1 Various decipherments of the Voynichese alphabet

Voynichese character	Decipherment										
	Tiltman (1967–1969)[a]	1st Friedman study group[a]	2nd Friedman study group[a]	Currier (1970–1976)[a]	D'Imperio (1978)	Comegys (2001: 53)	Sherwood (2008)[b]	Tucker and Talbert (2013)[c]	Bax (2014)	Voynichese character count and frequency	Count and frequency seventeenth century Classical Nahuatl
[glyph]	o	o	o	o	d	o	o	ā	a	25,468 = 15.9%	ā = 609 = 6.4%
[glyph]	c	c	c	c	c		e	a	o	20,227 = 12.6%	a = 753 = 8.0%
[glyph]	g	g	9	9	n	y	g	i/y	n	17,655 = 11.0%	i = 1318 + y = 160 = 1478 = 15.6%
[glyph]	d	d	v	f	w	tl	tl	} tl	k	16,020 = 10.0%	tl = 443 = 4.7%
[glyph]	h	h	b	p	a	ll	l				
[glyph]	a	a	a	a	h	a	a	o	ə/u/wa	14,281 = 8.9%	o = 685 = 7.2%
[glyph]	8	8	8	8	k	s	b	ch	t/d/ɵ/	12,973 = 8.1%	ch = 125 = 1.3%
[glyph]	t	t	s	s	b	t		m		11,008 = 6.9%	m = 285 = 3.0%
[glyph]	e	e	e	e	i	p	i	câ		10,471 = 6.5%	ca = 180 = 1.9%
[glyph]	r	r	u	r	o	x, ç	r	e	r	6716 = 4.2%	e = 449 = 4.8%
[glyph]	4	4	4	4	f	t final	v	qu/kw		5423 = 3.4%	qu = 214 = 2.3%
[glyph]	s	s	z	z	r	tl		ts/tz	k^h/x/tʃ	4501 = 2.8%	tz = 113 = 1.2%
[glyph]		m		m		m initial	m	ll	ur	4076 = 2.5%	ll = 53 = 0.6%
[glyph]	2	2	2	2	e	s	s	n	s	2886 = 1.8%	n = 741 = 7.8%

[illegible]	dz	dz	%	x	x	tl		} cu		1858 = 1.2%	cu = 85 = 0.9%
[illegible]	hz	hz	#	q	j	tl					
[illegible]	d	f	f	v	z	tl	p	} hu/gu		1844 = 1.1%	hu = 306 + uh = 155 = 461 = 4.9%
[illegible]	h	p	p	b	g	ll					
[illegible]		l	n	n		n initial	n	l	ir	1752 = 1.1%	l = 710 = 7.5%
[illegible]	e		h	j		k		yâ/hâ	r	1046 = 0.7%	ya = 13 + ha = 2 = 15 = 0.2%
[illegible]			k	t		n rare		c/k		591 = 0.4%	c = 496 = 4.9%
[illegible]		y						?		524 = 0.3% (listed by Knight as *?)	
[illegible]		l		i		i		sh/x		316 = 0.2%	x = 206 = 2.2%
[illegible]	dz	fz	@	y	t	tl		} p		291 = 0.2%	p = 148 = 1.6%
[illegible]	hz	pz	$	w	2	tl					
[illegible]	l	l	d	d	p	u	u	z/ç	r	157 = 0.1%	z = 351 = 3.7%
[illegible]			w	3				t		156 = 0.1%	t = 545 = 5.8%
[illegible]			g	u		m rare		?		148 = 0.1%	
[illegible]				6	y			h		96 = 0.1%	h = 306 = 3.2%
[illegible]			l	k				?		52 = <0.1%	
[illegible]			i	g				?		31 = <0.1%	
[illegible]			r	l				?		17 = <0.1%	

(continued)

Table 11.1 (continued)

Voynichese character	Decipherment									Voynichese character count and frequency	Count and frequency seventeenth century Classical Nahuatl
	Tiltman (1967–1969)[a]	1st Friedman study group[a]	2nd Friedman study group[a]	Currier (1970–1976)[a]	D'Imperio (1978)	Comegys (2001: 53)	Sherwood (2008)[b]	Tucker and Talbert (2013)[c]	Bax (2014)		
[illegible]			y	h				?		14 = <0.1%	
[illegible]			j	i				?		2 = <0.1%	
[illegible]			t	5				?		1 = <0.1%	
[illegible]			q	o				?		1 = <0.1%	
^		v									

Voynichese character counts and frequencies (out of 160,602 total characters) were calculated from Knight (2009). Classical Nahuatl frequencies (out of 93,451 total characters) were calculated from incantations of a *ticitl* (doctor or seer) from Alcarón (1987)

[a]D'Imperio (1978)

[b]http://www.edithsherwood.com/voynich_decoded/

[c]Decipherment not underlined is tentative

James C. Comegys (2001) concluded that the *Voynich Codex* was written in sixteenth century Mexico and assumed that the symbols represented Classical Nahuatl. He based his symbol decoding on the word frequency analysis of Mik Clark (cited as http://geocities.com/soho/café/2260/voynich.html but not found) who determined that the most common word in the *Voynich Codex* was EVA *daiin* = Voynichese [Voynichese symbol]. EVA stands for Extensible Voynich Alphabet (or EVA), which was formerly called the European Voynich Alphabet. Comegys simply asserted that EVA *daiin* could be translated as the very common Nahuatl word *mach,* meaning "*it is said*." Montemurro and Zanette (2013) do state that EVA *daiin* is the most common thematic word and Ruihang Feng, in a University of Adelaide (Australia) thesis submitted in 2016, stated that it was the most common word out of a total of 37,104 (2.17%.). However, Comegys considered that the Voynichese words are spelled backward, as in Hebrew. The Voynichese word [Voynichese symbol] spelled backward as [Voynichese symbol] suggests that [Voynichese symbol] = "m," [Voynichese symbol] = "a," and [Voynichese symbol] = "ch."

The same logic was used to decipher the Nahuatl verb *pac(a),* meaning "wash" (Karttunen 1983). Citing Takeshi Takahashi (www.voyihcom//opages?pagesh.tst); but again not located), there were 161 sequences of EVA *ral* (pronounced cap) = Voynichese [Voynichese symbol] (spelled backward is [Voynichese symbol]. It was also asserted that EVA *ral* = Nahuatl *pac*. From this exercise, it follows that the pronunciation of [Voynichese symbol] represents the sound of "p" and [Voynichese symbol] represents "c." Transforming *pac* with suffixes and stems leads to other forms of the verb, such as bathe, mothers' bath, belly bathe, and so forth. Although we concur with Comegys that the source of the *Voynich Codex* is Mexico, we do not agree that the Voynichese text is read backward or that the text is read from bottom to top. We find no rationale for associating Nahuatl *mach* with EVA *dain,* or Nahuatl *pac* with EVA *ral.* Furthermore, Comegys gives the same decoding for many terms, e.g., seven different Voynichese symbols [Voynichese symbols] = "tl," which is improbable. Our conclusion is that the symbol decoding of Comegys is without merit.

Stephen Bax (2014), an academic linguist, based his translation of Voynichese words by associating images in Voynich with their words in various languages. We agree that this approach is a correct one and we do not dispute his linguistic expertise, but the problem hinges on the correct identification of the images. One of the problems is that Bax, who is not a botanist, has relied on others, such as Edith and Erica Sherwood, also non-botanists, who have a priori decided that all the plants illustrated in the *Voynich Codex* are Mediterranean. For example, the plant of folio 93v, recognized as a sunflower by the botanist O'Neill (1944), was identified first as an artichoke (obviously incorrectly), and then as *Inula helenium* by the Sherwoods, also incorrectly (see Chap. 16).

Folio 68r(3) of the *Voynich Codex*, which includes various labeled stars, contains an image of the star cluster known as the Pleiades located in the constellation Taurus and labeled with Voynichese symbols as [Voynichese symbols]. In our opinion, this image is correctly identified. However, Bax then assumes that the Voynichese word refers not to a word for the Pleiades, of which there are many, but rather to the name of the well-known constellation Taurus (the bull). The Voynichese word has five letters and Taurus has six, which is only a minor problem if some of the letters are syllables.

Table 11.2 Vowel frequency in the *Voynich Codex*

Voynichese symbol	Pronunciation	Frequency in Voynich (%)	Sequence
o	ā	15.9	1
c	a	12.6	2
9	i/y	11.0	3
a	o	8.9	5
?	e	4.2	9

In any case, it might be assumed that the 8 symbol is pronounced as a "t" and the 9 symbol has a "s" sound. However, Bax proposes that the symbols spell out TAURN. Thus, 9 = "t," o = "a," ? = "r," and 9 = "n." To determine the last symbol as a "n" sound, Bax resorts to linguistic jujitsu. However, it appears that the real reason why the "n" sound is used is that Bax had previously asserted that the 9 has a "n" sound to conform to the identification of folio 2r as a species of *Centaurea* with the Turkish name *KANTARON* (Bax 2014: 26–29). He has assumed that the first Voynichese name in the illustration folio 2r represents the name of the plant ll98aw9, deciphered as KNTARN. Note that the 9, which appears twice, has the pronunciation of a "n" sound. This is ingenious, but we have problems as follows. First, the plant is misidentified, Second, it assumes that both ? and w have the same "r" sound, one "ir" and the other "ur." Folio 2r shows a plant that has linear leaves in groups and an asterid inflorescence. The bracts are prominently tipped in a brownish color, whereas the petals of the flowers are serrated and off-white. This is definitely not *Centaurea diffusa* (Asteraceae), as identified by Sherwood and Sherwood (2008). This species of *Centaurea* has spines on the bracts, and is not darkened at the tips. The petals of *C. diffusa* are linear, whereas the foliage is pinnatifid, often gray-green. A better match would be *Lactuca graminifolia* (Asteraceae), which has linear leaves, often in groups, and bracts are tipped in purple. The petals are serrated and vary from a very light blue-pink to white (see Chap. 4, Fig. 4.12).

Bax expands his decipherment (Table 11.2) using a European plant (folio 41v), which he asserts is "convincingly demonstrated" as coriander (*Coriandrum sativum*) by Sherwood and Sherwood (2008). Coriander in Greek is *korion*. Bax notes that the first name under the image is llcc?o8ax and is partially deciphered as KOORATƏ/U?. Again, the plant is misidentified. *Coriandrum sativum* (Apiaceae) has orbicular schizocarps on a tall plant with leaves gradually becoming linear near the apex, but does not bear tuberous roots, which are prominent in folio 41v (see Chap. 4, Fig. 4.5). The phytomorph in folio 41v bears a dense fruiting umbel with schizocarps that are flat, characteristic of the Apiaceae. These schizocarps have a dark center and white wings. The growth habit is a rosette bearing coppery tuberous roots. The leaves are deeply dissected and palmate. The best match would be biscuitroot, a *Lomatium* sp., most probably *L. dissectum* (Apiaceae). As a result of the plant misidentifications, Bax's decipherment cannot be correct.

Tucker and Talbert (2013), as previously noted, decoded the Voynichese symbols based on the identification of two labeled plants, one as prickly pear (*Opuntia ficus-indica*) = Nahuatl *nochtli*, and the other as *Agave atrovirens* = Taino *maguey*, as explained in Chap. 5, Table 5.1. On this basis, other plants were used to expand the symbol decoding. Some of the names chosen identify the plant, but none was found to be known Nahuatl names. However, following the practice of other Nahuatl names, many were descriptive. Additionally, the decoded symbols did identify the city of Huejotzingo, which confirmed a previous identification and gave us confidence that the decoding was correct. We admit to some weaknesses in this decipherment. For example, the assumption that was made that symbols with open and closed loops were the same pronunciation is a conjecture, e.g., = , = , = , = . Furthermore, we admit that the symbol decoding 0 has failed to decipher much of the text. However, it is intriguing that the symbol decipherments of Comegys (2001) come to the same conclusion as those of Tucker and Talbert (2013) for four Voynichese letters: = "i/y," = "ch," = "q," and = "tl."

The decoding of the alphabet by Tucker and Talbert, although tentative, is consistent in the examples presented. Furthermore, if we check the frequency of each symbol in Table 11.1, the most commonly used symbols are vowels, in 1st, 2nd, 3rd, 5th, and 9th place in frequency (Table 11.2). This confirmed that the Voynichese symbols consist of vowels and consonants and therefore comprise a phonetic language. It had predictive value for the identification of the city Huejotzingo (see Chap. 10) and led to subsequent identification of other cities and perhaps stars. We surmise that the failure to decipher the Voynich text has two explanations. It may be based on an unknown Mesoamerican language or dialect, or, as suggested by William Friedman, it is a mixed synthetic language, i.e., an invented language based on a number of native languages.

Decoding Test Based on Apothecary Jar Labels

To determine which of the decodings were likely to be correct, we examined 31 labeled apothecary jars in the Pharmaceutical section and the Voynichese symbols were transliterated based on the results of the nine decodings in Table 11.1, as shown in Table 11.3. With the exception of Tucker and Talbert's (2013) decipherment, all of the others appear to be gibberish. Furthermore, four of the deciphered Voynichese jar labels based on this decipherment and using Classical Nahuatl (CN) or Huastec Nahuatl (HN) cognates suggested "remedies," as shown below:

Folio 88r#1. *atlaematli: atl* = water, liquid (CN), *matli* = arm, (animal) front leg (CN): "Remedy for swollen limb."

Folio 88r#2. *atlocachi: atl* = water, liquid (CN), *cachi cualli* = better (CN): "Remedy to improve urination."

Folio 88r#3. *ahuintlichoca: a* = negative in compound word (CN), *huin* = to express pain (CN), *tli* =?, *choca* = weeps (HN): "Remedy to reduce intense pain, an analgesic."

Folio 88v(1)#2. *atlaeoha: atl* = water, liquid (CN), *eoh* = impers. of *eua* = *levantarse* (to raise) (CN): "Remedy to increase urination."

Table 11.3 Decipherments of Voynich apothecary jars (*majolica*)

	Voynich jar							
Decipherment	**101v(2)#2**	**99r#4**	**99r#2**	**99r#1**	**99v#4**	**99v#3**	**101v(2)#1**	**99v#2**
Tiltman[a] (1967–1969)	ahaeaeg	ghcoe?g	aharae	odaroeo?	araea2	8aroea8g	8acro8	odoe?o8g
1st Friedman study group[b]	opoeaeg	ghcoe?g	oporae	odaro?o?	oraea2	8aroea8g	8acro8	odoe?o8g
2nd Friedman study group[b]	opoeae9	9bcoe?9	opouae	ovauoho?	ouaea2	8auoea89	8acuo8	ovoe?o89
Currier[c] (1970–1976)	oboeae9	9pcoe69	oborae	ofarojo6	oraea2	8aroea89	8acro8	ofoe6o89
D'Imperio (1978)	dgdihin	nacdiyn	dgdohi	dwho?oy	dohihe	khoihkn	khcodk	dwdiydkn
Comegys (2001)	ollopapy	yll?op?y	olloxap	otlaxoko?	oxapas	saxopasy	sa?xos	otlop?osy
Sherwood (2008)	o?oiaig	gleoi?g	o?oaai	otlaro?o?	oraias	baroiabg	baerob	otloi?obg
Tucker and Talbert (2013)	āhuācâocâi	itlaacâhi	āhuāeocâ	ātloeāhāah	āeocâan	choeācâochi	choaeāch	ātlācâhāchi
Bax (2014)	a?a?ə?n	n?oa??n	a?arə?	akərara?	arə?əs	təra?ətn	təorat	aka??atn

	Voynich jar							
Decipherment	**99r#3**	**101v(3)#1**	**99v#1**	**102r#2**	**102r#1**	**88r#1**	**89r#3**	**88v(3)#4**
Tiltman[a] (1967–1969)	hsoe8g	ho8oa???	ahara?g	odoedg	ho2toe2 2occ8g	ohortchg	ohaeo?g	oda? gora?
1st Friedman study group[b]	hsoe8g	ho8oll?	ahara?g	odoedg	po2toe2 2occ8g	ohortchg	opae?g	odal goral
2nd Friedman study group[b]	bzoe89	bo8oa???	abaua?9	ovoev9	po2soe2 2occ89	obousb9	opaeow9	odan 9ouan
Currier[c] (1970–1976)	pzoe89	po8oaii6	apara69	ofoef9	bo2soe2 2occ89	oporscp9	obaeou9	ofan 9oran
D'Imperio (1978)	ardikn	adkdh??y	hahohyn	dwdiwn	gdebdie edcckn	dadobcan	dwhid?n	dwh? ndoh?
Comegys (2001)	lltlopsy	llosoaii?	allaxa?y	otloptly	llostops so??sy	olloxt?lly	ollapomy	otlan yoxan
Sherwood (2008)	l?oibg	loboa???	alara?g	otloitlg	?os?oig	olor?etlg	o?aio?g	otlan goran
Tucker and Talbert (2013)	tltzācâchi	tlāchāo?	otloeohi	ātlācâtli	huānmācân naaachi	ātlāematli	āhuocâāshshei	ātlol iāeol
Bax (2014)	?ka?tn	?ataə???	ə?ərə?n	aka?kn	?as?a?s saootn	a?ar?o?n	a?ə?a?n	akəir narəir

(continued)

Table 11.3 (continued)

	Voynich jar							

Decipherment	**88v(3)#2**	**88v(3)#1**	**88v(2)#2**	**89v#1**	**89r#2**	**88r#2**	**88v(1)#2**	**88v(2)#3**
Tiltman[a] (1967–1969)	ohoe8g	a8arg	gdg8	odaroe8g	odarg	ohae8g	ohorae	gdodzg
1st Friedman study group[b]	ohoe8g	o8org	gdg8	odaroe8g	odarg	ohae8g	ohora?	gdofzg
2nd Friedman study group[b]	oboe89	o8ou9	9v98	ovauoe89	odau9	obae89	obouah	9vo@9
Currier[c] (1970–1976)	opoe89	o8or9	9f98	ofaroe89	ofar9	opae89	oporaj	9foy9
D'Imperio (1978)	dadikn	dkdon	nwnk	dwhoikn	dwhon	dahikn	dadoh?	nwdtn
Comegys (2001)	ollopsy	osoxy	ytlys	otlaxopsy	otlaxy	ollapsy	olloxak	ytlotly
Sherwood (2008)	oloibg	obrg	gtlgb	otlaroibg	otlarg	olaibg	olora?	gtloe?g
Tucker and Talbert (2013)	ātlācâchi	āchāoei	itlich	ātloeācâi	ātloei	ātlocâchi	ātlāeoha	itlāpi
Bax (2014)	a?a?tn	atarn	nknt	akəra?tn	akərn	o?ə?tn	a?arər	nka?n

	Voynich jar						
Decipherment	88v(2)#1	88v(1)#3	89r#1	88v(3)#3	88r#3	88v(1)#1	89v#2
Tiltman[a] (1967–1969)	odtsg	8arae?ae	tocc2g	dora?g	adg2dg8ae	ohoe8g	doccora?
1st Friedman study group[b]	odtsg	8ara??ae	tocc2g	doralg	ofg2dg8ae	ohoe8g	doccoral
2nd Friedman study group[b]	ovsz9	8auah?ae	socc29	vouam9	of92v98ae	oboe89	voccouan
Currier[c] (1970–1976)	ofsz9	8araj6ae	socc29	foran9	ov92f98ae	opoe89	foccoran
D'Imperio (1978)	dwbrn	khoh?yhi	bdccen	wdoh?n	dznewnkhi	dadikn	wdccdoh?
J.C. Comegys (2001)	otlttly	saxak?ap	to??sy	tloxany	otlystlysap	ollopsy	tlo??oxan
Sherwood (2008)	otl??g	bara??ai	?oeesg	tlorang	opgstlgbai	oloibg	tloeeoran
Tucker and Talbert (2013)	ātlmtzi	choeohahocâ	māaani	tlāeoli	ahuintlichocâ	ātlācâchi	tlāaaāeol
Bax (2014)	ak?kn	tərər?ə?	?aoosn	karəirn	a?nskntə?	a?a?tn	kaooarəir

? = no decipherment given

[a]John H. Tiltman (1967, 1968) from D'Imperio (1978)

[b]D'Imperio (1978)

[c]Prescott Currier (1970–1976) from D'Imperio (1978)

Conclusion

Decipherment of the *Voynich Codex* depends on two interrelated issues: decoding the Voynichese symbols and determining the language (or languages) involved. One cannot be done without the other. It is truly amazing that despite extensive cryptological analysis, the manuscript has remained undeciphered based on the general assumption that the basic language was Latin.

Our interpretation of the Voynichese symbols is that they are a phonetic alphabet of a real or invented language.

The decoding of the Voynichese symbols by Tucker and Talbert (2013) was predicated on the botanical evidence of the presence in the codex, without exception, of species indigenous to New Spain. This suggested that the language of the *Voynich Codex* was the lingua franca of New Spain, probably a version or combination of Nahuatl, related dialects or language, and some Spanish. The botanical evidence was confirmed by finding illustrations of indigenous animals (alligator gar, armadillo, jaguarundi, and ocelot). Two plants, prickly pear cactus with the Nahuatl name *nochtli* and an agave with the Taino name *maguey*, allowed decipherment of nine Voynichese symbols ([Voynichese symbols]), which were then expanded to 15 symbols using other labeled plants, as discussed in Chap. 5. This decoding system allowed decipherment of Nahuatl cognates, and some cities and words, including dialects of Nahuatl, Spanish, Mixtec, Taino, and perhaps Arabic (see Chap. 13) that would have been known to a Nahua educated by the Spanish in sixteenth century New Spain. However, much to our chagrin, it failed to translate the text, leading to the conclusion that it was definitely not Classical Nahuatl and may be a synthetic language (as suggested by William Friedman) composed of various indigenous languages. We are convinced that a Nahuatl dialect is involved (see Chaps. 1 and 5).

There are at least nine groups that claim to have deciphered the Voynichese symbols and all their results are different, indicating that all but one or none is correct. The decoding of Voynichese symbols based on plant identification (Tucker and Talbert 2013; see Chap. 5) indicated that the alphabet is phonetic and that the most common symbols are vowels. Labeled apothecary jars were examined to compare the validity of various Voynichese symbol decoding systems. Only the decoding of Tucker and Talbert (2013) appeared to resemble words and four decipherments were successful in making sense of the script.

Literature Cited

Ackerson, T. 2012. Cited in Dragoni (2012).

Amancio, D.R., W.G. Altmann, E. Rybski, O.N. Oliveira Jr., and L. da F. Costa. 2013. Probing the statistical properties of unknown texts: Application to the Voynich manuscript. *PLoS/ONE* 8 (7): e67310. https://doi.org/10.1371/jounakl.pone.0067310.

Bax, S. 2014. A proposed partial decoding of the Voynich script. www.stephenbax.net Version January 2014.

Brumbaugh, R.S. 1978. *The most mysterious manuscript*. Carbondale: Southern Illinois Press.

Burisch, D. 2012. Cited in Dragoni (2012).
Casanova, A. 1999. Méthodes d'analyse du langage crypté: Une contribution à l'étude du manuscript du Voynich. Docteur thesis. Université Paris.
Child, J.R. 1976/2002. The Voynich Manuscript revisited. National Security Agency/Central Security Service, Fort George G. Meade, Maryland. DOCID: 636899.
———. 2007. Again, the Voynich Manuscript. Unpubl. pdf. http://web.archive.org/web/20090616205410/http://voynichmanuscript.net/voynichpaper.pdf. 3 Feb 2017.
Comegys, J.C. 2001. Keys for the Voynich Scholar: Necessary clues for the decipherment and reading of the world's most mysterious manuscript which is a medical text in Nahuatl attributable to Francisco Hernández and his Aztec Ticiti collaborators. J.C. Comegys, Madera, CA. Library of Congress MLCM 2006/02284 (R) FT MEADE.
Comegys, J.D. 2013. The Voynich Manuscript: Aztec herbal from New Spain. voynichms.com/wp-content/uploads/2014/03/VMSAztecHerbal-JDComegys.pdf. Accessed 5 June 2017.
Currier, P. 1970–1976, Voynich MS transcription alphabet. Plans for computer studies: transcribed text of Herbal A and B material: Notes and Observations. Unpublished communication to John H. Tiltman and M. D'Imperio, Damariscotta, Maine. Cited in D'Imperio.
D'Imperio, M.E. 1978. The Voynich manuscript: An elegant enigma. In *National Security Agency/central security Service*. Meade: Fort George G.
Dakin, K. 1981. The characteristics of a Nahuatl *lingua franca,* 55–67. In: *Nahuatl studies in memory of Fernando Horcasitas*, ed. Frances Karttunen. Texas Linguistic Forum 18.
De Alarcón, H.R. 1987. *Treatise on the heathen superstitions that today live among the Indians native to This New Spain, 1629.* Trans. & ed. J.R. Andrews and R. Hassig. Norman: University of Oklahoma Press.
Dragoni, L. 2012. Il manoscritto Voynich: Un po' di chiarezza sul libro più misterioso del mondo. Unpubl. pdf. http://www.cropfiles.it/altrimisteri/Voynich.pdf. 1 Sept 2014.
Feely, J.M. 1943. *Roger Bacon's cipher: The right key found.* New York: Rochester.
Feng, R. 2016. 141: Cracking the Voynich manuscript code (2nd thesis draft). In *School of electrical and electronic engineering*. Adelaide: Ther University of Adelaide.
Finn, J.E. 2004. *Pandora's hope: Humanity's call to adventure*. Baltimore: PublishAmerica.
Friedman, E.S. 1962. 'The most mysterious ms' still an enigma. *The Washington Post*, Times Herald (1959–1963), August 5.
Gibson, D. 2003. The Voynich question: A possible Middle East connection. Unpubl. web page. http://nabatea.net/vconnect.html. 3 Feb 2017.
Guy, J.B.M. 1991. On Levitov's decipherment of the Voynich manuscript. Cited in Kennedy and Churchill (2006).
Gwynn, B. 2006. Cited in Kennedy and Churchill (2006).
Hoschel, 2012. Cited in Dragoni (2012).
Huth, V. 2012. Cited in Dragoni (2012).
Karttunen, F. 1983. *An analytical dictionary of Nahuatl*. Norman: University of Oklahoma Press.
Kennedy, G., and R. Churchill. 2006. *The Voynich manuscript: the mysterious code that has defied interpretations for centuries*. Rochester, Vermont: Inner Traditions. (First published in 2004 by Orion press, London).
Kircher, F., and D. Becker. 2012. *Le manuscrit Voynich décodé*. Agnières: SARL JMG editions.
Knight, K. 2009. The Voynich Manuscript. MIT. http://www.isi.edu/natural-language/people/voynich.pdf. 1 May 2014.
Landsmann, E. 2008. The Voynich-Manuscript. Unpubl. pdf. http://elifonaot.q32.de/cms/lib/exe/fetch.php?media=de:pub:2007:20071231e:20071231e_dasvoynichmanuskript.pdf. 4 Feb 2017.
Levitov, L.E. 1987. *Solution of the Voynich manuscript: A liturgical manual for the Endura rite of the Cathari heresy, the cult of Isis*. Laguna Park: Aegean Park Press.
Maat, J. 2012. Cited in Dragoni (2012).
Manly, J.M. 1931. Roger Bacon and the Voynich Ms. *Speculum* 6: 345–391.
Montemurro, M.A., and D.H. Zanette. 2013. Keywords and co-occurrence patterns in the Voynich Manuscript: An information-theoretic analysis. *PLos ONE* 8 (6): e66344. https://doi.org/10.1371/journal.pone0066344.

Newbold, W.N., and K.R. Grubb. 1928. *Cipher of Roger Bacon*. Philadelphia: University of Pennsylvania.

O'Neill, H. 1944. Botanical observations on the Voynich MS. *Speculum* 19: 126.

Packheiser, U. 2013. *Das Voynich manuscript*. FP Web & Publishing. http://www.utep.de.

Rivero, H. 2012. Cited in Dragoni (2012).

Rugg, G. 2004a. An elegant hoax? A possible solution to the Voynich manuscript. *Cryptologia* 28: 31–46.

———. 2004b. The mystery of the Voynich manuscript: New analysis of a famously cryptic medieval document suggests that it contains nothing but gibberish. *Scientific American* 291 (1): 104–109.

Sherwood, E. and E. Sherwood. 2008. The Voynich botanical plants. http://edithsherwood.com/voynich_botanical_plants/index.php. 5 June 2017.

Stojko, J. 1978. *Letter to God's eye: The Voynich manuscript for the first time deciphered and translated into English*. Vantage Press. New York.

Stolfi, J. 2006. Cited in Kennedy and Churchill (2006).

Strong, L.C. 1945. Anthony Askham, the author of the Voynich manuscript. *Science* 191: 608–609.

Sullivan, M. 2012. Cited in Dragoni (2012).

Tucker, A.O., and R.H. Talbert. 2013. A preliminary analysis of the botany, zoology, and mineralogy of the Voynich Manuscript. *Herbalgram* 100: 70–85.

Vonfelt, S. 2014. The strange resonances of the Voynich manuscript. Unpubl. pdf. http://graphoscopie.free.fr/publications/Voynich_en.pdf. 23 Jan 2017.

Williams, R.L. 1999. A note on the Voynich manuscript. *Cryptologia* 23: 305–309.

Zandbergen, R. 2004–2017. The Voynich Manuscript. www.voynich.nu/. 6 June 2017.

Chapter 12
Mesoamerican Languages and the Voynich Codex

Fernando A. Moreira

Voynichese and Mesoamerican Languages

Our general hypothesis is that Voynichese is composed of some combination of Mesoamerican languages that includes Classical Nahuatl, its dialects, other central Mexican languages, and Spanish loan words. We determined that Voynichese is not Classical Nahuatl, but instead many Nahuatl type matches have been identified based on the decoding of Voynichese symbols by Tucker and Talbert (2013). We assume that Voynichese is either a dialect, perhaps an extinct form of Nahuatl, or more likely a constructed (fictional) language made up of a number of Mesoamerican languages. In this chapter, we review the pre-contact (with Europe) languages of central Mexico, with special emphasis on Chichimeca Jonaz and Acolhuacatlatolli. We consider common words that we assume would be in the *Voynich Codex* and compare them with their counterparts from various languages. In addition, Voynichese common transliterated words (Knight 2009) are compared with Voynichese optimal and thematic words (Montemurro and Zanette 2013). The Nahuatl meanings of these words are also considered.

Language Diversity in Mexico

Language diversity in Mexico was nothing short of a New World marvel, from language families that dominated the Mexican landscape and beyond to the isolates found in explicit zones. There were more than 200 distinct languages in Mesoamerica alone, and this number can increase depending on classification. The following is a broad overview of languages found in what was roughly central Mexico at the time of European contact. First and foremost, we need to have a basic understanding of what language families, languages, isolates, and dialects actually are and how they are classified. Language relationships are usually described metaphorically by the

J. Janick, A. O. Tucker, *Unraveling the Voynich Codex*, Fascinating Life Sciences, https://doi.org/10.1007/978-3-319-77294-3_12

use of a phylogenetic tree and classified via comparative linguistics. This classification is based on various features, such as the use of sounds, morphology, and order of elements in utterances. Languages can have multiple tones or stresses (essentially pitch patterns versus the force in which sounds are produced) that change the meaning of a word.

Isolating languages consist mostly of single morphemes (e.g., Classical Chinese). A morpheme is the smallest grammatical unit of a language that cannot be divided, say, for example: cat, dance, and -ing (a suffix, say from singing). Analytic languages convey grammatical relationships without using inflectional morphemes. Synthetic languages consist of several morphemes. There are three accepted subtypes, which can be exaggerated. The first subtype, polysynthetic languages, includes those where one word from one language can equate to an entire sentence in another language. The second synthetic subtype is agglutinating languages, where each morpheme usually has a single function and words can consist of many morphemes. The third subtype is fusional languages, in which affixes can combine functions. Please note that a "synthetic language" in this chapter refers to the above, and should not be confused with Arika Okrent's mixed artificial languages (Okrent 2010), which shares the same name. Basic word order (subject, object, verb), which can influence basic meaning, also suggest language relationships.

Language families (a monophyletic unit) are groups of languages that are related by descent from a common ancestor. These can be further divided into smaller phylogenetic elements denoted as "branches" of a family. Language, for the purposes of this summary, is considered a natural communication system that is arbitrary, symbolic, and creative, consisting of sounds, words, and grammar, and is used by people in particular regions. An isolate is a language that has no demonstrable genetic affiliation with another, and essentially does not share a common ancestor. They are language families of a single language, such as Basque. A dialect, simply put, is a variety of a language. The latter is attributed to changes in vocabulary, grammar, and pronunciation and often how a language is spoken in different regions, or by different social classes. There is a point when a dialect can differ substantially that it is no longer mutually intelligible, meaning that it becomes its own unique language, but this is vague. The language families found in central Mexico are the following: Uto-Aztecan, Otomanguean, Mixe-Zoquean, and Purépechan. It should be noted that the identification of a language group does not entail the identification of a particular cultural group, instead native groups themselves use language, geographic region, and membership of particular communities for self-classification.

Uto-Aztecan

Ancient Uto-Aztecan is a geographically broad language family that extends from Oregon to Panama. This family contains 58 languages (however, colonial sources do advocate a larger size). Linguistic studies suggest that its origins stem from southeast California, Arizona, and northwest Mexico. The family is divided into two generally recognized groups, Northern and Southern Uto-Aztecan. The latter group is the one of interest and is further broken down into Pimic, Taracahitic, and Corachol-Aztecan (Campbell 1997: 134). From Pimic, only Southern Tepehuán is spoken in our area of interest and is found in Sonora, Durango, and Jalisco states. Taracahitic languages are beyond the scope of this book. Corachol-Aztecan is the family that contains Nahuatl (Classical Nahuatl), along with Pipil (Nawat), Pochutec (extinct), Cora, and Huichol. The former is well known for the lateral affricate "tl" /t͡ɬ/. Nahuatl was spoken primarily in central Mexico, notably by members of the Aztec empire. It also has various dialects. Absolutive noun suffixes, with the exception of Cora/Huichol, are typical in Uto-Aztecan grammar. These suffixes have no apparent semantic value and appear in nouns as heads of lexical entries and, depending on the linguistic environment, can be dropped. A noteworthy absolutive noun suffix is in Classical Nahuatl and takes the form "-tl" after vowels and "-tli" after consonants: *a:-tl* (water) and *no:ch-tli* (cactus fruit) respectively. Note that the colon denotes a long vowel. These sounds are distinguished in Nahuatl and can affect meaning. They are rarely or hardly ever included in older dictionaries.

Classical Nahuatl, for example, is an extinct agglutinative language from central Mexico, and was considered the lingua franca in the sixteenth century. This particular form survived in colonial texts and is likely a sociolect, a language variety associated with a specific social group, in this case the aristocracy. Pajapan Nahuatl, also known as Isthmus Nahua, has various lexical and phonological distinctions compared with Classical Nahuatl. One case, for example, is the simplification of the lateral fricative to a dental stop, illustrated in the word for "person," *tla:catl* versus *ta:gat*. Pajapan, conversely, has close affinities to Pipil. Phonological changes like these have been attributed to linguistic contact with Mixe-Zoquean languages (Ramírez 2005: 8).

Mixe-Zoquean

The Mixe-Zoquean language family was attributed to the early Olmec culture and inconclusively linked to the Isthmian script. Mixe-Zoquean comprises two groups, Mixean and Zoquean, and contains a total of 12 distinct languages and 11 dialects (Wichmann 1995). Of these, only the Mixean line falls into our area of interest, specifically Oaxaca Mixean, which breaks down to Lowland, Midland, South, and North Highland Mixe. Although Sayultec and Olutec, along with two Zoquean languages (Sierra Popoluca and Texistepec Popoluca) have been shown to have played a role in influencing Nahuan tongues, they may have extended into our area of interest before European contact.

Mixe-Zoquean has complicated morphophonology, that is, complex phonetic processes, such as assimilation, metathesis, palatalization, laryngeal gain/loss, and vowel lengthening/shortening (Longacre 1977: 102). The family shares morphosyntactic (morphology and syntax) traits found in polysynthetic languages and has a general object-verb (OV) configuration. Its expressions and clause formations are syntactically complex. The languages are head-marking, which means grammatical marks causing dependency changes between different words are placed at the nuclei of phrases, among other traits. An interesting note is that Nahua borrowed significant lexical and grammatical forms from Mixe-Zoquean languages (Kaufman and Justeson 2006). In 1524, Hernán Cortés, in his fourth letter to Emperor Charles V, admitted that the Mixe people (likely Oaxaca Mixean speakers) had yet to be conquered.

Otomanguean

Considered the oldest language family in Mesoamerica by most linguists, Otomanguean is highly diverse, possessing eight sub-groups and including some 40 languages (176 according to Simons and Fennig 2017). The family is geographically extensive, and was found as far away as Nicaragua. The branches of Oto-Pamean, Chinantecan, Popolocan, Zapotecan, Amuzgo, and Mixtecan are all located in our area of interest, mainly in the states of Oaxaca, Guerrero, Mexico, Hidalgo, and Querétaro.

All Otomanguean languages are tonal, either having register or contour tones systems. These tones can range from two to five. Otomanguean, comparatively speaking, has fewer morphemes per word than surrounding languages and thus is more analytic. Most Otomanguean languages have phonemic vowel nasalization and open syllables, in addition to limited syllable-initial consonant clusters, and are oddly lacking labial consonants, such as the "p" sound (Campbell 1997: 157). Otomanguean is head-initial, that is, it has right-branching word order and is generally verb-subject-object (VSO). Reconstructed words of cultigens such as maize, beans, and squash indicate this family's prominent role in the origin and diffusion of agriculture. Lexical influences from Mixe-Zoquean and Mayan can also be traced.

Purépechan (Tarascan)

Purépechan is an agglutinative isolate spoken in the state of Michoacán, and the people who spoke it were the builders of the city of Tzintzuntzan. Nahuatl speakers referred to the Purépechan people as Michhuàquê, meaning "those who have fish." Tarascan has no laterals, that is, it lacks an "l" sound and also has no phonemic glottal stop. In Tarascan, a distinction exists between non-aspirated and aspirated plosives and affricates. The stress accent is phonemic and can change the meaning of word, for example *karáni* versus *károni*, meaning "write" and "fly," respectively (Zerbe 2013: 2, 43–44). Its complex morphology and long words make it very synthetic; yet, noun incorporation, that is, when a noun forms a compound with its direct object/adverbial modifier, is not present. The language is described as either subject-verb-object (SVO) or subject-object-verb (SOV) in nature, although not arbitrarily so, as word order is used rationally to convey contrastive information.

Totonac-Tepehua (Totonacan)

Totonacan is a relatively small language family with two branches, Totonac and Tepehua, amounting to 12 languages. These languages are spoken in the states of Veracruz, Puebla, and Hidalgo. These speakers have been associated with the builders of the ancient city of Teotihuacán, in the present-day state of Mexico. The conjecture is supported by Totonacan loanwords found in Lowland Mayan and Nahuan languages, although genetic affinities have been associated with the Mixe-Zoquean language family. Totonacan has a complicated word formation. These languages tend to possess a three-vowel system, with each vowel containing length and creaky voice (laryngealization) variances; for example, the phoneme /u/ can also be /uː/, /ṵ/, or /ṵː/. Totonacan is the only Mesoamerican language with a lateral obstruent /ɬ/. The lateral affricate /t͡ɬ/ is also present in this family.

Totonacan is very agglutinative and polysynthetic. A prominent and interesting feature of Totonac-Tepehua languages is the use of sound symbolism (phonosemantic). This is when the meaning of a word is partially represented by its phonemes. Phonemic changes such as /s/ → /š/ → /ɬ/ indicate grade. For example, *spipispipi* means "something trembling," whereas *ɬpipiɬpipi* means "something having convulsions" (Childs 2015). A system of causatives, applicatives, and object/body parts prefixes are used in lieu of morphological case on nouns. Most Totonacan languages lack prepositions.

Chichimecan Languages

The Nahuatl term *chīchīmēcah* is a pejorative ethnonym and a political term used to describe uncivilized nations, essentially barbarians. The term itself was derived from the "inhabitants of Chichiman," where */chi:chi:/* + */–ma:-n/* means "place of milk" (Morritt 2011: 23). In the past, the Nahua reserved the term Chichimeca to describe themselves in origin stories, before becoming "refined." When describing the Chichimeca, one must note that it is not merely a particular culture or language, it is a general term that often engulfs Otomi, Chichimeca Jonaz, Cora, Huichol, Pame, and Tepehuán speakers, among others. These typically belong to the Otomanguean and Uto-Aztecan language families. Nigel Davies argues convincingly that most all the Chichimec groups that migrated into central Mexico in the Nahua histories spoke Otomi, and it has been suggested that at least some of the tribes might have spoken Pame as well (Smith 1984: 158).

For purposes of this chapter, we will be strictly talking about Chichimecan as the Chichimeca Jonaz language. This language belongs to the Otomanguean (also called Oto-Manguean) language family and is from the Oto-Pamean branch, the northernmost branch of this family. The language area was in the states of Querétaro and Guanajuato, just north of Mexico state, essentially nested in between the Aztec Empire and the Tarascan state.

Oto-Pamean languages, in general, have relatively basic tones systems compared with other Otomanguean languages (Palancar 2017: 6). Pamean languages, broadly speaking, have low/high opposition. For example, most words in Chichimeca Jonaz are disyllabic, that is, words made up of two syllables. These are organized suprasegmentally into "four melodies," and minimal pairs are rare. Like its family, Oto-Pamean is verb-initial, with the exception of Chichimeca Jonaz, which is SOV (Palancar 2017: 4). Chichimeca Jonaz has the following consonants and vowels: /p t t͡s t͡ʃ k ʔ b d d͡ʒ g s ʃ h z l r w/ with fortis and lenis nasals /m m̲ n n̲/, and spoken vowels /i ɪ u e o æ ɑ/ and nasal vowels /ĩ ɪ̃ ũ ẽ õ œ̃ ɑ̃/.

Two major classes of nouns exist in Chichimec Jonaz: variable nouns (which have internal possession inflection) and invariable nouns (which have invariable stems). These arguably mirror a division between inalienable and alienable nouns. According to Jaime de Angulo (1933), possession in Chichimec variable nouns lacks an absolute form, basically when the noun is not possessed, no form exists. This suggests obligatorily possession, as exists on nouns that refer to kinship terms, cultural artifacts, body parts, and food. As a rule, verbs in Oto-Pamean languages have inflectional formatives. These are markers placed in front of a verbal stem (forms of the verb). In Chichimec there are at least 14 different stem patterns (Palancar 2017: 22). Verbal inflection is toneless and monosyllabic.

Chichimeca Jonaz tense-aspect-mood (TAM) marking is rich, having present, future, potential, and sequential tenses, in addition to three different past tenses. Conversely, other Oto-Pamean languages can have up to a dozen values. Chichimeca Jonaz also has seven formative-based conjugation classes.

Acolhuacatlatolli

Acolhuacatlatolli is one of the pre-contact languages being considered. Acolhuacan was a province just east of the Valley of Mexico. The root for Acolhuacan is argued to come from the Nahuatl *ahcol(−li)*, meaning "shoulder," combined with "-hua" (possessive suffix) and "-can" (locative suffix), giving a meaning equivalent to "place of strong men." Hanns J. Prem (Lee 2008) argues correctly that the initial vowel in Acolhuacan is long, not short, as in *ahcol(−li)*, and is not followed by a glottal stop /h/. Given one version of the glyph, Jongsoo Lee states that the roots may actually be a(−tl) + col(−li), that is, "water" + "bent," meaning something like "place of those who live near bent water" (Lee 2008: 255).

What was the nature of Acolhuacatlatolli in sixteenth century Nueva España (New Spain)? In pertinent references (Carrasco 1963; Davies 1980; de Alva Ixtlilxóchitl 2012; Garcia Icazbalceta and Pomar 1891; Gerste 1891; Gibson 1964; Gradie 1994; Kirchhoff et al. 1976; Orozco 1864; Rossell 2006; Smith 1984; Stampa 1971; Swanton 2001; Thomas and Swanton 1909), there are very few data on Acolhuacatlatolli, and there is no agreement on the nature of the language. It may have been:

1. A Chichimec language distinct from Uto-Aztecan and the Oto-Pamean families.
2. A language related to Otomi (from the Oto-Pamean branch).
3. An Uto-Aztecan language.

Most modern, peripheral references only mention hypothesis no. 2, but nobody knows the vocabulary or grammar of this language. Certainly, if the Acolhua people transitioned to speaking Nahuatl, as commanded by Techotlalatzin (*tlatoani,* or king, from ca. 1357 to 1409 (Lee 2008: 83) of Texcoco, in the present-day Mexico state), some remnants of Acolhuacatlatolli would have survived in their Nahuatl dialect (herein Texcocan Nahuatl). Conversely, any surviving Acolhuacatlatolli would have had extensive influence from Nahuatl.

Gerste (1891) stated that Texcocan Nahuatl frequently substituted "i" and "o" for the vowels "e" and "u." Writings from Texcoco, in addition to the letters of Cortés, clearly show these substitutions, i.e., Otumis versus Otomis or Mutezuma versus Moctezuma, to compare a few. Gordon Whittaker (1988) notes that Texcocan Nahuatl also lacks "a-raising" that occurs in the /iya/ sequence. It lacks the palatalization from "a" to "e" after the /y/. Texcocan Nahuatl also developed an on-glide, that is, a semi-vowel (glide) that immediately precedes the vowel "e." For example, the compound word for "three years" in Texcocan Nahuatl was *yexhuitl* versus the standard Classical Nahuatl *ēxihuitl* (McAfee and Barlow 1946). It is conceivable that these phonetic differences are artifacts of their rejected Chichimec tongue (although the latter example may be an exception).

Texcocan Nahuatl was considered the most refined speech in the Triple Alliance. Although Juan Bautista de Pomar makes it clear that there are many Chichimec borrowings that defy understanding; therefore, the difference is not limited to simply phonetics, but is also in their lexicon. Texcocan Nahuatl surrounded the capital, and was spoken as near as Otumba (Mexico state), to as far east as Cempoala (Veracruz state), and to the north in Tulancingo (Hidalgo state). We know that the base language in the *Voynich Codex* is neither Nahuatl nor Otomi. References do point to the ambiguous "Populuca" language, another epithet describing an ethno-lingual group. The term itself comes from the Nahuatl "to babble" (Hill 2006). It has been attributed to Otomanguean and Mixe-Zoquean languages. In the case of Texcocan Nahuatl, it is likely related to the Popolocan or another in the Oto-Pamean branch (such as Otomi).

The Popolocan branch includes Mazatec, Ixcatec, Chocho, and Popoloca (Nginua), and is typically VSO word order (Veerman-Leichsenring 2001). The *Historia Tolteca-Chichimeca*, a Cuauhtinchan bilingual document, was confirmed to have been written in both Nahuatl and Popoloca (Swanton 2001: 134). Although conjecture, could this actually have been Acolhuacatlatolli? The Popoloca spoken in Tlacoyalco has the following consonants and vowels: /p β k t̪ ð t t͡s s l ɾ t͡ʃ t͡ʃ̡ ʃ ʃ̡ j m n ɲ ɣ ʔ h/ and /a e i o u/ (including nasalized versions), where "o" and "u" are a single phoneme (Stark 2011). Popoloca has three to four tones depending on the language (San Marcos Tlacoyalco versus San Juan Atzingo) (Stark 2011; Krumholz et al. 1995).

Analysis

As part of our efforts to determine a base language for the *Voynich Codex*, we used a portion of the Swadesh list (Swadesh 1971: 283). The ranked word list, named after the linguist Morris Swadesh, is often used in glottochronology and includes assumed stable and loan-resistant words (i.e., they have increased retention rates). Relative stability is a statistical measure of the semantic meaning of lexical items (i.e., essentially how often a particular word – for example, "dog" – changes in a language), where a higher number indicates longer retention. In addition, we included some obvious extra words, based on the assumed context of the *Voynich Codex*, such as those subjects related to the illustrations in the text. We examined connections with Nahuatl, Otomi, Northern Pame, and Popoloca languages. Unfortunately, no Chichimeca Jonaz comparisons were available to the authors. If the base is Nahuatl or another central Mexican language, then we should see retention of lexical items in that language, if any.

Comparing the list of lexical items in Table 12.1, such as colors, body parts, states of matter, etc., with Table 12.2, compiled by Knight (2009), along with Table 12.3, compiled by Montemurro and Zanette (2013), we did not obtain any clear matches. The closest we got to a match was the Nahuatl word for "tree," *cuahuitl*, with "tree trunk," quātlaachi/*cuauhtlactli* (4olfc89). This match is problematic. Assuming that these are correctly transliterated, it is difficult to demonstrate a case when the absolutive suffix *-tli* (−t͡ɬi) in Nahuatl can transform into *-chi* (−t͡ʃi), a voiceless postalveolar affricate. Most Nahuatl nouns are bounded,

Table 12.1 Selected words from the Swadesh 100 list expected to be found in the *Voynich Codex*, with additions ([a]) and their equivalents in central Mexican languages

			Valle del Mezquital	Xi'iuy	San Juan Atzingo
Word (English)	Stability	Nahuatl	Otomi (Hñähñu)	Northern Pame	Popoloca (Nginua)
Water	37.4	a:tl	dehe	kante	ìntā
Tree	33.6	cuahuitl	za	nkuãng	inta
Fish	33.4	michin	huä	xikiau	coche
Stone	32.1	tetl	do	gatu	ixrō
Bone	30.1	omitl, pitztli	ndo'yo	n'kiuãng	–
Night	29.6	yohualli	xui	dagũjũn'	tie
Leaf	29.4	izhuatl, quilmaitl	xi	n'xĩ	ica, chaon ca
Rain	29.3	quiahuitl	'ye	mikie	ichrin
Blood	29	eztli	ji	kakji	ijnī
Person	28.7	tlacatl	jä'i	–	chojni
Star	26.6	citlalin, centli	tso̱	gank'u'	conotsé
Mountain	26.2	tepe:tl	t'o̱ho̱	nt'ue	ijnā
Fire	25.7	tletl, tlatla	nzät'i, nzo̱	nkiue	xrohi
Drink	25	a:tli	tsi	juau	its'ini
Dog	24.2	itzcui:ntli, chichi	tsat'yo, 'yo	ndru	comīā
Sun	24.2	to:natiuh	hyadi	ganu'	yāon
Moon	23.4	me:tztli	zänä, zinänä	nm'ãu	nitjó
Heart	23.2	yollotl, yolli	mu̱i, dämu̱i, koraso	n'uà	āsénni

(continued)

Table 12.1 (continued)

Word (English)	Stability	Nahuatl	Valle del Mezquital Otomi (Hñähñu)	Xi'iuy Northern Pame	San Juan Atzingo Popoloca (Nginua)
Feather	23.1	ihhuitl, atlapalli	xi	–	cānēnēe ca, nēnēe ca
White	22.7	iztac, iztactic	nt'axi, t'axi	danua	tjóá
Yellow	22.5	cozauhqui, coztic	k'ast'i	dajuajan	sine
Bird	21.8	tototl	ts'ints'u̱	xiljiañ	coxroxe
Earth	21.7	tla:lli, tla:lticpac	xi'mhai	npu'	inche, nonte
Black	21.6	capoltic	mboi, boi, 'bo	dampu	tié
Foot	21.6	icxitl	ua	makua	tōténi
Green	21.1	xo:tl	'ñixki, k'angi, nk'angi, nk'ant'i	nklɨjɨyl	yoa
Sleep	21	cochiliztli, cochi	ähä	a'ũɨ̃jɨ̃yl	ts'ajuani
Root	20.5	nelhuatl, nelhuayotl, tlanelhuayotl	'yu̱	skamu	nuèe nta
Red	20.2	chiltic, eztic, tlatlatic	the̱ni, nthe̱ni	da'uã	cjátse
Eat	20	tlacua, tlacuah	tsi	asɨjɨñ	sineni
Dry	18.9	hua:qui	'yoni, 'yot'i	m'u	xámá, tsíxámá
Smoke	18.5	poctli, ayahuitl, pocyotl	'bifi	xkl'ĩ	ihnchi
Seed	18.2	achtli, xina:chtli	nda	gusi	itje
Woman	17.9	cihua:tl, zohuatl	'be̱hñä	ntjuy	ichjin, chojni chjin, tjachjin
Long	17.4	hueyac, melactic	ma	npa'au	cjínjīnxīn
Man	16.7	oquichtli	'ño̱ho̱, däme	na'u	ch'ín, ch'ín ntoa, táda
Cold	16.6	cecuiztli	tse̱	matse	iquin
Burn	15.5	chichinoa, (i) hchinoa:	tsät'i	juɨñ	tsjacani
Big	13.4	hue:intin	dängi, däta, nda, ndä	ntɨ-	ijié, ntájín, ntájní
Hot	11.6	totoctic	pa	mpa	sóa
Small	6.3	tepito:n	ts'u̱, notsi, zi	–	intsí, ihntē, –xjan
Bathe[a]	–	altia:	hi, hñi, nxaha	–	sinchec'óntini, ts'óntini
City/town[a]	–	a:ltepe:tl	dähni, dänga hnini	kam'us	chjasin
Bark (tree)[a]	–	–	–	–	tjōé
Flower[a]	–	xo:chitl	do̱ni	ntung	itsjo
Plant/herb[a]	–	xihuitl	t'axi	ntjũ'ũ	ica, ca
Temple[a]	–	teocalli	nijä	–	nīnco

Table 12.2 The 20 most frequent words in the *Voynich Manuscript* (Knight 2009) and their equivalents

Count/8116	Transliteration	Meaning in Nahuatl	Others
863	choll	–	Spanish: *cholla* = skull, cactus
537	aca	*aca* = someone	–
501	machi	*mache* = mainly, on the whole, particularly, especially	–
469	oll	–	–
426	tsachi	–	–
396	maca	*Maca* = "give," or the imperative "no"	–
363	or	–	–
350	ar	–	–
344	mai	–	Otomi: *mai* = shrimp, vine
318	choe	–	Popoloca: *cho* = deceased
308	quatlaai	–	–
305	quatlachi	*cuauhtlactli* = tree trunk	–
283	tsai	–	–
279	quatlol	–	–
272	quatlachi	*cuauhtlactli* = tree trunk	–
270	chi	*-chi* = toward	–
262	quatholl	–	–
260	oca	–	Otomi: *ǫke* = take off
253	choca	*choca* = to weep, cry; for animals to make various sounds	Popoloca: *chóc'á* = in front of
243	n	–	–
219	mae	–	Otomi: *mai* = shrimp, vine

that is, non-possessed nouns must take a suffix called the absolutive. The portion that is far harder to explain is the deletion of the voiceless velar, the "k" sound: *cuauhtla**c**tli* /kʷaht͡ɬa**k**tkʷ/ versus quātlachi /kuat͡ɬat͡ʃi/. Based on the context of the Voynich, *cuauhtlactli* is a logical term to find, but why is it more common than the root words: "tree" or "leaf" or "plant"?

Frequent words (Table 12.2), unfortunately, are not an effective way to determine the context of a text, although they are a very good way of identifying pronouns, articles, conjunctions, and prepositions. For example, the top five words in the works of Shakespeare are "the," "and," "I," "to," and "of." Suffice to say, these say nothing of romance or betrayal, key Shakespearean themes. This is a fair reason why there are virtually no matches between Tables 12.1 and 12.2. That being said, we fail to show that the frequent words in Table 12.2, with the exception of a few, have semantic meaning.

The transliterated word chi (89) has a Nahuatl match, but it is not without issue. This relational noun stem "-chi" is a preposition, meaning "toward," and is rarely

Table 12.3 The most informative optimal and thematic words in the *Voynich Manuscript* (Montemurro and Zanette 2013) and their equivalents

Voynichese	Transliteration	Nahuatl meaning	Others
	oll	–	–
	ocâ	–	–
	oe	–	–
	machi	–	–
	mācâ	*maca* = "give," or the imperative "no"	–
	māe	–	Otomi: *mai* = shrimp, vine
	mi	–	Otomi: *mǐ* = to settle; *mí* = 3rd person imperfect
	cuācâ	*cua:cuah* = a possible truncation of *cua:cuahueh*, "animal with horns"	–
	cuāe	*cua* = eat	Popoloca: *cuaye* = abundant
	cuy	–	Otomi: *ku̱i* = to take out
	choll	–	Spanish: *cholla* = skull, cactus
	chol	–	–
	chocâ	*choca* = to weep, cry; for animals to make various sounds	Popoloca: *chóc'á* = in front of
	chi	*-chi* = toward	–
	câmachy	*camohtli* = yam	–
	ātlosh?z	–	–
	ātlachi	–	–
	ātlaay	–	–
	ācâ	*aca* = someone	–
	ātlachi	–	–
	quātloll	–	–
	quātlosh?z	–	–
	quātlol	–	–
	quātlocâ	*cuauhtloc* = next to tree	–
	quātloe	–	–
	quātlachi	*cuauhtlactli* = tree trunk	–
	quātlaachi	*cuauhtlactli* = tree trunk	–
	quātlaai	–	–
	quācâ	–	Otomi: *kueke* = separate
	quātlmi	–	–
	n	–	–
	tsachi	–	–
	tsai	–	–
	tsā	–	Otomi: *tsa* = bite, power, shame
	tsāe	–	Popoloca: *tsāa* = stomach, to steal
	i	*i-* = possessive singular prefix	Popoloca: *í* = other, this

found embedded with anything other than the stem *(tla:l)-li-*, “ground,” as in the following form *tla:lchi* (Richard Andrews 2003: 473). Chi is presented as a stand-alone and not a stem, questioning the validity of the match. The indefinite pronoun *aca* (Voynichese [Voynich glyphs]), Nahuatl for “someone,” is a term one would expect to find in a frequent words list. It so happens to be the second most common term. The remaining Nahuatl matches are verbs and adverbs.

The Popolocan preposition *chóc’á*, meaning “in front of,” may be a match for the transliterated chocâ ([Voynich glyphs]). If so, this could be an indication that the *Voynich Codex* is a multilingual document. A major setback for advocation of a multilingual text is the orthography. In the case of Otomanguean languages, there would likely be a lack of representation of tones and nasalized vowels, which would make it difficult to distinguish lexical or grammatical meaning. Glottal stops are rarely represented, if at all, in colonial dictionaries; thus, the expectation that they would be in the *Voynich Codex* is low. The Popoloca word for “deceased,” *cho*, could be related to *choe*. It is not uncommon for diphthongs (double vowel sounds) to simplify into single vowel sounds. Unfortunately, there are not enough examples to demonstrate this. The Otomi verb *o̱ke* is an approximation of the transliteration oca, meaning “take off.” In this particular case, it is difficult to explain the vowel raising, a → e from this example alone.

The optimal and thematic words tell us very little about the *Voynich Codex*. What they can possibly tell us, given the illustrations and words mentioned, are its relation to trees. “Tree trunk” has already been discussed, but not the Nahuatl word *cuauhtloc*, meaning “next to tree,” which is comparable with the Voynichese, quātlocâ ([Voynich glyphs]). Here, we witness epenthesis, that is, adding a sound to a word, in this case, an “a” at the end of the word. This typically occurs to break up unwanted sequences; here, it appears to be /–t͡ɬok/ to /–t͡ɬoka/, the former is permitted in Nahuatl. Once again, the sample size is far too small to conclude that such a change is systematic, assuming that a derivation of Nahuatl is used in the *Voynich Codex*.

The transliteration of [Voynich glyphs] yields câmachy, which could be *camohtli*, Nahuatl for “yam.” Similar to *cuauhtlactli* (tree trunk), the absolutive suffix “-tli” is transformed into the syllable /–t͡ʃi/. If a true cognate, it would strongly suggest that this could be a systematic sound change. “Yam” as a theme does fit the context of the Voynich. The transliterated cuācâ could refer to the Nahuatl *cua:cuah,* which may be a truncation of *cua:cuahueh*, “animal with horns.” There are plenty of quadrupeds with horns illustrated in the *Voynich Codex* and they appear predominantly in the zodiac illustrations. As mentioned earlier, diphthongs can be simplified: ua → a.

Another term that fits the overall Voynich theme is the Otomi word *mai*. This can mean both “shrimp” and “vine” and corresponds to the Voynichese chor. Spanish *cholla* meaning “skull,” or “cactus,” from the transliterated [Voynich glyphs] (choll), demonstrates deletion of a vowel, in this case, the “a” sound. This calls into question the mentioned epenthesis. Nonetheless, this sound deletion could be limited to a number of specific cases. This is difficult to demonstrate given our sample size. The Popolocan *tsāa* means both “stomach” and “to steal,” although the former use seems more likely and it corresponds to the Voynichese [Voynich glyphs]. It makes perfect sense that an herbal mentions the words “stomach,” and “eat,” as it does with the Nahuatl *cua* from [Voynich glyphs].

Conclusion

It could not be determined if a Mesoamerican language was inscribed in the *Voynich Codex*. Furthermore, based on the analysis of common words, it was difficult to find evidence of a constructed language from the languages surveyed. Using this transliteration, the document does have strong affinities with Nahuatl, but these appear to be superficial, although our analyzed sample size was relatively small. There are sound changes that are outside convention, but also hints of repetition worth investigating. However, the potential author, Gaspar de Torres, born in Santo Domingo, does hint at the possible inclusion of Taíno, an Arawakan tongue that was once spoken on the island, in the *Voynich Codex*. Furthermore, Torres' half-uncle was a premier slaver and contact with an African slave, perhaps as a nanny, opens up new venues.

Although Montemurro and Zanette (2013) demonstrate connections between sections of the *Voynich Codex* via linguistic structure, suggesting that the manuscript actually may be an intelligible document, this analysis demonstrates that our approach or assumptions may be faulty. Nevertheless, according to Montemurro and Zanette's list, the *Voynich Codex* seems to revolve around trees, yams, vines/shrimp, and horned quadrupeds.

Literature Cited

Campbell, L. 1997. *American Indian languages: The historical linguistics of native America. Oxford studies in anthropological linguistics*. Oxford, UK: Oxford University Press.

Carrasco, P. 1963. Los caciques chichimecas de Tulancingo. *Estudios de Cultura Náhuatl* 4: 85–91.

Childs, G.T. 2015. Sound symbolism. In *The Oxford handbook of the word*, ed. J.R. Taylor. Oxford: Oxford University Press.

Cortés, H. 1986. *Letters from Mexico*. Trans. and ed. A. Pagden. New Haven: Yale University Press.

Davies, N. 1980. *The Toltec heritage*. Norman: University of Oklahoma Press.

De Alva Ixtlilxóchitl, F. 2012. Historia de la Nación Chichimeca. www.linkgua-digital.com. Barcelona.

De Angulo, Jaime. 1933. The Chichimeco language. *International Journal of American Linguistics* 7: 152–194.

García Icazbalceta, J., and J. B. Pomar. 1891. Pomar y Zurita: Pomar, Relación de Tezcoco; Zurita, Breve relación de los señores de la Nueva España. Varias relaciones antiguas. (Siglo XVI).

Gerste, R.P. 1891. La langue des Chichimèques. Comte Rendu du Congrès Scientifique International es Catholiques. *Philologie* 6: 42–57.

Gibson, C. 1964. *The Aztecs under Spanish rule: A history of the Indians of the valley of Mexico, 1519–1810*. Stanford: Stanford University Press.

Gradie, C.M. 1994. Discovering the Chichimecas. *The Americas* 51: 67–88.

Hill, J. 2006. The languages of the Aztec empire. In *The Oxford Handbook of the Aztecs*, ed. D.L. Nichols and E. Rodríguez-Alegría, 129–131. New York: Oxford University Press.

Kaufman, T., and J. Justeson. 2006. The history of the word for "cacao" and related terms in ancient Meso-America. In *Chocolate in Mesoamerica: A cultural history of cacao*, ed. C.L. McNeil, 118–139. Gainesville: University of Florida Press.

Kirchhoff, P., L.O. Gümes, and L.R. Garcia. 1976. *Historia Tolteca-Chichimeca*. México: Instituto Nacional de Antropología e Historia.

Knight, K. 2009. The Voynich Manuscript. MIT. http://www.isi.edu/natural-language/people/voynich.pdf.

Krumholz, J. A., M. K. Dolson, and M. H. Ayuso. 1995. *Diccionario Popoloca de San Juan Atzingo Puebla*. No. 33. Ed. 1. Tucson: The Summer Institute of Linguistics.

Lee, J. 2008. *The allure of Nezahualcoyotl: Pre-Hispanic history, religion, and Nahua poetics*. Albuquerque: University of New Mexico Press.

Longacre, R.E. 1977. Comparative reconstruction of indigenous languages. In *Native languages of the Americas*, ed. T.A. Sebeok, 99–139. New York: Plenum Press.

McAfee, B., and R.H. Barlow. 1946. The Titles of Tetzcotzinco (Santa Maria Nativitas). *Tlalocan* 2: 110–127.

Montemurro, M.A., and D.H. Zanette. 2013. Keywords and Co-Occurrence Patterns in the Voynich Manuscript: An Information-Theoretic Analysis. *PLoS ONE* 8 (6): e66344. https://doi.org/10.1371/journal.pone.0066344.

Morritt, R.D. 2011. *Olde New Mexico*. Newcastle: Cambridge Scholars Publishing.

Okrent, A. 2010. *In the land of invented languages*. New York: Spiegel & Grau Trade Paperbacks.

Orozco y Berra, M. 1864. *Geografía de las lenguas y carta etnográfica de México*. México: Imprenta de J.M. Andrade y F. Escalante.

Palancar, E.L. 2017. Oto-Pamean. In *The languages and linguistics of middle and central America: A comprehensive guide*, ed. S. Wichmann. Berlin: Mouton de Gruyter.

Ramírez, V.P. 2005. *El Nawat de la costa del golfo*. México: Algunas semejanzas y diferencias Estructurales con el Náhuatl Central. Escuela Nacional de Antropología e Historia-FFyL/IIA-UNAM.

Richard Andrews, J. 2003. *Introduction to classical Nahuatl, volume 1*. Norman: University of Oklahoma Press.

Rossell, C. 2006. Estilo y escritura en la Historia Tolteca Chichimeca. *Desacatos* 22: 65–92.

Simons, G.F., and C.D. Fennig. 2017. *Ethnologue: Languages of the world*. 20th ed. Dallas: SIL International. Online version http://www.ethnologue.com.

Smith, M.E. 1984. The Aztlan migrations of the Nahuatl chronicles: Myth or history? *Ethnohistory* 31: 152–186.

Stampa, M.C. 1971. Historiadores indígenas y mestizos novohispanos. *Siglo XVI-XVII. Revista Española de Antropología Americana* 6: 206–243.

Stark, S. L. 2011. Ngigua (Popoloca) Pronouns. SIL-Mexico Branch Electronic Working Papers #012. http://www.sil.org/mexico/workpapers/WP012i-PopolocaPronouns-pls.pdf.

Swadesh, M. 1971. In *The origin and diversification of language*, ed. Joel Sherzer. Chicago: Aldine.

Swanton, M.W. 2001. El texto Popoloca de la Historia Tolteca-Chichimeca. *Relaciones. Estudios de Historia y Sociedad* 22: 116–140.

Thomas, C., and J.R. Swanton. 1909. *Indian languages of Mexico and central America*. Washington, DC: Government Printing Office.

Tucker, A.O., and R.H. Talbert. 2013. A preliminary analysis of the botany, zoology, and mineralogy of the Voynich Manuscript. *Herbalgram* 70 (100). American Botanical Council. www.herbalgram.org.

Veerman-Leichsenring, A. 2001. Changes in Popolocan word order and clause structure. In *Grammatical relations in change*, ed. J.T. Faarlund, 303–322. Amsterdam: John Benjamins Publishing.

Whittaker, G. 1988. Aztec dialectology and the Nahuatl of the friars. In *The work of Bernardino de Sahagun, pioneer ethnographer of sixteenth-century Aztec Mexico*, ed. J.J. Klor de Alva, H.B. Nicholson, and E.Q. Keber, 321–330. Albany: Institute for Mesoamerican Studies, University at Albany, State University of New York.

Wichmann, S. 1995. The *Relationship Among the Mixe–Zoquean Languages of Mexico*. Salt Lake City: University of Utah Press.

Zerbe, A. 2013. P'urépecha fortis v. lenis consonants. University of Washington, working papers. In *Linguistics (UWWPL)*, ed. S. Song and J. Crowgey, vol. 31 https://depts.washington.edu/uwwpl/vol31/zerbe_2013.pdf.

Chapter 13
Progress and Problems in Decipherment

Jules Janick and Arthur O. Tucker

Deciphering the Text

The *Voynich Codex* is rich in illustrations, but without doubt the most important part is the text. Clearly, full understanding of the Voynich depends on decipherment. Our efforts have achieved a few tentative successes, but our inability to translate the great bulk of the text has led us to first analyze the illustrations, and to understand it we have employed the technology and techniques of botanists and zoologists, along with art historians. Decipherment, maddeningly difficult, has been the downfall of countless workers. It has been considered the kiss of death to academics seeking tenure, and the most famous cryptologists in the world have admitted defeat (see Chap. 11). There is no doubt that the decipherment of the *Voynich Codex* has been a field of broken dreams, including our own.

We have made the assumption that the most promising route to decipherment is to use the illustrations as the key. Some success has been achieved with analysis of the plants, animals, a mineral, a few cities, and apothecary jars. We are not alone on this path. Other researchers have mined this field, but in our opinion, their efforts have been limited by errors in plant identifications and assumptions that the manuscript is a fifteenth century European text, a fallacy that has been explained throughout this book. We aspire to walk in the footsteps of the great decipherments of world history, including Egyptian hieroglyphics, Sumerian cuneiform, Linear B of Crete, and the decipherment of Mayan logograms. It may be that our skills, experience, and intellect are insufficient for the task, but we persevere.

J. Janick, A. O. Tucker, *Unraveling the Voynich Codex*, Fascinating Life Sciences, https://doi.org/10.1007/978-3-319-77294-3_13

Progress in Decipherment

The decoding of Voynichese symbols (see Chap. 5) based on plant names and the conjecture that the languages were those known to New Spain in the sixteenth century, including Nahuatl, Mixtec, Taino, and Spanish, has led to the decipherments mentioned above, but we admit to its failure to decipher most of the text of the *Voynich Codex*. In this chapter, we review our attempts at decipherment. Many are speculative, but hopefully they will suggest the proper path to proceed. We have demonstrated repeatedly that Classical Nahuatl is not the language of the *Voynich Codex*, but we know that it is a part of it.

In the 12 pages of the Pharmaceutical section of the *Voynich Codex*, plants or plant parts, apothecary jars, and a mineral crystal are illustrated and many are labeled with Voynichese symbols (folios 99v–102v; see Chap. 2, Fig. 2.3). The presence of labeled plants in this section has made it possible for Tucker and Talbert (2013) to propose a decipherment of alphabetic symbols of the *Voynich Codex*, as discussed in Chap. 5. Using this syllabary, some success in decipherment has been achieved and is summarized in Table 13.1.

Plants, Animal, Mineral, and Apothecary Jars The decipherment of the alphabetic symbols was made possible by the identification of two plants (prickly pear cactus and agave) in the Pharmaceutical section (see Chap. 2, Fig. 2.2) that were labeled in Voynichese symbols. The assumption made was that the alphabet was phonetic. In the case of prickly pear cactus, the five letters were assumed to be pronounced *nochtl*, the Nahuatl name, and the six letters of agave were pronounced *maguey,* the sixteenth century Spanish Taino name. This process was extended to: 12 other plants, as described in Table 5.1 of Chap. 5; the mineral boleite, also found in the Pharmaceutical section; the fish, alligator gar, labeled in the Zodiac section (see Chaps. 6 and 8); and four apothecary jars (see Chap. 11). The alphabet decipherment is presented in Table 5.3 of Chap. 5. The successes are summarized in Table 13.1.

Cities The use of plant names to develop the syllabary was confirmed by the serendipitous identification of the name of a city in circle 1 of folio 86v (see Chap. 10). Tucker had independently proposed that circle 2 of folio 86v represented the city Huejotzingo, which in Nahuatl means "the place of small willows." In circle 2, there was a picture of a fortress that we identified as the monastery of San Miguel Arcángel and above the name in Voynichese were the words . Using the symbol decoding determined by the labeled plants, this name, to our astonishment and glee, was deciphered: *Huoxeatlcapi* or "willow capital" (*huexötl* = "willow" in Nahuatl, *capi* = "capital" in sixteenth century Latin American Spanish. This confirmed our expectations of the city designation, the fact that the kabbalah-like folio 86v was a map, and that the alphabet decoding was on the right track.

Based on this evidence, we assumed that other cities could be translated. We had one success. In the drawing of the passageway from circle 2 to circle 3 in folio 86v,

Table 13.1 Decipherment of plants, an animal, a mineral, cities, and apothecary jars in the *Voynich Codex*

Class and folio location	Voynichese name	Translation	Cognate translation and identification	Comments
Plants				
100r#83	[illegible]	*nāshtli*	*nochtli* (Nahuatl), *Opuntia ficus-indica*	Phytomorph resembles a rooted cladode of prickly pear
100r#4	[illegible]	*māguoey*	*maguey* (Spanish from Taino), *Agave atrovirens*	Phytomorph resembles agave plants, perhaps dried
88r#100	[illegible]	*āguocâchā*	*agua* (Sp.) + *cacha* (Nahuatl for callus) *Lupinus* species	Name is descriptive of callus in root common in Fabaceae species
99r#16	[illegible]	*mātli*	*matli* (Nahuatl) animal front leg, branches	No evidence for name and not good description of phytomorph
99r#28	[illegible]	*tlācânoni*	*tlacanoni* (Nahuatl) = bat or paddle, *Dioscorea remotifolia*	Name descriptive of rhizome
99v#10	[illegible]	cholla	*cholla* (Spanish = skull, cactus	Name does not describe phytomorph (only tuber shown)
99v#18	[illegible]	*ācâtli*	*acatli* (Nahuatl) = reed	Name does not describe phytomorph
100r#1	[illegible]	*mānoetzācâ*	mano (Spanish) = hand + tzacua (Nahuatl) to close	Phytomorph shows five organs, perhaps descriptive of a closed fist
100r#2	[illegible]	*nāmāepi*	*nama* (Mixtec) = plant that produces soap *Philodendron mexicanum*	Unclear if name is appropriate
100r#5	[illegible]	*noe, moe-choll-chi*	*cholla* (Spanish) = skull + *chi* (Nahuatl) = owl, *Scutellaria mexicana*	Flower resembles bird skull
100r#7	[illegible]	*mācanol*	*macana* (Taino) = obsidian and wooden word, similar to *macuahuitl* (Nahuatl), *Philodendron goeldii*	Name describes phytomorph

(continued)

Table 13.1 (continued)

Class and folio location	Voynichese name	Translation	Cognate translation and identification	Comments
100r#9	[illegible]	*nānoya*	*nanoya* (Spanish) = grandmother (*Taxodium huegelii*) with Spanish moss (*Tillandsia usneoides*)	Name nondescriptive of phytomorph
100r#14	[illegible]	*ācâmaāya*	acamaya (Nahuatl) = crab or crayfish, *Gonolobus chloranthus* fruit	Name nondescriptive of phytomorph
100r#15	[illegible]	*ātlācân*	*atlacaneci* (Nahuatl) = bestial man; Mexican or Montezuma cypress (*Taxodium huegelii*)	Name nondescriptive of phytomorph
Animal				
70r	[illegible]	atlacaoca	*atlaca* (Nahuatl) = fishing + *oc*(*a*) = still, still fished	Name suggestive of fish
Mineral				
102r#4	[illegible]	*ātlaān*	*atlan* (Nahuatl) = in or under water, **boleite**	Boleite found under water
Cities				
86v circle 2	[illegible]	*huoxeātlocapi*	*huexötl* (willow) + *capi* Spanish (capital)	Huejotzingo (Nahuatl) = "city of small willows"
86v circle 8	[illegible]	*āltmchonon*	*atl* = water m chon = seller	"Water seller" is the nickname of Vera Cruz
86v circle 2 passageway	[illegible]	*ātlaachi* *ātlācachi*	*atla* = water achi = a small amount, atlacachi = fishermen, thus lake fishing	City or village not identified
Villages			*Similar cities (states) in Mexico*	
86v circle 9	[illegible]	*āhuol*	ahualulca, ahualulco (Guanajuato)	
86v circle 9	[illegible]	*ātlā*	Atla (Puebla)	Perfect fit
86v circle 5	[illegible]	*ātlachi*	Atlaco (Hildago)	Atlachi may refer to "small amount of water" or lake

86v circle 7	[illegible]	*ātlahocâ*	Atlatlahuaca, Atlatlahuca (Mexico)	
86v circle 5	[illegible]	*ātlmchi*	Atlan (Puebla, Hildago)	
86v circle 5	[illegible]	*ātlichi*	Atlaxco, Atlixco (Puebla), Atlicos (Veracruz)	
86v circle 5	[illegible]	*ātlmchoya*	Atlmozoyahua (Veracruz)	
86v circle 9	[illegible]	*chālli*	Chalma (Puebla)	
86v circle 9	[illegible]	*ch?an*	Chicayan (Veracruz)	
86v circle 7	[illegible]	*mtlaachi*	M. Avila Camache, Machanche (Queretaro)	
Apothecary jars				
88r#1	[illegible]	*atlaematli*	*atl* = water, liquid (Classical Nahuatl), *matli* = arm, (animal) front leg (Classical Nahuatl)	Remedy for swollen arm or leg
88r#2	[illegible]	*atlocachi*	*atl* = water, liquid (Classical Nahuat) *cachi cualli* = better (Classical Nahuatl)	Remedy to improve urination
88r#3	[illegible]	*ahuintlichoca*	*huin* = to express pain (Classical Nahuatl), *tli* =?, *choca* = weeps (Huastec Nahuatl)	Remedy to reduce intense pain, an analgesic
88v(1)#2	[illegible]	*atlaeoha*	*atl* = water, liquid (Classical Nahuatl), *eoh* = impersonal of *eua* = *levantarse* (to raise) (Classical Nahuatl)	Remedy to increase urination

there was a picture of three towers with the name *ātlaachi ātlācâchi*. The word *ātlaachi (atla* = "water," *achi* = small "amount") might mean a "lake," whereas the word *ātlācâchi* could be translated as "spear throwers" or "fishermen." Note that *Atlacachichimeca* has been translated as *los chichimeca con atlatl* (*chimeca* with "spear-throwers") by Staedtler and Hernández (2006). Thus, the two words *ātlaachi ātlācâchi* could be translated as "lake fishermen," but we failed to locate any city or town of that name. In circles 5, 7, and 9 of folio 86v there are eight, eight, and 10 words respectively (see Chap. 10, and Table 10.2) that we suspected might refer to names of villages. Ten of them (see Appendix 3) did bear some resemblance to present towns in Mexico that appeared to be derived from Nahuatl names. Circles 5 and 9 contained the word *ātlachi* (in circle 5 and in circle 9). Thus, the decipherment of the symbols indicated that city names could be deciphered.

Star Names

Astronomy was an important component of Aztec culture as demonstrated by their complex solar calendar (see Chap. 9). Although many constellations were known, individual Aztec star names appeared to be lost, probably as a result of the destruction of Aztec books by the Spanish in New Spain. In Book 7 of the *Florentine Codex*, Bernardino de Sahagún (1953, Fig. 21) referred to "morning star," the great star (undoubtedly Venus), and showed pictures of shooting stars, a number of stars of the constellation Scorpio, and the star cluster known as the Pleiades.

Three folios of the *Voynich Codex* – 68r(1), 68r(2), and 68r(3) – contained stars labeled in Voynichese. There are 29 labeled stars in folio 68r(1), 24 in 68r(2), and 10 in 68r(3), for a total of 63 individually named stars (Figs. 13.1, 13.2, and 13.3). The 63 stars have unique names that may be:

1. Lost Aztec names in Nahuatl or some other Mesoamerican language.
2. Spanish names.
3. Arabic names, as Arabic astronomy was widely developed and used in Spain and may have been known by the author of the *Voynich Codex*. None of the names resembled Spanish names. In addition, in folio 68r(3) there was a cluster of seven stars assumed to be the Pleiades. Here, we examine the supposition that some of the star names in the *Voynich Codex* are based on Arabic names.

Arabic Association Of the 64 transliterated star names in folios in the Cosmological section, 43 (67.2%) begin with the letter "A." Examination of star names in Arabic shows that 124 of 166 (74.7%) in a Google list, and 256 of 1039 (24.6%) in the Arabic index from Allen (1899) begin with the letter "A." Many Arabic star names begin with the prefix "al" (meaning "the" in Arabic) suggesting that the star names in the *Voynich Codex* might be related to Arabic names. A famous example is Aldebaran, the brightest star in the constellation Taurus, which derives from the Arabic *Al Dabaran*, meaning the "the follower," probably of the Pleiades. We suggest

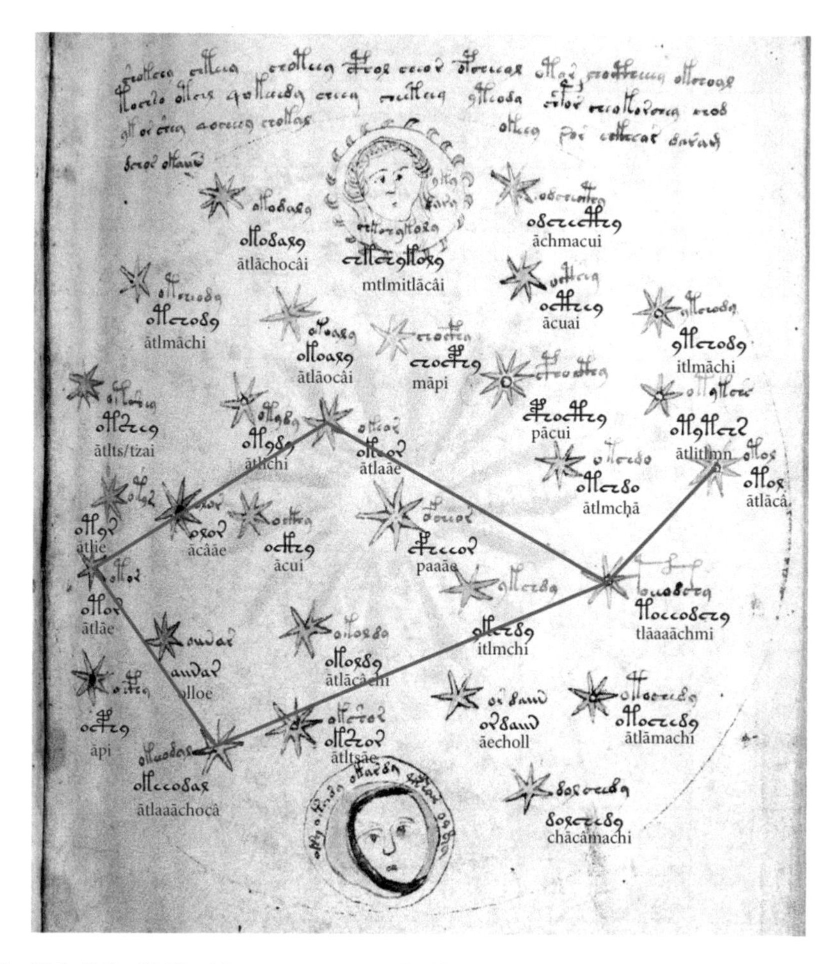

Fig. 13.1 Folio 68r(1) with star names transcribed in Voynichese font and transliterated based on the symbol decipherment of Tucker and Talbert (2013). The connected stars are assumed to represent the constellation Virgo (Al'Awwa or Al Awwā). Tabular explanation below

Voynichese	Translation	Similar Arabic star name from Allen (1899)	Comments
[illegible]	ācâāe		
[illegible]	āchmacui	Acamar (page 219)	Eridanus (constellation in the Southern Hemisphere)
[illegible]	ācuai		
[illegible]	ācui		
[illegible]	āecholl		
[illegible]	āpi		
[illegible]	ātlaaāchocâ	Al Awwā (barker or barking dog) + *choca* (howl in Nahuatl)	Virgo, perhaps star Spica

Voynichese	Translation	Similar Arabic star name from Allen (1899)	Comments
[illegible]	ātlaāe	Al Awwā (barking dog)	Virgo
[illegible]	ātlācâ	Alacel (page 467)	Virgo
[illegible]	ātlācâchi		Spear thrower = fisherman (Nahuatl)
[illegible]	ātlāchocâi	Alasch'a (γ Scorpio, page 370)	
[illegible]	ātlāe	Al Awwā (barking dog)??	Virgo
[illegible]	ātlāmachi	Alamac, Alamak, Alamech (γ) Andromedae (page 36)	Andromeda
[illegible]	ātlāocâi		
[illegible]	ātlichi	Alasch'a (γ Scorpio, page 370)	Scorpio
[illegible]	ātlie		
[illegible]	ātlitlmn		
[illegible]	ātlmāchi	Alamac, Alamak, Alamech (γ) Andromedae (page 36)	Andromeda
[illegible]	ātlmchā	Alamac, Alamak, Alamech (γ) Andromedae (page 36)	Andromeda
[illegible]	ātltsai		
[illegible]	ātltsāe		
[illegible]	chācâmachi		
[illegible]	itlmāchi	Alamac, Alamak, Alamech (γ Andromedae (page 36)	Andromeda
[illegible]	itlmchi	Alamac, Alamak, Alamech (γ Andromedae (page 36)	Andromeda
[illegible]	māpi		
[illegible]	mtlmitlācâi [under the image of the sun]	????	Sun = *Shams* (Arabic) *Tonaltzintli* (Nahuatl), *Mamalhuaztli* = fire sticks for the New Fire ceremony
[illegible]	olloe		
[illegible]	paaāe		
[illegible]	pācui		
[illegible]	tlāaaāchmi	Al Awwā (barking dog) + *chmi*	Virgo

? = no decipherment given

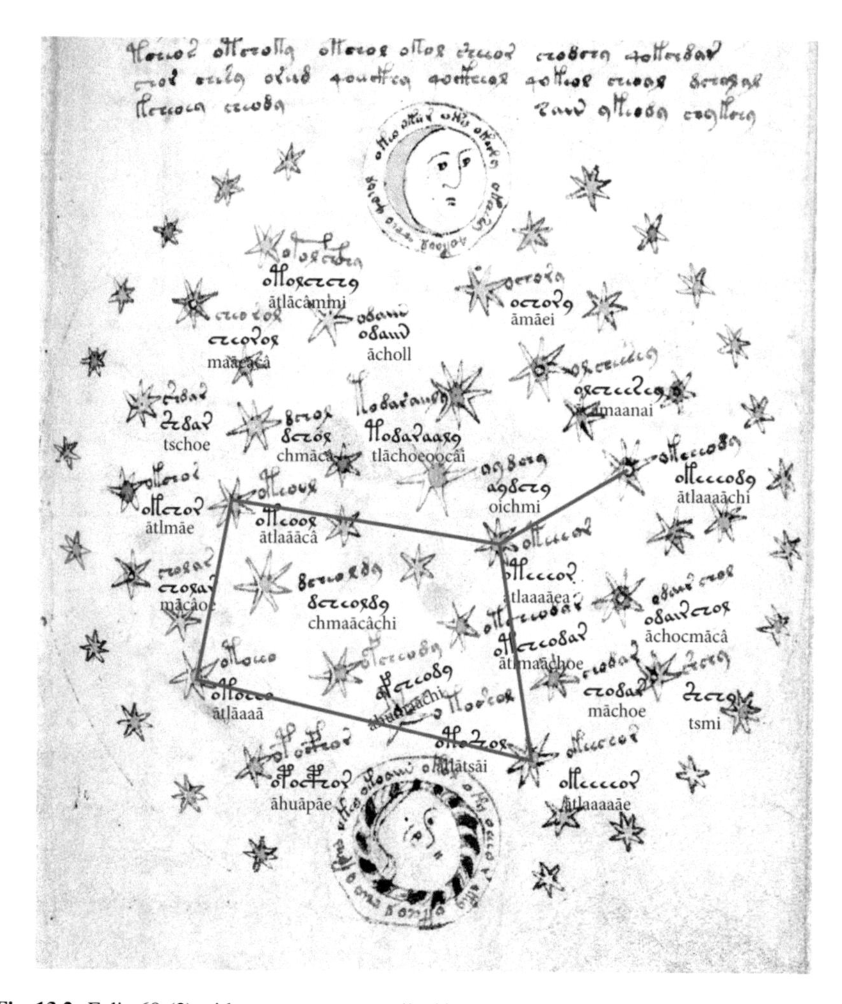

Fig. 13.2 Folio 68r(2) with star names transcribed in Voynichese font and transliterated, based on the symbol decipherment of Tucker and Talbert (2013). The connected stars are assumed to represent the constellation Virgo (Al'Awwa or Al Awwā). Tabular explanation below

Voynichese	Translation	Similar Arabic star name from Allen (1899)	Comments
	ācâmaanai		
	āchocmācâ		
	ācholl		
	āhuāpāe	(Al) Agribah page 130 (= the Ravens)	Canis Minor
	āhumaāchi		
	āmāei		
	ātlāaaā	Al Awwā (barking dog)	Virgo
	ātlaaaaāe	Al Awwā (barking dog)	Virgo
	ātlaaaāchi	Al Awwā + *chi* (barking dog)	Virgo

Voynichese	Translation	Similar Arabic star name from Allen (1899)	Comments
[illegible]	ātlaaaāe	Al Awwā (barking dog)	Virgo
[illegible]	ātlaāācâ	Al Awwā + *ca* (barking dog)	Virgo, perhaps star Spica
[illegible]	ātlācâmmi	Alacast and Alcalst (ε Virginis, page 471)	Virgo
[illegible]	ātlātsāi		
[illegible]	ātlmaāchoe		
[illegible]	ātlmāe		
[illegible]	chmaācâchi		
[illegible]	chmācâ		
[illegible]	maāeācâ		
[illegible]	mācâoe		
[illegible]	māchoe		
[illegible]	oichmi		
[illegible]	tlāchoeoocâi		
[illegible]	tschoe		
[illegible]	tsmi		

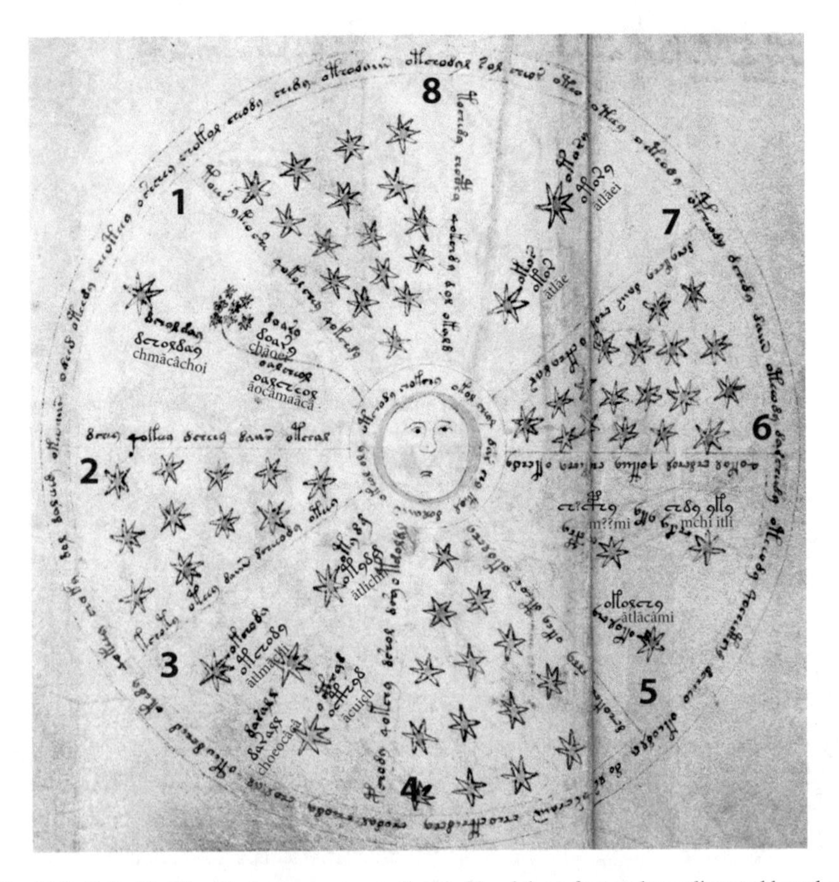

Fig. 13.3 Folio 68r(3) with star names transcribed in Voynichese font and transliterated based on symbol decipherment of Tucker and Talbert (2013). Tabular explanation below

Voynich	Translation	Similar Arabic star name, Allen (1899)	Comments
Segment 1			
8oa?9	chāoei	(Al) Thurayya	Pleiades (*Chow* [Egyptian])
8czo88a9	chmācâchoi		*chamahu(a)* in Karttunen (1983: 450) = a swelling up (i.e., large star) = Aldebaran
oa8czco8	āocâmaācâ		*aoccampa* in Karttunen (1983: 11) = no longer anywhere or disappears, suggesting that the Pleiades might be able to disappear
Segment 3			
ocHz98	ācuich		
oH988	ātlichh		
ollczo89	ātlmāchi	Alamac, Alamak, Alamech (γ) Andromedae, page 36)	Andromeda
8a?a88	choeocâcâ		Similar to *chaoei* + *caca* (frog) in Nahuatl (*ad-difta* = frog in Arabic)
Segment 5			
ollo8cz9	ātlācâmi		
cz??cHz9	mpi		
cz89 9ll9	mchi itli		
Segment 7			
ollo2	ātlān	Alanac, Alanak, Alioc (α Aurigae, pages 85, 87)	Auriga
		Alanin (Draco, page 205)	Draco
		Alanac, Alanak, Alioc (α Aurigae, pages 85, 87)	Auriga
		Alanin (Draco, page 205)	Draco
ollo?9	ātlāei		

that in some cases the letters "atl" at the beginning of many star names in Voynich might refer to the Arabic *al* ("the") rather than the Nahuatl *atl* (for water, i.e., pronounced as *al* rather than the glottal stop "atl"). However, there is an area in the sky on the edge of the Milky Way that Aratos of Solos, a Greek poet who flourished in Macedonia in the third century BCE, designated as "water" (Allen 1899: 337).

Star Names in Folios 68r(1), 68r(2), and 68r(3) In the tables associated with Figs. 13.1, 13.2, and 13.3, the stars for each of the three folios are presented in Voynichese font and listed alphabetically using the symbol decoding of Tucker and

Table 13.2 The ten deciphered names that phonetically resemble *Al'Awwa* in folios 68r(1) and 68r(2), and with similar sounding nymph names in the constellation Virgo, identified by ring and position

Folio	Voynichese name	Voynichese name deciphered	Similar nymph name in Virgo zodiac folio 72r(2) (ring and position)
68r(1)	[illegible]	ātlaaāchocâ (ātlaaā + *chocâ*)	āaaācholl (2-18)
	[illegible]	ātlaāe	ilaaāe (2-14)
	[illegible]	ātlāe	itlaaāe (2-14)
	[illegible]	ātlācâ (similar to alacel; also in Virgo)	āaachocây (4-6)
	[illegible]	tlāaaāchmi (tlāaaā + *chmi*)	
68r(2)	[illegible]	ātlāaaā	ātlaaāha (4-4)
	[illegible]	ātlaaaaāe	itlaaāe (2-14)
	[illegible]	ātlaaaāchi (ātlaaaā + *chi*)	āaachy (4-2) ātlaāchy (2-4 and 2-12) ātlaaāch (2-10) ilaāch (2-11) āaachay (2-17) āaachocây (4-6)
	[illegible]	ātlaaaāe	itlaaāe (2-14)
	[illegible]	ātlaāācâ ātlaāācâ ātlaāā + *câ*	

Talbert (2013), along with similar Arabic names from the 1024 entries in the Arabic index of Allen's *Star Names: Their Lore and Meaning* (1899, Dover ed., 1963), in addition to a few Nahuatl words from Karttunen's *An Analytical Dictionary of Nahuatl* (1983). Twenty-three of the 64 transliterated names – 12 in folio 68r(1), seven in 68r(2), and three in 68r(3) – show some similarity to Arabic star names, suggesting that some Arabic names of stars were transliterated into Voynichese symbols.

It is assumed that if Arabic star names are referred to in the *Voynich Codex*, the pronunciation of the transliterated Voynichese names would correspond to the pronunciation of the Arabic names. For example, ten of the deciphered Voynichese star names – five in folio 68r(1) and five in folio 68r(2) – are phonetically similar to the Arabic *Al'Awwa* (or *Al Awwā*; the "barker" or "barking dog"), which refers to the constellation Virgo (Table 13.2). In addition, there are similar names associated with 30 nude nymphs in two concentric rings around Voynich folio 72r(2), the constellation Virgo in the Zodiac section (Fig. 13.4), which may also indicate that some nymph names are based on Arabic star names.

One star in Virgo known in Arabic as *Rigl al'Awwa* ("foot of the barker" or "barking dog") has been identified as *mu virginnis,* one of the many visible stars in Virgo (Fig. 13.5). Thus, we assume that the other name variants could refer to other stars in Virgo. It is intriguing that one star was deciphered as *ātlaaā* + *chocâ* and

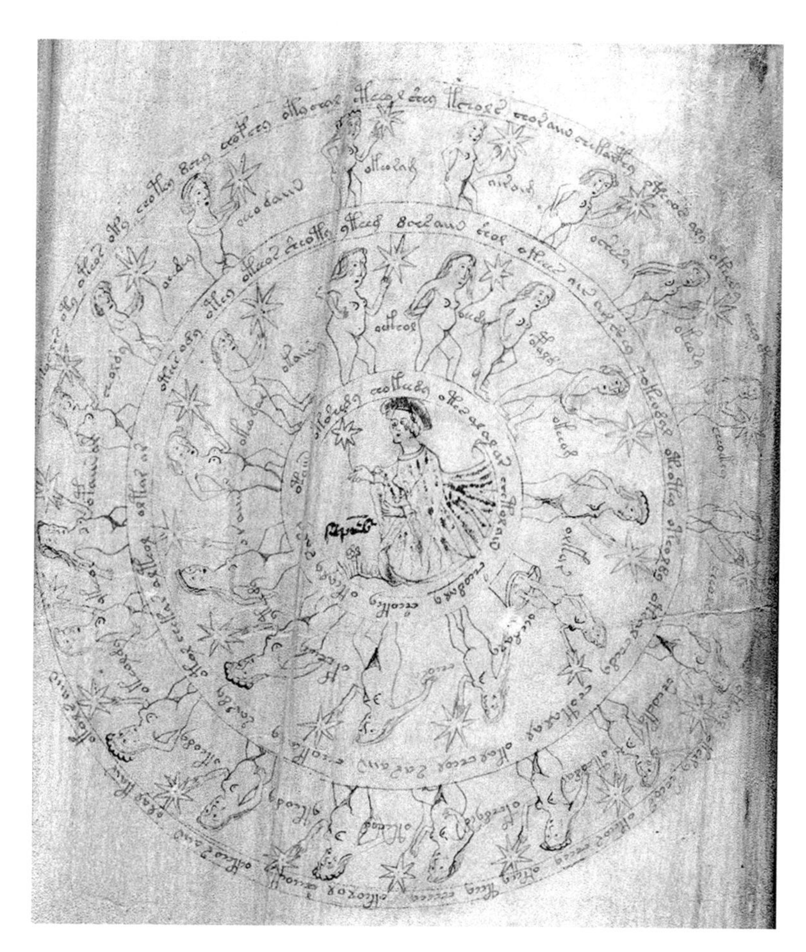

Fig. 13.4 The 30 nymphs "names" in Virgo, folio 72r(2). Tabular explanation below

Voynichese "names"	Latin	Ring (outer ring = 1)	Order (12 o'clock = 1, clockwise)
[illegible]	ātlaānoyâ	2	1
[illegible]	oco?	2	2
[illegible]	ātsany	2	3
[illegible]	ātlaāchy	2	4
[illegible]	maāatlay	2	5
[illegible]	maācâchy	2	6
[illegible]	tsaāatly	2	7
[illegible]	ātlāochon	2	8
[illegible]	āhumchynch	2	9
[illegible]	ātlaaāch	2	10

Voynichese "names"	Latin	Ring (outer ring = 1)	Order (12 o'clock = 1, clockwise)
[Voynich script]	itlaāch	2	11
[Voynich script]	ātlaāchy	2	12
[Voynich script]	ātlaocâchy	2	13
[Voynich script]	itlaaāe	2	14
[Voynich script]	āhuolloca̋	2	15
[Voynich script]	mācâchy	2	16
[Voynich script]	āaachay	2	17
[Voynich script]	āaaācholl	2	18
[Voynich script]	ācuācâ	4	1
[Voynich script]	āaachy	4	2
[Voynich script]	āhucâh	4	3
[Voynich script]	ātlaaāha	4	4
[Voynich script]	ācâtloe	4	5
[Voynich script]	āaachocây	4	6
[Voynich script]?	maān?	4	7
[Voynich script]?[Voynich script]	āhum?y	4	8
[Voynich script]	ytlaachy	4	9
[Voynich script]	āeoll	4	10
[Voynich script]	ātlaānon	4	11
[Voynich script]	ātlo?	4	12

? = no decipherment given

that *choca* is the Nahuatl word for "howling." One very bright star in Virgo has the current name of Spica, supposedly representing an ear of wheat. Perhaps this star is represented by either *atlaaachoca* in folio 68r(1) or *atlaaaca* in folio 68r(2).

It should be noted that folios 68r(1) (Fig. 13.1) and 68r(2) (Fig. 13.2) are similar, except that the position of the sun and the moon are reversed, suggesting that one might be the evening sky and the other the morning sky. It should be noted that although the sun and the moon appear in a north–south orientation, this position would be east–west in Aztec culture. Connecting the five stars in folio 68r(1) and the five stars in folio 68r(2) suggests a constellation shape that resembles Virgo (Figs. 13.1 and 13.2).

The Pleiades

Segment 1 of folio 68r(3) (Fig. 13.3) shows three items:

1. A seven-star cluster assumed to be the Pleiades labeled [Voynich script] (*chāoei*).

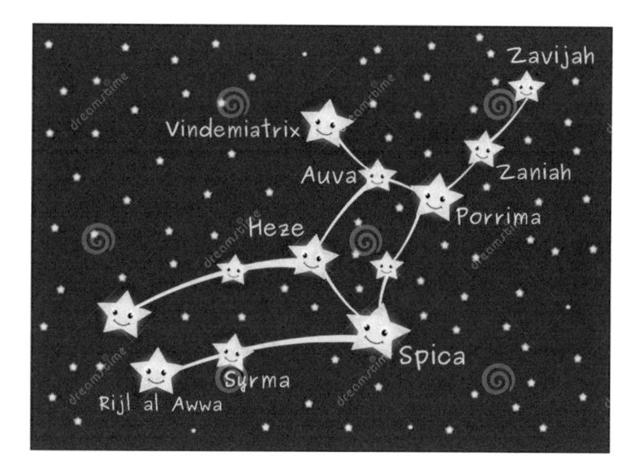

Fig. 13.5 Stars in the constellation Virgo

2. A curved line under the seven stars, suggesting movement underscored with the name oaꝛcꞇcoꝛ (*āocâmaācâ*).
3. A large star labeled 8cꞇoꝛ8aꝗ (*chmācâhoi*). The four segments in this folio with named stars may represent the overhead sky (perhaps Puebla or vicinity) in the four seasons of winter, spring, summer, and fall: perhaps 15 December (segment 1), 15 March (segment 3), 15 June (segment 5), and 15 September (segment 7).

The Pleiades are overhead in the winter season (e.g., 15 December). Thus, segment 1 is suggested to represent the night sky over central Mexico, say Puebla on a winter day, perhaps 15 December. It indicates a very large star to the west, most likely to be the enormous red star Aldebaran in the shoulder of Taurus. The sky overhead in the summer (opposite segment 3) would not show the Pleiades, which would only be apparent overhead in the day sky. In the summer, the prominent constellation overhead would be Aquarius. Thus, the Pleiades disappears from the night sky in the summer. The overhead sky in the spring (segment 3) and fall (segment 7) would show the Pleiades on the horizon and would essentially be invisible on 15 June 15 and 15 September.

The name *chāoei* under the seven stars representing the Pleiades in folio 68r(3) is very similar to the name *chow*. According to Allen (1899: 399), the Pleiades was a marked object on the Nile, at one time probably called *chu* or *chow* and supposed to represent the goddess Nit (or Neith). Clearly, the word *chow* is close to *chaoei.* The Arabic name for the Pleiades is *Al-Thurayya* (related now to the woman's name Saraya or Thoraya). The Arabic name *Al Thurayya,* or "the many little ones," bears a remarkable similarity to the Egyptian name Athurai or Acauria, also known as *Chu* or *Chow*. According to Mokhles Elsysy, an Egyptian colleague at Purdue University in Indiana, *chāoeāi* is close "in tone" to *Thurayya.*

The Voynichese name of the large star labeled *chmacahoi* is close to the Nahuatl word *chamahu(a)* (Karttunen 1983: 45) meaning "a swelling up," i.e., a large star. Furthermore, the word *āocâmaācâ* means "under the line," suggesting that the Pleiades moves, and can be interpreted as being close to *aoccampa* (Karttunen 1983: 11), which is defined as "no longer anywhere" or "disappears." If this is correct, the

Voynichese names in segment 1 can be interpreted to mean: the Pleiades disappears in the night sky in different seasons and is next to a large swelling star (i.e., Aldebaran, which means "the follower"). This suggests that the star names in the tables in Fig. 13.1 and Fig. 13.2 may be a combination of Arabic names, Nahuatl names, and perhaps names in another Mesoamerican language.

Star Names Conclusions The conjecture that some of the star names in three Voynich folios, 68r(1), 68r(2), and 68r(3), are derived from Arabic names appears possible and perhaps likely, but cannot be considered proven beyond a shadow of a doubt. If correct, it provides evidence that the decipherment of the Voynichese alphabet has predictive value, not only in Nahuatl, but in other languages. It suggests that the text of the *Voynich Codex* is phonetic and may be used for various languages including Nahuatl or other Mesoamerican languages (some perhaps lost), Spanish, and Arabic. We know from the letters of Francisco Hernández, who was sent to New Spain by King Philip II to study medicinal plants, that he interviewed indigenous people through interpreters, and transcribed their words using the Latin alphabet. Fifty dictionaries of Mesoamerican languages were examined to decipher transliterated words found in the *Voynich Codex*, so far without success. The proximity of some star names to Arabic names can be explained by the wide use of these names by European astronomers, as exemplified by the star name Aldebaran, derived from the Arabic *Al Dabaran* ("the follower"), for a very bright star in the constellation Taurus. We consider the closeness of names of some nymphs in the Virgo zodiac circle a potential breakthrough in understanding nymph names in the *Voynich Codex*.

Problems in Decipherment

Although, as already mentioned, some success has been achieved in deciphering names of plants, an animal, a mineral, and some cities, most of the Voynichese text defies decipherment and we conclude that the text is not Classical Nahuatl, despite its use of some Nahuatl words and cognates. Some examples where illustrations may give a clue to decipherment are listed, as follows:

Examples from Folio 86v

Circle 1 (Fig. 13.6) This circle is a T-O map that we believe represents the Earth and each segment contains a name: (*ānocâ*), (*ātlae*), and most likely (pai) or perhaps (āhuaai). In many medieval texts, the words in these three segments usually represent Asia, Africa, and Europe (see Chap. 9, and Fig. 9.4). If our conjecture is correct, the names in this sphere should probably be geographic. Anoca is a tiny town located in the municipality of Techaluta de Montenegros in the Mexican

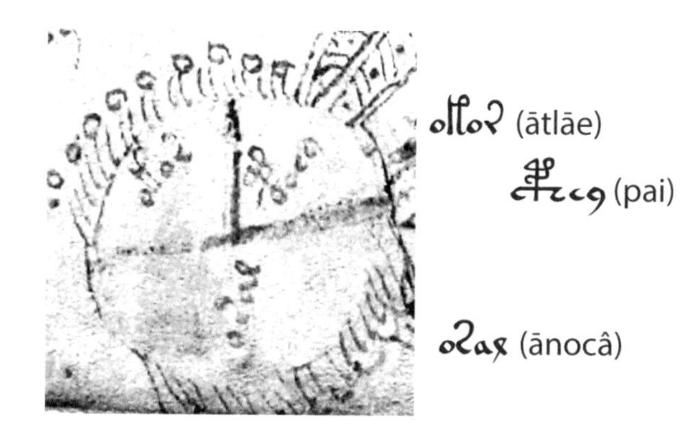

Fig. 13.6 Circle 1 of folio 86v

state of Jalisco; *atlae* may refer to water; and *pai* or *āhuaai* are not recognizable words in Nahuatl. But "-pa" can mean "in," "on," "toward," or "away from." Although all these words resemble Nahuatl, we cannot come up with a logical meaning. But it is intriguing that *ānoca* is the name of a place.

Circle 2 (Fig. 13.7b) The single word coming out of the volcano is assumed to relate to material emanating from the vent: (*ocho ts ya o n o l*). However, we make no sense of the meaning from Nahuatl cognates, but *tsoncoyal* in Huasteca is close and means *ensuciar* in Spanish or "to dirty" in English. As *ocho* means eight in Spanish, it is possible that the word means "eight eruptions," and in fact there were eight eruptions of Popocatepetl from 1518 to 1571 (Volcano Discovery 2017).

In the center of circle 2, which we have assumed represents the city of Huejotzingo, there is a spiral of 14 words (Fig. 13.7c) embedded in a matrix of 118 stars that should have some meaning appropriate to this city. These words, listed in Fig. 13.7d from outer to inner, defy decipherment.

Text from Folio 1v of the Voynich Codex Describing **Ipomoea arborensis**

The ten Voynichese lines describing folio 1v (Fig. 13.8) and found word by word in sequence in Table 13.3, list the Voynichese names, deciphered names, and potential meanings based on Nahuatl dictionaries. There is no obvious relationship between the possible meanings of words and plant descriptions. However, the many short words indicate that the text refers to prose. Some of the words may suggest a description of some sort, such as "a unit of measure," "to sell," "an animal leg," "have an odor," or "frog."

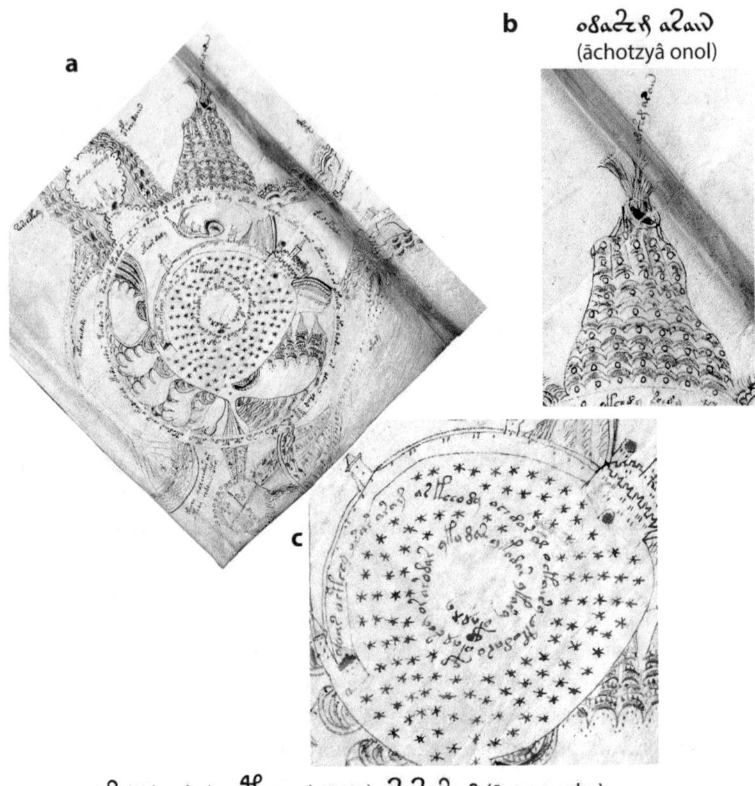

d (āshoshe) (ocuay) (ānonoeoha)
(ontlaachi) (āmchāe) (ācâ) (āatloshni)
(itlāchoeā) (āhuocâtzi) (āeoeāchoe)
(itlochoe) (itlachoe) (itloei) (ahuocâ?i)

Fig. 13.7 (**a**) Circle 2 of folio 86v representing Huejotzingo. (**b**) Enlargement of spewing volcano. (**c**) Enlargement of spiraling words. (**d**) Deciphered spiraling words

Lines 1–4 of Folio 103r in the Recipe Section

The Recipe section consists of 23 pages (12 folios) of text in which each sentence is separated by a star. There are 324 stars and 325 paragraphs. These are presumed to be either prescriptions or possible poetry, incantations, or sayings. The first sentence (Fig. 13.9) has been transliterated in Table 13.4. However, attempts at decipherment based on Nahuatl words offer no clue as to its meaning, although there are many Nahuatl cognates.

Fig. 13.8 Folio 1v from the *Voynich Codex* showing the plant *Ipomoea murucoides* surrounded by Voynichese text. Decipherment in the text

Lists of Nouns in the Voynich

There are at least five lists of what should be considered nouns in the *Voynich Codex*, as follows:

1. 26 names in circles 5, 7, and 9 in folio 86v, presumed to be cities or towns (see Appendix 1).
2. 288 names associated with nymphs in the zodiac section, folios 70r–73r (see Appendix 2).
3. 64 star names: 29 in 68r(1), 24 in 68r(2), and 11 in 68r(3) (see Chap. 13, Figs. 13.1, 13.2, and 13.3, and Appendix 3).
4. 31 names of apothecary jars (see Chap. 11, Table 11.3, and Appendix 4).
5. 189 plant names alphabetized from the Pharmaceutical section (see Appendix 5).

Three sets of names (nymphs, cities, and stars) were alphabetized based on their deciphered names and the lists were examined to determine if there were names in common within each list and among the three lists (Table 13.5). The results indicated

Table 13.3 Transliteration and attempted decipherment of folio 1v text (*Ipomoea arborensis* description). Note the presence of short words composed of only two syllables that may be non-nouns

Voynichese words	Transliteration	Similar Nahuatl cognates
	tlmni	
	mācholl	
	ācâ	aca = someone (Wood 2000–2016)
	ācâtlmai	
	moe	mo = negative particle on its own in questions expressing doubt (Karttunen 1983)
	poe	
	ohâ	
	itlaai	
	moe	mo = negative particle on its own in questions expressing doubt (Kartunnen 1983)
	āe	
	āmi	ami = *see* an; *plural* amique; *see* quemmach; *perfect* oan: *noun* = accommodation; an or am before a vocal m *pronoun* of the second person plural you; quemmach? *adverb* is it possible? (Siméon 2010)
	chmā	
	câtlāchi	
	ātlāchoe	
	māchi	mache = mainly, on the whole, particularly, especially (Siméon 2010)
	chā	
	cui	cui = to take something or someone (Karttunen 1983)
	cuācui	cui = to take something or someone (Karttunen 1983)
	tzi	
	chtltzaai	
	cui	cui = to take something or someone (Karttunen 1983)
	tlātlmāchi	mache = mainly, on the whole, particularly, especially (Siméon 2010)
	chocâ	choca = to weep, cry; for animals to make various sounds, e.g., to roar, to bray, or for birds to sing (Wood 2000–2016)
	chācâ	choca = to weep, cry; for animals to make various sounds (e.g., to roar, to bray, or for birds to sing) (Wood 2000–2016)
	mātlaā	matl = unit of measure, about 6 feet (Wood 2000–2016)
	choc	
	choyâ	

(continued)

Table 13.3 (continued)

Voynichese words	Transliteration	Similar Nahuatl cognates
	nāmai	namaca = to sell (Wood 2000–2006)
	mā	ma = *participle* preceding the imperative and optional (Siméon 2010)
	tlāchi	tlachia = to look, see, or observe from a watchtower (Wood 2000–2016)
	huātlāi	
	tzācâ	tzaca = to enclose, lock up (Karttunen 1983)
	choc	
	pāocâ	
	choe	
	mi	
	tlāchi	tlachia = to look or see (Wood 2000–2016)
	ātlāoll	
	tzātzi	
	mātli	matli = animal front leg (Wiktionary. Index: Nahuatl)
	mācâ	maca = singular of macamo, no (before the imperative) (Siméon 2010); maca = let no, let no, be not, don't, no (Wood 2000–2016)
	cuā	cuac = the end, at the top, after (Dictionnaire de la langue nahuatl classique); qua = *perfect* oqua: nite = biting, eating someone (Siméon 2010); aca = someone (Wood 2000–2016)
	câ	
	tzācâ	tzaca = to enclose, lock up (Karttunen 1983)
	otlocânācâmai	
	māchā	macha = stink, have an odor (Walters et al. 2002)
	câācâ	caca = toad, frog (Karttunen 1983)
	mi	
	cui	
	quā	
	ācâ	aca = someone (Wood 2000–2016)
	māaan	
	maācâ	maca = singular of macamo, no before the imperative (Siméon 2010); maca = let no, let no, be not, don't, no (Wood 2000–2016)
	chācâ	choca = to weep, cry; for animals to make various sounds, e.g., to roar, to bray, or for birds to sing (Wood 2000–2016)
	cuai	cui = to take something or someone (Karttunen 1983)

(continued)

Table 13.3 (continued)

Voynichese words	Transliteration	Similar Nahuatl cognates
	itlācâ	
	chācâ	choca = to weep, cry; for animals to make various sounds (e.g., to roar, to bray, or for birds to sing) (Wood 2000–2016)
	chācââ	choca = to weep, cry; for animals to make various sounds (e.g., to roar, to bray, or for birds to sing) (Wood 2000–2016)
	itlācâ	
	chācâmāchi	choca = to weep, cry; for animals to make various sounds (e.g., to roar, to bray, or for birds to sing (Wood 2000–2016) mache = mainly, on the whole, particularly, especially (Siméon 2010)
	ātlācâtzācâ	
	tlācâ	
	tlami	
	mācâ	maca = sing. of macamo, no before the imperative (Siméon 2010); maca = let no, let no, be not, don't, no (Wood 2000–2016)
	tli	
	mācâ	maca = singular of macamo, no before the imperative (Siméon, 2010); maca = let no, let no, be not, don't, no (Wood 2000–2016)
	cuācâ	cuac = the end, at the top, after (Dictionnaire de la langue nahuatl classique)
	māchi	Mach = even, ever, however, indeed (Dictionnaire de la langue nahuatl classique)
	mācâ	maca = singular of macamo, no (before the imperative) (Siméon 2010); maca = let no, let no, be not, don't, no (Wood 2000–2016)
	choll	
	tzāe	
	ātlocâ	
	mācâ	maca = sing. of macamo, no before the imperative (Siméon 2010); maca = let no, let no, be not, don't, no (Wood 2000–2016)
	chācâtli	
	choe	
	tzācâ	tzaca = to enclose, lock up (Karttunen 1983)
	chmāe	
	ātlmā	

(continued)

Table 13.3 (continued)

Voynichese words	Transliteration	Similar Nahuatl cognates
	choe	
	tzāchi	
	tlāāe	tla = *pronoun relative indeterminate* for things (Siméon 2010); *pronoun* something; *conjunction* if (Herrera 2004)
	mātlmai	
	chocâ	choca = to weep, cry; for animals to make various sounds, e.g., to roar, to bray, or for birds to sing (Wood 2000–2016)
	māchi	mache = mainly, on the whole, particularly, especially (Siméon 2010)
	nmāchi	mache = mainly, on the whole, particularly, especially (Siméon 2010)
	huācâ	
	māchoe	mache = mainly, on the whole, particularly, especially (Siméon 2010)

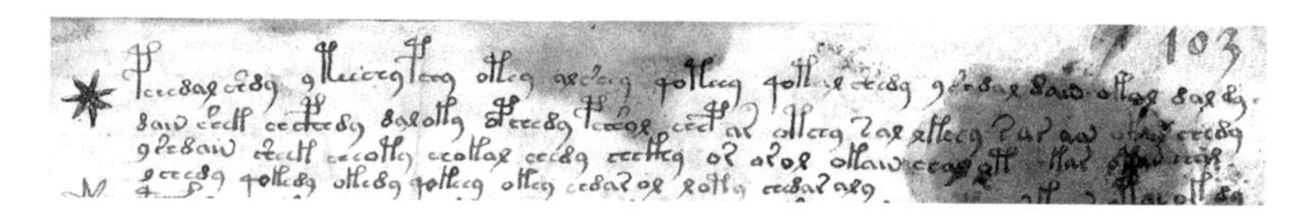

Fig. 13.9 Folio 103r, lines 1–4, of the Recipe section. Decipherment in the text

many duplications within lists and a few duplications between lists. For example, two nymph names, *ātlmāchi* and *ātlychi,* were the same as star names. Three nymph names, *ātlaachi*, *ātlācâoe,* and *ātlych,* were the same as cities. Two cities, *ātlāei* and *ātlychi,* had the same names as stars. Finally, one name, *ātlchi,* was a nymph, city, and star name.

Conclusion

Our present hypothesis is that Voynichese might be based on a phonetic alphabet. It may have been developed as a means of recording a number of extant languages that were in use in colonial New Spain. Similarly, Francisco Hernández also had native interpreters to record the names of plants throughout the region. Voynichese includes some Nahuatl or Taino words, Spanish loan words, and apparently Arabic in the star names. Although the main text is not Classical Nahuatl, there were enough Nahuatl cognates to enable a number of words to be deciphered. We surmise that the other languages involved include a dialect or a related Aztec language as there were

Table 13.4 Folio 103r, lines 1–4, decipherment

Voynichese	Transliteration	Compound words?	Similar Nahuatl cognates
Line 1			
[illegible]	humachocâtzchi	huma + chocâ + tzchi	humo = (*Spanish,* smoke) + choca, (*Nahuatl)* "to cry"
[illegible]	itlaamihumi	itlaa + mi + humi	humo = (*Spanish*, smoke)
[illegible]	ātlaiocâtzai	ātlaio + catzai	atlei (*Nahuatl*) = nothing (de Molina 1970)
[illegible]	quātlaai	quātla + ai	quatlaca (*Nahuatl*) = head with presumption of vanity
[illegible]	quatlocâtzachi	quatlo + câtz + chi	quatlaca (*Nahuatl)* = head with presumption of vanity + catzahuac = something
[illegible]	itzchocā	itz + choca	itz = to look at oneself + choca = crying
[illegible]	chol	chol	choloa to flee
Line 2			
[illegible]	ātlocâ	ātl + oca	water + aca = someone (Wood 2000–2016)
[illegible]	chocâchi	choca + câchi	choca = crying
line 2			
[illegible]	chol	chol	choloa to flee
[illegible]	tzatl	tzatl	
[illegible]	aapachi	aap + achi	apachoa = to inundate, to soak
[illegible]	chocâātli	chocâ + ātli	choca = crying
[illegible]	āhumachi	āhu + machi	ah = negative prefix + machia = to be known
[illegible]	huatzācâ	hua + tzācâ	tzacu = to close
[illegible]	aaahu	aaahu	huah = someone who possesses
[illegible]	oe	oe	
[illegible]	ātlmi	ātlmi	
[illegible]	nocâ	nocâ	
[illegible]	câtlaai	câtlaai	
[illegible]	noe	noe	
[illegible]	al	al	
[illegible]??	ātl??	ātl??	water
[illegible]	machi	machi	
Line 3			
[illegible]	itzchol	itz + chol	itz = to look at oneself + choloa = to flee

(continued)

Table 13.4 (continued)

Voynichese	Transliteration	Compound words?	Similar Nahuatl cognates
	tzaatl	tzaatl	
	maātli	maātli	
	aaātlocâ	aaātl + ocâ	water + aca = someone (Wood 2000–2016)
	aaachi	aaachi	
	mcui	mcui	
	ān	ān	
	oeācâ	oeācâ	
	ātlol	ātlol	
	aaocâ	aaocâ	
	ātl	ātl	water
	tloe	tloe	
	ātlos	ātlos	
	aaiyâ	aaiyâ	
Line 4			
	câmachi	câmachi	
	quātlachi	quātl + achi	
	ātlachi	ātla + chi	water
	quātlaai	quātlaai	quatlaca = head (*Nahuatl*) with presumption of vanity
	ātlai	ātlai	atlei = nothing
	aachoe	aachoe	
	ācâ	ācâ	aca = someone (Wood 2000–2016)
	câātli	câātli	
	machoe	machoe	mache = mainly, on the whole, particularly, especially (Siméon 2010)
	ocâi	ocâi	

? = no decipherment given

many. It clearly is not Totonac or Upper Necaxa Totonac (Beck 2011). It may be extinct. One possibility is Acolhuacatlatolli, but there is no agreement on the nature of the language. It may have been:

1. A Chichimec language allied, but distinct from Nahuatl and Otomi.
2. A language similar to Otomi.
3. A language similar to Nahuatl, perhaps extinct.

Table 13.5 Commonality of noun names within and among nymphs, cities, and stars

Comparison	288 Nymph names		26 Cities		Star names	
Among lists	[illegible]	ātlaachi (1x)	[illegible]	ātlaachi (1x)		
	[illegible]	ātlācâoe (1x)	[illegible]	ātlācâoe (1x)		
	[illegible]	ātlachi (3x)	[illegible]	ātlachi (2x)		
			[illegible]	ātlāei (2x)	[illegible]	ātlāei (1x)
	[illegible]	ātlmāchi (1x)			[illegible]	ātlmāchi (2x)
	[illegible]	ātlichi (2x)	[illegible]	ātlichi (1x)	[illegible]	ātlchi (1x)
Within lists	[illegible]	āaaācholl (2x)	[illegible]	ātlachi (2x)	[illegible]	ātlmāchi (2x)
	[illegible]	āaachi (2x)	[illegible]	ātlāei (2x)		
	[illegible]	ātlaācâi (4x)	[illegible]	ātlmchi (2x)		
	[illegible]	ātlaāchi (7x)				
	[illegible]	ātlaān (3x)				
	[illegible]	ātlaay (2x)				
	[illegible]	ātlachi (3x)				
	[illegible]	ātlāchi (2x)				
	[illegible]	ātlaocâ (2x)				
	[illegible]	ātlay (2x)				
	[illegible]	ātlocâ (6x)				
	[illegible]	ātlocâchi (2x)				
	[illegible]	ātlocâoe (2x)				
	[illegible]	ātlocâoll (2x)				
	[illegible]	ātlocâoyâ (2x)				
	[illegible]	ātlocây (6x)				
	[illegible]	ātli (4x)				
	[illegible]	ātlychi (2x)				
	[illegible]	mcuai (2x)				
	[illegible]	ytlaaāchi (2x)				
	[illegible]	ytlaāch (2x)				

Most modern peripheral references only mention hypothesis no. 2, but nobody knows the vocabulary or grammar of this extinct language, except for references that Juan Bautista de Pomar and other period writers made to vowel substitutions. Certainly, if the Acolhua people were transitioned to speaking Nahuatl by Techotlalatzin, the king, any surviving Acolhuacatlatolli would have had an extensive influence from Nahuatl (and Spanish nouns) by the mid-sixteenth century (Orozco y Berra 1864; Gerste 1891; García Icazbalceta and Pomar 1891; Thomas and Swanton 1909; Carrasco 1963; Gibson 1964; Stampa 1971; Kirchhoff et al. 1976; Davies 1980; Smith 1984; Swanton 2001; Rossell 2006; de Alva Ixtlilxóchitl 2012).

Based on the evidence uncovered, we infer, as suggested by the renowned cryptologist William Friedman, that the text of the *Voynich Codex* is probably a mixed synthetic language developed from a number of languages most likely invented from various contemporary languages extant in sixteenth century New Spain. This would include Nahuatl, Taino, and other languages of central Mexico, and also include some words of Spanish, and perhaps Arabic for star names. A modern example would be Esperanto, defined as an a posteriori synthetic language. One of the best hypotheses to date claims that Voynichese might be based on an ancient language of communication between speakers whose native language was different, i.e., a Nahuatl lingua franca that had been established between traders of the Aztec empire (Dakin 1981). It was a phonetic rendition of Nahuatl dialects that extended from Durango and Jalisco Mexico to as far south as areas of Nicaragua. Although related to Classical Nahuatl, it included dialects of central Mexico and, later, Spanish nouns. It was separate from the dictionaries of Classical Nahuatl compiled by Luis de Molina and other Spanish priests. This language was not based on a tribe or area, but varied with use and was replaced after the sixteenth century by Spanish.

We admit that our success in decipherment is only partial, but we hope that it is a foundation for others to build on. We are convinced that the translation of the main text of the *Voynich Codex* requires a linguistic analysis based on Mesoamerican languages. If the language is extinct, some way of resurrecting it will be needed. We still maintain that images in the *Voynich Codex* that are associated with names will provide the key.

Literature Cited

Allen, R.H. 1899. *Star names: Their lore and meaning.* New York: G.E. Stechert. (Dover edition 1963).

Beck, D. 2011. *Upper Necaxa Totonac dictionary*. Berlin: De Gruyter Mouton.

Carrasco, P. 1963. Los caciques chichimecas de Tulancingo. *Estudios de Cultura Náhuatl* 4: 85–91.

Dakin, K. 1981. The characteristics of a Nahuatl lingua franca, 55–67. In *Nahuatl studies in memory of Fernando Horcasitas,* ed. Frances Karttunen. Texas Linguistic Forum 18.

Davies, N. 1980. *The Toltec heritage*. Norman: University of Oklahoma Press.

De Alva Ixtlilxóchitl, F. 2012. Historia de la Nación Chichimeca. www.linkgua-digital.com. Barcelona.

de Molina, A.. 1970. *Vocabulario en lengua Castellana y Mexicana, y Mexicana y Castellana.* (Originally written in 1555–1571.) Mexico City: Porrua.

de Sahagún, B. 1953. Florentine Codex. *General history of the things of New Spain. Book 7—The sun, moon, and stars, and the binding of the years*. Transl. A.J.O. Anderson and C.E. Dibble. Salt Lake City: University Utah Press.

García Icazbalceta, J., and J.B. Pomar. 1891. Pomar y Zurita: Pomar, Relación de Tezcoco; Zurita, Breve relación de lose señores de la Nueva España. Varias relaciones antiguas. (Siglo XVI).

Gerste, R.P. 1891. La langue des Chichimèques. Pages 42–57 in Comte Rendu du Congrès Scientifique International es Catholiques. Sciences Historiques.

Gibson, C. 1964. *The Aztecs under Spanish rule: A history of the Indians of the valley of Mexico, 1519–1810.* Stanford: Stanford University Press.

Herrera, F. 2004. *Hippocrene concise dictionary. Nahuatl-English, English-Nahuatl (Aztec).* New York: Hippocrene Books.

Karttunen, F. 1983. *An analytical dictionary of Nahuatl*. Norman: University of Oklahoma Press.

Kirchhoff, P., L.O. Gümes, and L.R. Garcia. 1976. *Historia Tolteca-Chichimeca.* México DF: Instituto Nacional de Antropologia e Historia.

Orozco y Berra, M. 1864. *Geografía de las lenguas y carta etnografica de México*. México.

Rossell, C. 2006. Estilo y escritura en la Historia Tolteca Chichimeca. *Desacatos* 22: 65–92.

Siméon, R. 2010. *Diccionario de la lengua Nahuatl o Mexicana*. México: Siblo Veintiuno.

Smith, M.E. 1984. The Aztlan migrations of the Nahuatl chronicles: Myth or history? *Ethnohistory* 31: 152–186.

Staedtler, M.C., and M.F. Hernández. 2006. Chapter 10: Hydraulic elements at the Mexico-Texcoco lakes during the postclassic period. In *Water management: Ideology, ritual, and power*, ed L.J. Lucero and B.W. Fash. Tucson: The University of Arizona Press.

Stampa, M.C. 1971. Historiadores indigenas y mestizos novohispanos. Siglo XVI–XVII. *Revista Española de Antropología Americana* 6: 206–243.

Swanton, M.W. 2001. El texto Popoloca de la Historia Tolteca-Chichimeca. *Relaciones. Estudios de Historia y Sociedad* 22: 116–140.

Thomas, C., and J.R. Swanton. 1909. *Indian languages of Mexico and central America*. Washington, DC: Government Printing Office.

Tucker, A.O., and R.H. Talbert. 2013. A preliminary analysis of the botany, zoology, and mineralogy of the Voynich manuscript. *HerbalGram* 100: 70–85.

Volcano Discovery. 2017. *Popocatepetl volcano news & eruption update*. https://www.volcanodiscovery.com/popocatepetl/news.html. 22 May 2017.

Walters, J.C.W., M.M. de Wolgemuth, P.H. Pérez, E.P. Ramírez, and C.H. Upton. 2002. *Dicconario Náhuatl de los municipios de Mecayapan y Tatahuicapan de Juárez.* Veracruz: Instituto Lingüístico de Verano.

Wood, S. (ed.) 2000–2016. Online Nahuatl Dictionary http://whp.uoregon.edu/dictionaries/nahuatl/index.lasso.

Part IV
The Author and the Artist

Chapter 14
The Author/Artist of the Voynich Codex

Jules Janick

Profiling the Author/Artist

The evidence presented in this book presupposes that the *Voynich Codex* is a sixteenth century Mesoamerican work. This premise made it possible to create a psychological profile and historical portrait of the author/artist. Although there is evidence that there may be multiple hands in some of the drawings and that the author and artist are likely different persons, we assumed here that one principal person was responsible for the manuscript, even if the illustrator was different from the author. The portrait presented here would have to account for the parts of the manuscript (Herbal, Cosmological, Balneological, Pharmaceutical, and Recipe) and include the author's education and training, artistic ability, spiritual and religious feelings, facility with languages, scientific training, sexuality, and personality.

This chapter was composed early in our investigations of the *Voynich Codex* and represented our hypothesis based largely on the artwork. The profile was a working hypothesis based on the images. The aim and approach were to discover if the multifaceted aspects of the manuscript were compatible with the time, place, and history associated with sixteenth century New Spain. The manuscript is a long, complex work, but we were convinced that the personality and psychological profile of the principal author/artist could be revealed.

Education of the Author/Artist

Education

We conjectured that the author of the manuscript was a member of the Aztec nobility trained at one of several colleges set up by the Franciscan friars who came to convert the Indians and protect them from the rapacious conquistadors. There were two

J. Janick, A. O. Tucker, *Unraveling the Voynich Codex*, Fascinating Life Sciences, https://doi.org/10.1007/978-3-319-77294-3_14

famous schools. The first was La Escuela de San José de los Naturales, established in 1526 by the Belgian Franciscan missionary Peeter Van Der Moers (1486–1572), known as Pedro de Gante, which included instruction in the plastic arts (e.g., sculpture and ceramics). The other, more famous institution was El Colegio de Santa Cruz de Tlatelolco (in present-day Mexico City; also called Colegio de Santa Cruz), established by Fray Juan de Zumárraga between 1533 and 1536.

The education of some of the sons of Aztec chiefs at Tlatelolco has been summarized by Emmart (1940) and SilverMoon (2007). In 1526, Pope Paul III (Alessandro Farnese) gave the order for 20 sons of the Mexican nobility to be educated in Spain and then return to the New World to act as instructors for their own people. This was not accomplished, but the early Franciscan friars, including Toribio de Benavente, known as Motolinía, established schools and began instruction in various convents. In 1533, the Franciscans gathered 50 sons of *caciques* (chiefs) of the neighboring towns of Mexico City and instructed them in Christian theology and elementary education. Charles I of Spain (also known as Charles V of the Holy Roman Empire), who ruled Spain from 1516 to 1556, contributed funds to support the convent schools. Native boys were trained to read, write, chant, and play church instruments, whereas those who had acquired skills in drawing and writing were given the task of illuminating manuscripts. In 1536, as a result of the success of these schools, the Colegio de Santa Cruz (Fig. 14.1), the first institution of "higher" education, was established at the convent of the Church of Santiago in the district of Tlatelolco on the outskirts of Mexico City. One of the founders was the Franciscan friar, Bernardino de Sahagún (ca. 1499–1590), who compiled the famous *General*

Fig. 14.1 Colegio de Santa Cruz chapel and convent of Tlatelolco, Mexico. (Source: Emmart 1940)

History of the Things of New Spain, a bilingual work in Spanish and Nahuatl (also called the *Florentine Codex*). This great work in ethnology consisted of about 2400 pages organized into 12 books, including 2468 illustrations drawn by native artists (Sahagún 1950–1982). This celebrated work was the most important work produced at the Colegio de Santa Cruz, and although attributed to Sahagún, it represents a collaboration by Nahua students to memorialize the pre-conquest culture of the Nahua people, who we remember as Aztecs (Silvermoon 2007). No doubt some of the students were among the original 50 who had attended elementary school, but the student body also included 100 recruited from the Indian nobility of principal towns and districts near Mexico City. The first students of the Colegio were children aged 10–12 years, but some were described as young men and were selected on the basis of special aptitudes. We conjectured that one of these students might have been the primary author of the *Voynich Codex* and that it was likely that he was one of the young men, probably 15–20 years of age in 1536, putting the date of birth of the author/artist of the *Voynich Codex* at between approximately 1521 and 1526. The date of the *Voynich Codex* has been estimated to be around 1565 to 1572, which would make the author 39–51 years old at that time.

The students were instructed in philosophy, logic, arithmetic, music, and in native Indian medicine because the Spanish believed that the remedies of the New World were superior to those of Europe for native diseases. Indian physicians were brought in as instructors. The authors of the renowned Nahuan herbal now known as the *Codex Cruz-Badianus* were two members of the Colegio: Martinus de la Cruz, the school's indigenous doctor who also gave instruction in medicine (SilverMoon 2007), and Juannes Badianus, a native student who translated the de la Cruz text into Latin, and was born and reared in the village of Xochimilco, the site of the famous "floating" gardens, or chinampas. Their Spanish names, Martin de la Cruz and Juan Badiano, were conferred upon their baptisms.

The preface to the *Codex Cruz-Badianus* contains an obsequious, self-deprecating, but revealing sentence by de la Cruz (Emmart 1940): "*But you will recollect that we poor unhappy Indians are inferior to all mortals, and for that reason our poverty and insignificance implanted in us by nature merit your indulgence.*" We surmise that the primary author of the *Voynich Codex* was of a very different personality: strong, self-confident, and clearly very proud of his heritage. We suggest he might have been one of the older boys, likely not a long-time student, and perhaps not as inculcated in religion, as only a few Christian symbols were found in the manuscript. It is clear that he had a natural ability for sketching. His talents were completely original and unique.

Many of the indigenous students, such as Antonio Valeriano, Martin Jacobita, and Pedro de San Buenaventura, had extraordinary talents and great intellectual ability. For example, San Buenaventura was expert in Latin, Spanish, and the native Nahuan language. He was a historian and author of *Anales de Cuauhtitlan* and wrote medicinal and zoological texts. He was knowledgeable about the Aztec calendar, explaining it to Sahagún, who had difficulty in understanding it. Furthermore, many of the indigenous people had great artistic talents, as exemplified by the sculpture, murals, and illustrated codices (both pre- and post-Columbian) that have survived.

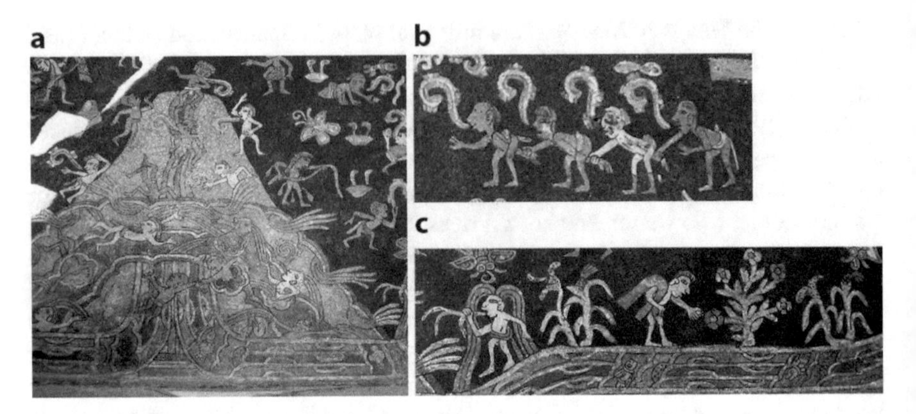

Fig. 14.2 Tepantitla mural, Teotihuacan. (**a**) Bathing in mountain streams; (**b**) a daisy-chain of men; and (**c**) tending and tasting herbs

The Aztec language basically consisted of pictograms; thus, illustrations and drawing were part of their culture. A pre-Columbian mural known as the *Great Goddesses of Teotihuacan,* executed between 100 BCE and 700 CE, shows details reminiscent of the *Voynich Codex* (Fig. 14.2) where almost a dozen males, most wearing only a loincloth, are swimming and cavorting in crisscrossed rivers associated with a mountain. Beside the mountain are many figures with speech emanating from their mouths, and there are four crouching male figures with one hand under their legs attached to the hand of the figure behind in a ritual daisy chain that is redolent of the nymphs touching or holding hands in pools (Fig. 14.5). Next to the river are various plants, including maize being tasted or examined, suggesting they might have been part of a medicinal garden.

There are two famous post-Columbian examples in sixteenth century New Spain of paintings and murals made by Nahuan artists known as *tlacuiloque* that include features similar to the *Voynich Codex* (see Chap. 16). The first example comprises ceiling paintings on paper made in 1562 in a Franciscan church in Tecamachalco, Puebla (Azpeitia 1972). The paintings of the Apocalypse are in a Northern Renaissance style by an indigenous artist baptized as Juan Gerson. They include many elements found in the *Voynich Codex* that are discussed in the Chap. 15. There are many scenes from the Hebrew Bible, such as a panel referred to as the *City of God* (Jerusalem) painting, in addition to a view of the Celestial City of Jerusalem. The second example consists of the murals at the Casa del Deán in Puebla, completed in 1584 (Morrill 2014), created by unknown artists in the home of Don Tomás de la Plaza, dean of the Cathedral of Puebla from 1553 to 1589. They include a parade of sibyls, female prophets from Greek mythology, in sixteenth century clothing (Fig. 14.3). The beautifully painted figures show some resemblance to the nymphs in the Balneological section of the *Voynich Codex*. Below the sibyls, there are nude female figures on horseback in a band that resembles a frieze.

The artist of the *Voynich Codex* shows great facility for sketching in pen and ink. Very few of the drawings show corrections. They are swiftly executed with amazing dexterity and wit. The botanical drawings in the *Voynich Codex* are quite different

Fig. 14.3 Murals in the Salon of the Sibyls at the Casa del Deán in Puebla, Mexico. (Source: Morrill 2014)

from the drawings of the *Codex Cruz-Badianus*. They are much more free-spirited and in many aspects bizarre and whimsical, but with much botanical detail (see Chap. 4, Fig. 4.16). It would appear that the Voynich artist, in contrast to Martin de la Cruz, did not have formal training in botanical illustration. In Chap. 16, we show that many of the illustrations in the *Voynich Codex* were consistent with other artwork made by *tlacuiloque* (indigenous artists) in sixteenth century Mexico.

Sexuality

The artist displays a clear sexual interest in the female body. There are about 500 women, referred to as nymphs, in the *Voynich Codex*, all but 30 unclothed in full frontal view, and all appearing to be the same person (Fig. 14.4). (A few have been suggested to be males, but this is doubtful.) The absence of a back view suggests Aztec sensibility against revealing the female buttocks. The figures, mostly of blonde, curvy, voluptuous women, show that the artist clearly had a sexual interest in the female form. The fact that they are mostly blonde indicates that he was imbued with the Spanish concept of beauty. Blonde women were the Spanish ideal (da Soller 2005), despite many Spanish people being dark-haired because of their Moorish heritage. There are only two obvious males in the *Voynich Codex*: the male

Fig. 14.4 Nymphs bathing in pools in the *Voynich Codex*

"twin" in the sign Gemini and the archer (Sagittarius) with a crossbow in the zodiac images (see Chap. 8).

The nymphs are mostly found in two locations of the *Voynich Codex*: associated with the zodiac (see Chaps. 6 and 8) and associated with the Balneological section (see Chap. 7). In the zodiac folios, they appear in two or three rings around each zodiac sign and seem to represent degrees or days in the zodiac calendar (e.g., folios 70r–73r). With the exception of 30 clothed nymphs associated with the two Aries zodiacs, one with a ram and the other a ewe, all the other 301 nymphs are nude. The zodiac concept is clearly Western, but the artist has modified the signs by replacing some of them with animals found in Mexico in the sixteenth century: fish (Pisces) with alligator gar, sheep (Aries) with either Spanish imports or a native ram and an ewe, the bull (Taurus) with a cow and a bull breed imported by the Spanish, the crab (Cancer) with two crayfish, the lion (Leo) with an ocelot, and the scorpion (Scorpio) with a jaguarundi (see Chap. 6). Clearly, the artist combined the Spanish zodiac with Aztec influences and further incorporated gender complementarity, a fundamental part of the Aztec religion, by including both male and female forms in Pisces, Aries, Taurus, Gemini, and Cancer.

There are at least 202 nymphs in the Balneological section (see Chap. 7), most of them cavorting in green pools connected by tubes or plumbing. Many are touching each other or holding hands (Fig. 14.5). They appear to be partaking in ritual bathing and it is well known that the Aztecs were devotees of daily bathing (von Hagen 1956). The Voynich images are reminiscent of the bathing scene in the pre-Columbian mural associated with the Great Goddesses of Teotihuacan (Fig. 14.2).

Religion and Spirituality

The *Voynich Codex* incorporates Christian, Jewish, Moorish, and Aztec themes. The evidence of a Christian influence is twofold. The first is the presence of nine crosses (Fig. 14.5): one on a nymph crown on folio 72v, one being held by a nymph

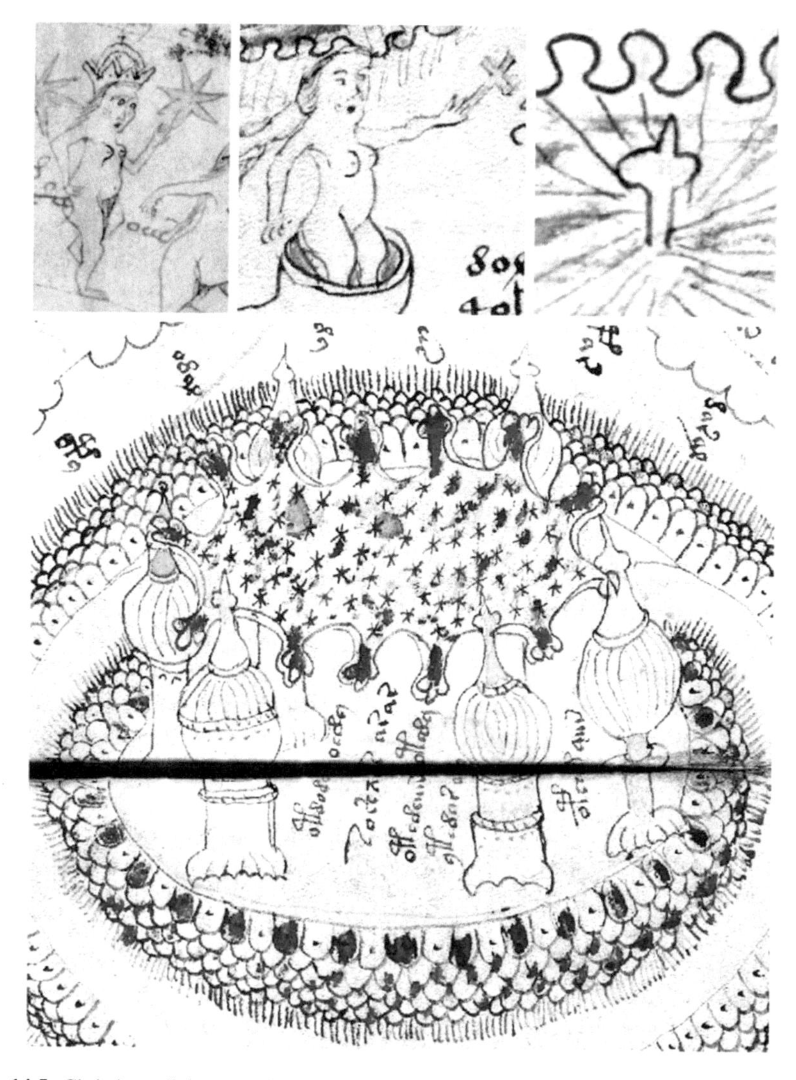

Fig. 14.5 Christian religious symbolism in the *Voynich Codex*

in folio 79v, one standing alone in folio 75v, and six on qubba (domes) or possibly ciboria (goblet-shaped vessels for holding Holy Communion wafers) in circle A of folio 86v. The second is the association of circle A in folio 86v with the New Jerusalem (also referred to as the Celestial City of Jerusalem, City of the Angels, Puebla de los Angeles or Angelopolis), founded by the Franciscan friars, including Motolinía, in 1530. The association of the artist with Christianity may be explained by his connection to a Franciscan school or *colegio*. However, the relatively scarcity of Christian symbols suggest that the artist might not have considered himself a devout Catholic. Furthermore, the presence of individual nymphs holding what appears to be either an iron collar or the pear of anguish, torture devices used by the

Spanish Inquisition, may indicate that the artist was critical of devices used to subjugate the Indians (see Chap. 7). The interpretation of these tiny, crude drawings, by their very nature, is controversial, and alternative interpretations should be considered.

The Jewish influence is based on the kabbalah-like arrangement of circles of folio 86v (see Chap. 10). The Moorish (Islamic) influence is found in the arches of circle 2, and in the domes (or ciboria) in circle A of folio 86v that hold up a starry mantle. In addition, some of the star names in the *Voynich Codex* appear to be Arabic (see Chap. 13).

The author was also influenced by Aztec culture and customs such as ritual bathing, which had religious overtones (see Chap. 7), and medicinal gardens. In a study of the ancient gardens of Huaxtepec, Granziera (2005) described the natural pools in the pre-Hispanic gardens as being part of a ritual landscape and ceremonial center dedicated to the cult of water and fertility: "*the earth is like a vessel containing water, it is like a womb filled with amniotic fluid.*" It is clear that the ceremonial combination of bathing and pools in the Balneological section is based on the religious sensibility of the Aztecs.

The Cosmological section (see Chap. 9) indicates that the author was imbued with Aztec influences. Common cosmological themes in the circles of this section include the sun, the moon, the planet Venus, and stars, including the presence of the star cluster known as the Pleiades. The prominence of sun images, e.g., folios 67r(1), 67v(2), and 68v(3), clearly indicates Aztec cosmology. We consider the Cosmological section strong evidence that the *Voynich Codex* represents a Mesoamerican manuscript.

Herbalist

The Herbal section makes up a major part of the *Voynich Codex*. It includes 131 pages of plants (usually a single plant per page) and 20 pages with roots, leaves, or shoots of various plants. The Recipe section (folios 103r–116r) appears to be concerned with the medical uses of plants (see Chap. 4). This emphasis on herbal botany suggests that the author/artist was the Aztec version of the Roman physician Pedanius Dioscorides. Sixty-one of the 359 phytomorphs of plants were identified and all are Mesoamerican (Tucker and Talbert 2013; Tucker and Janick 2016) (see Chap. 4). Identifications of the Voynich plants as Old World plants by non-botanists such as Sherwood and Sherwood (2008) and Velinska (2013) are not credible (see Chap. 16). The identification of medicinal plants and presumably their culture and uses underscores pharmacology as an Aztec science. It appears that the artist would have been considered a medical doctor.

Conclusion

A summing up of the profiles of the primary author/artist indicates that he was an intellect and talented native son of Mexico with wide interests and many talents, in short, a polymath. He was a brilliant Nahua, much like Leonardo da Vinci, with the artistic, surrealistic talents of Salvador Dali. He was probably born in the 1520s and likely had some education at or connection to the Colegio de Santa Cruz. He was knowledgeable about botany, herbal medicine, and cosmology, and skilled in sketching and calligraphy. He was knowledgeable about various religions, and clearly fluent in Spanish, Nahuatl, and other native languages. He was an excellent scribe and he seems to have perfected a new script we called Voynichese. He had a healthy interest in women, particularly one voluptuous blonde, who he drew more than 500 times.

Although knowledgeable about Christianity, he was clearly a supporter of Aztec culture, and there was a suggestion that he protested against some of the cruel torture devices of the conquistadors. He was an expert in languages and could speak Spanish, Latin, Nahuatl, other Mesoamerican languages, and perhaps Arabic. He must have been a prodigious worker because the manuscript was long and complex. He was obviously familiar with Motolinía and symbolically illustrated the Celestial City of Jerusalem, established in Puebla in 1530. He was knowledgeable about the kabbalah in its Christian formulation. It would appear that the author/artist of the *Voynich Codex* was a key intellectual of sixteenth century colonial New Spain. Although various names have been identified in the Colegio de Santa Cruz, the author/artist, and the accuracy of the predictions, are revealed in Chap. 15.

Literature Cited

Azpeitia, R.C. 1972. *Juan Gerson, Pintor indigena del siglo XVI – simbolo del mestizage – Tecamachalco Puebla*. Mexico: Fondo Editorial de la Plastica Mexicana.

Da Soller, C. 2005. *The beautiful woman in medieval Iberia: Rhetoric, cosmetics, and evolution*. Columbia: University of Missouri-Columbia.

De Sahagún, B. 1951–1982. *Florentine Codex. General history of the things of New Spain*. [1540–1585] 12 vol. Trans. A.J.O. and Anderson C.E. Dibble. Salt Lake City: Univ. Utah Press.

Emmart, E.W. 1940. *The Badianus manuscript (Codex Barberini, Latin 141) Vatican library. An Aztec herbal of 1552*. Baltimore: Johns Hopkins Press.

Granziera, P. 2005. Huaxtepec: The sacred garden of an Aztec Emperor. *Landscape Research* 30 (1): 81–107.

Morrill, P.C. 2014. *The casa del Deán: New world imagery in a sixteenth century Mexicmural cycle*. Austin: University of Texas Press.

Sherwood, E., and E. Sherwood. 2008. The Voynich botanical plants. http://www.edithsherwood.com/voynich_botanical_plants/index.php. 26 July 2012.

SilverMoon, 2007. The imperial college of Tlatelolco and the emergence of a new Nahua intellectual elite in New Spain (1500–1760). Durhan: Department of History, Duke University.

Tucker, A.O., and R.H. Talbert. 2013. A preliminary analysis of the botanical, zoological and minerals of the Voynich manuscript. *HerbalGram* 100: 70–85.

Tucker, A.O., and J. Janick. 2016. Identification of phytomorphs in the Voynich codex. *Horticultural Reviews* 44: 1–64.

Velinska, E. 2013. The Voynich manuscript: Plant ID list. ellievelinska.blogspot.com/2013/07/the-voynich-manuscript-plant-id-list.html. 5 June 2017.

Von Hagen. V.W. 1956. The ancient sun kingdoms of the Americas. Cleveland: The World Publishing Company.

Chapter 15
Voynich Codex Claimants

Arthur O. Tucker and Jules Janick

Potential Authors

The personality profile of the author/artist of the *Voynich Codex* was hypothesized by an examination of the illustrations (see Chap. 14) at an early stage of our interest in this work. The author/artist was assumed to be an intellect and native son of Mexico with wide interests and many talents; in short, a polymath and a brilliant Indian, much like Leonardo da Vinci. He was probably born in the 1520s and likely had some education at El Colegio de Santa Cruz de Tlatelolco (in present-day Mexico City; also called Colegio de Santa Cruz). He was knowledgeable about botany, herbal medicine, and cosmology, and skilled in sketching and calligraphy. He was familiar with various religions and had expertise in and was clearly fluent in Spanish, Nahuatl, and other native languages. He was an excellent scribe and seemed to have perfected a new script we called Voynichese. He had a healthy interest in women, particularly one shapely blonde who he drew more than 500 times (and all were "nudes," except for 30). Although knowledgeable about Christianity, he was clearly a supporter of Aztec culture, and there was a suggestion that he might have protested some of the cruel torture devices of the Spanish Inquisition. He must have been a prodigious worker because the manuscript was long and complex. He was obviously familiar with Motolinía and symbolically illustrated the New Celestial City of Jerusalem established in Puebla in 1530. He was aware of the kabbalah in its Christian formulation. Claimants were listed chronologically in the *Voynich Codex* hypotheses, 1921–2017, in Table 1.1 in Chap. 1.

There have been various potential authors of the *Voynich Codex*. The list of proposed claimants was first reviewed and then expanded. Later in the chapter, we present new discoveries that indicate that the author and artist identified themselves in the manuscript.

J. Janick, A. O. Tucker, *Unraveling the Voynich Codex*, Fascinating Life Sciences, https://doi.org/10.1007/978-3-319-77294-3_15

European Claimants

Roger Bacon (1214–1294)

Wilfrid Voynich wanted to believe that the Franciscan friar Roger Bacon (English philosopher, scientist, and often referred to as Doctor Mirabilis) was the author of his manuscript, which he entitled the *Roger Bacon cipher MS*. This theory was formally written up by William R. Newbold, a professor of moral and intellectual philosophy at the University of Pennsylvania in a manuscript entitled *The Cipher of Roger Bacon,* published posthumously in 1936. The theory was discredited by John M. Manly in an article entitled "Roger Bacon and the Voynich Ms.," (Manly 1931). The dating of the vellum (see Chap. 2) as early fifteenth century would not be compatible with Bacon authorship. The claim of Bacon as the author of the *Voynich Codex* was universally discredited.

Antonio Averlino (1400–1469)

Nicolas Pelling (2006) proposed that the Italian architect Antonio di Pietro Averlino was the author of the *Voynich Codex*, and that the manuscript could be his lost book of secrets. Pelling dated the *Voynich Codex* from 1450 to 1500, based on the swallowtail merlons in circle 2 of folio 86v, which were found in Milan after 1450. Circle A of folio 86v is considered a view of St. Mark's Basilica, as seen from the campanile (bell tower). Pelling offered as "evidence" Averlino's interest in cryptology, water, agriculture (including herbals and grafting), engines (which he detected in plant images), recipes, glass-making, and bees (which he found in folio 86v). We did not find this claimant to be reasonable. Buonafalce (2007) stated: "Some of his conclusions are rather extravagant. Despite his efforts, he does not arrive at any convincing decipherment of the encoded text."

Nicolas Pelling and Antonio di Pietro Averlino

Nicholas Pelling, who blogs as Nick Pelling, self-published a free-spirited book entitled *The Curse of the Voynich: The Secret History of The World's Most Mysterious Manuscript* in 2006. The work is rich in extravagant hypotheses, but also contains valuable information about the physical makeup of the manuscript into quires, their labeling, and misbinding. Many sections are obviously out of sequence, such as the Herbal section, which is found in three locations (see Chap. 2, Table 2.1). Pelling's assumption that the Voynich author is the Renaissance architect Antonio Averlino (ca. 1400–1469) is quite fantastic and frankly hilarious. We provide one example. He explains that the ligated initials on folio 1v (which we identified as ligated initials, "JGT") to be the numeral 5. They are not and can be compared with a perfectly clear numeral 5 found on the bottom right of folio 57v (see Chap. 2, Fig. 2.4). Pelling suggests that numeral 5 might be a code for information five pages later in which he finds a plant with the letter "F" embedded in one bud and the letters "TOA" embedded in the stem. The "F" is assumed to refer to Averlino's nom de plume, Filarete (lover of virtue), and the "TOA" is somehow a partial copy of [ONI]-TOA-[N], the personal cipher signature of Antonio Averlino. However, his identification of the letter "A" in the stem of folio 57v is spurious.

The choice of Averlino by Pelling was based in part on the identification of circle 2 of folio 86v as Milan, because at one time a circular wall surrounded that city. In addition, both the fortresses in Milan and in circle 2 have swallowtail merlons. However, there is no evidence that the city in circle 2 is completely encircled, despite the presence of part of a curved wall, and the merlons from the fortress do not quite resemble the curved merlons found in Milan's Castello Sforzesco. Pelling ignores the spewing volcano in circle 2. He includes extensive and inventive discussions on the Voynichese symbols, but in the end he confesses that he has not translated a single word. We should not be hard on Pelling, whom we do not know, but he goes out of his way to be contemptuous of others. He of course does not accept either the identification of the sunflower or the armadillo in the *Voynich Codex* and this lapse is fully discussed in Chap. 16. Of course, to accept the sunflower and armadillo would require him to disavow his entire thesis.

Leonardo da Vinci (1452–1519)

Young Leonardo da Vinci was proposed as a claimant by Edith Sherwood (2002) based on handwriting and her identification of European plants. To support her conjecture, she found hidden signatures of da Vinci. For example, in Voynich folio 70v, the ewe represents Aries. By the ewe's legs is the Occitan word *aberil*, for April, the date of the Aries sign (see Chap. 8, Table 8.8). Sherwood shows it in mirror image and comes up with the name Leonardo, which is ludicrous.

John Dee (1527–1608) and Edward Kelley (1555–1597)

The relation of John Dee, mathematician, astrologer, kabbalist, alchemist, and adviser to Queen Elizabeth I, is discussed in Chap. 1. The connection to the *Voynich Codex* is based on the visit of Dee to Rudolf II in Prague in 1583. A 1694 letter from

Sir Thomas Brown to Elias Ashmole states that Dee's son Arthur had seen a book containing "*hieroglyphicks*." The rumor that a book was sold to Rudolf II for 600 ducats conforms with Dee's diary entry that he left Prague with 630 gold ducats. Edward Kelley (1515–1597), an alchemist, occultist, and spirit medium who accompanied Dee to Prague, was considered to be a potential author alone or in collaboration with Dee. The Dee–Kelley conjecture as author (or authors) of the Voynich is supported by Brumbaugh (1978), Rugg (2004a, b), Dos Santos (2005), and Carthago (2006, 2015). The authorship of the manuscript by Dee and Kelley is based on hearsay and circumstantial historical evidence, but botanical and zoological evidence belies this claim. René Zandbergen (2004–2017) [listed as 2016 in literature cited] and Prinke (n.d.) are skeptical regarding the veracity of Dee being the author of the *Voynich Codex*, but we do not rule out him being involved with conveying the manuscript to Rudolf II in some way.

Other Claimants

There have been a number of claimants, including: Michel de Nostredame (Nostradamus) 1502–1566; or his son Cesar (Kennedy and Churchill 2008); Georg Handsch von Limuz 1529–1595, proposed by Karel Dudek (Dragoni 2012); Anthony Ascham, an English astronomer, fl. 1553 (Kennedy and Churchill 2008); Cornelius Drebbel 1572–1633 (Dragoni 2012); and Francis Bacon 1561–1626, proposed by Richard Santa Coloma (Dragoni 2012). Finally, Wilfrid Voynich 1865–1930 has himself been suggested to be the forger of the document (Kennedy and Churchill 2008: 232–236). We regard these suppositions as foolish.

New World Claimants

Francisco Hernández de Toledo (1514–1587)

James Comegys, a middle school teacher in Madera, California, carried out an analysis of the *Voynich Codex* in collaboration with his Spanish-speaking, Mexican students, and his twin brother, John Comegys. Their analysis was archived in a unique, difficult-to-access manuscript (James C. Comegys 2001). They reported, presciently in our view, that the *Voynich Codex* had its origin in Mexico based on "*hieroglyphs, plants, architecture etc.*" They believed that it was consistent with the themes and methods employed by Francisco Hernández on his 1570 expedition to New Spain in search of herbal medicines and the discovery of longitude. They inferred that Hernández was the author. The assumption was made that the grammar, syntax, lexicon, and orthography of the document were consistent with Nahuatl. James Comegys assumed that the text was written from right to left, as in Hebrew (he assumed Hernández was a *converso,* or convert), and that the page was read from the bottom to the top. However, all pages of the text are justified left and we disagree with both these assertions. James Comegys' attempts at decipherment of the *Voynich Codex* are discussed in Chap. 11.

Francisco Hernández was a Spanish naturalist and in 1567 became court physician to Philip II, who ruled Spain as king from 1556–1596. In 1570, the king appointed Hernández "General protomedico of our Indies, islands and mainland of the Ocean Sea" on a seminal scientific expedition to the New World to study medicinal plants. Accompanied by his son Juan, Hernández traveled in New Spain for 7 years collecting and classifying specimens, interviewing the indigenous people through interpreters, and conducting medical studies. He was assisted by three indigenous painters (baptized Pedro Vázquez, and Antón and Baltasar Elías). In his will, Hernández sought compensation for the artists and for Indian doctors of Mexico who had been engaged in bringing him herbs. Yet, in a 1580 poem to his friend Arias Montano, he complained of "*perverse Indian guides...their fraudulence... terrible tricks...and the mistakes of the artists.*" The extensive descriptions of Francisco Hernández' findings (the 2500-page manuscript covered 3000 plants) were published in Mexico in 1615 entitled, *Plantas y Animales de la Nueva Espana, y sus virtudes por Francisco Hernández, y de Latin en Romance por Fr. Francisco Ximenez* (Varey 2000).

There is a plausible connection between Hernández and Rudolf II of the Holy Roman Empire. Young Rudolf was sent to Spain in 1564 under the tutelage of his maternal uncle Philip II at the age of 11 and remained there until 1571. Rudolf must have been in contact with Hernández, who was the court physician before he left for New Spain in 1570. Hernández returned to Spain in 1577, 5 years after Rudolf had returned to Vienna. Hernández was unlikely to have been the author of the *Voynich Codex,* but it could have come into his possession from various indigenous translators. Although there were various letters between Hernández and Philip II while Hernández was in Mexico, no manuscripts were sent, but all his manuscripts and books were willed to Phillip II (Varey 2000). Thus, there was a possible path of the manuscript from the court of Philip II to his nephew Rudolf II after the death of Hernández.

An Unidentified Spanish-Educated Aztec

This claimant was proposed by Tucker and Talbert (2013) based on the identification of plants, animals, and a mineral as indigenous to Mexico. The assumption was made that a likely author of the *Voynich Codex* could have been a student at the Colegio de Santa Cruz. The careers of prominent students and other Nahua elites are summarized below.

Antonio Valeriano (1531–1605) was born in Azcapotzalco, a present-day municipality of Mexico City, and was one of the most prominent and accomplished graduates of the Colegio. He rose to be the governor of Indians for 8 years in Azcapotzalco, and then for 23 years (until his death) in Mexico-Tenochtitlan (its ruins are now located in present-day Mexico City). He was considered to be eloquent, a skilled Latinist, and well-instructed on Christian doctrine. He was a collaborator on Sahagún's magnum opus, *The General History of the Things of New Spain* (also called the *Florentine Codex*), and a number of other works.

Martin Jacobita, from Tlatelolco, was a professor from 1567 to 1570, and rector in 1572. He was also one of Sahagún's collaborators.

Martin de la Cruz (also called by his Latin name, Martinus de la Cruz) was an Aztec physician and a native teacher of medicinal knowledge at the Colegio. He wrote the text and was assumed to be the painter of thc illustrations of what is now known as the *Codex Cruz-Badianus* of 1552. It was written in Nahuatl, which was translated into Latin by the student **Juan Badiano** (also called by his Latin name, Juannes Badianus) 1484–1560 (Emmart 1940). This manuscript was probably sent to Spain soon after its completion as it was not mentioned by Sahagún or Hernández.

Pedro de San Buenaventura, an indigenous Indian from Cuauhtitlan, was educated at the Colegio de Santa Cruz. An expert in Latin, Spanish, and the language of the Nahuas, he was also an historian and author of *Anales de Cuauhtitlan*. He had great knowledge of the Aztec calendar, explaining it to Sahagún between 1575 and 1580 (SilverMoon 2007: 38), and as a scribe wrote about medicine and zoology. Buenaventura clearly had the intellectual capacity to write the *Voynich Codex*, but there was no evidence of any connection.

Juan Bautista (de) Pomar (1535–1601), a writer interested in pre-Columbian Aztec history, was the great-grandson of Nezahualcoyotl, ruler of the Alcohuan city-state of Texcoco, in the present-day state of Mexico. His father was half Spanish and his mother taught him Aztec traditions, although he was raised Christian. He spoke and wrote in both Spanish and Nahuatl. He was considered noble by the Spaniards and was able to obtain one of his great-grandfather Nezahualcoyotl's royal houses in Texcoco. Pomar's major work, *Relación de Juan Bautista Pomar,* completed in 1582, includes an account of the Aztecs and Tlatelolcas (inhabitants of Tlatelolco). Through interviews, Pomar recorded recollections of the older and lost customs of Aztec traditions, including herbs. His account, written at the suggestion of Francisco Hernández, complements the works of Bernardino de Sahagún and Fernando de Alva Cortés Ixtlilxochitl. He is also considered to be the author of *Romances de los señores de Nueva España*, a compilation of Nahuatl poetry.

Juan Gerson and Gaspar de Torres: Proposed Illustrator and Author of the Voynich Codex

The plant illustrations of the *Voynich Codex* have secrets to reveal beyond their identification (Janick and Tucker 2017). A close examination of the phytomorph of *Ipomoea arborescens* in folio 1v (Fig. 15.1a) reveals a symbol or glyph (Fig. 15.1b) resembling a typographic ligature that appears to be made up of three letters, "JGT," and resembles horse brands used in New Spain (Fig. 15.2). The "G" in the ligature resembles the "G" for *Guerra* (Fig. 15.1c), a face brand used on enslaved Indians

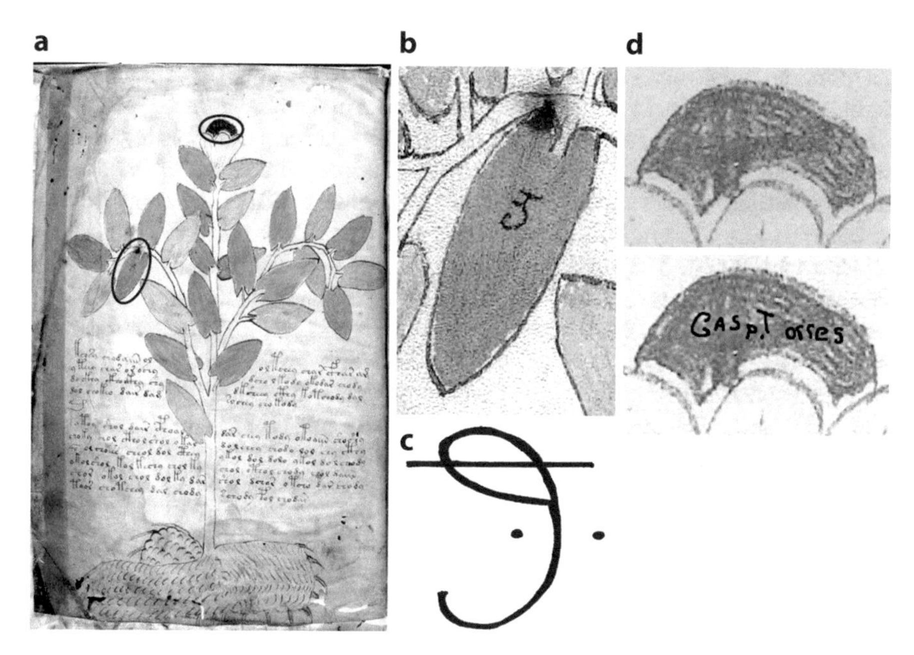

Fig. 15.1 (**a**) Putative initials of Juan Gerson, *Tlacuilo* (indigenous artist), and the name Gasp. Torres (Gaspar de Torres) embedded in folio 1v of the *Voynich Codex*, and (**b**) a leaf with ligated initials "JGT." (**c**) Letter "G" for *Guerra*, an Indian slave brand. (Source: Resendez 2016). (**d**) Bud with embedded name, Gasp.Torres, in the *Voynich Codex* folio 1v

Fig. 15.2 Horse brands in Michoacán, Mexico, 1597. Note the similarity of the brand for El Castaño (The Bay) with the ligated initials in folio 1v of the *Voynich Codex* in Fig. 15.1 and that the paragraph markers resemble the so-called "bird glyphs" in folio 1r (see Chap. 2, Fig. 2.7). (Source: Du Bron 2010)

captured in warfare to distinguish them from Indian slaves purchased from other Indians and branded with the letter "R" (*Rescae* for ransom) (Resendez 2016). Of all the artists in sixteenth century New Spain, the only one with the initials "JGT" was Juan Gerson, *Tlacuilo* (indigenous artist).

With the identification of Gerson's initials in folio 1v of the *Voynich Codex*, the phytomorph in folio 1v was carefully examined for the presence of other embedded initials or names (Fig. 15.1b). Alternating contrasts in the brown strokes of the bud of the tiff file of *Ipomoea arborescens* (folio 1v) by Arthur O. Tucker revealed the name "Gasp.Torres" (Fig. 15.1d), but a full multi-spectral analysis of the original codex is needed. We are convinced that the name refers to Gaspar de Torres, known to be *maestre de niños* (master of students) at the Colegio de Santa Cruz in Tlatelolco (SilverMoon 2007). The remarkable evidence that two names, Juan Gerson and Gaspar de Torres, are on the first botanical drawing on the verso side of the first folio led us to conjecture that Gerson, the native-born artist (*tlacuilo*), was the illustrator of the *Voynich Codex* and Torres was the author. If this conjecture is correct, it is not only plausible but reasonable that they would affix their signatures to the first botanical illustration. This conjecture was examined by biographic and iconographic analyses and is presented below.

Signature of Gaspar de Torres

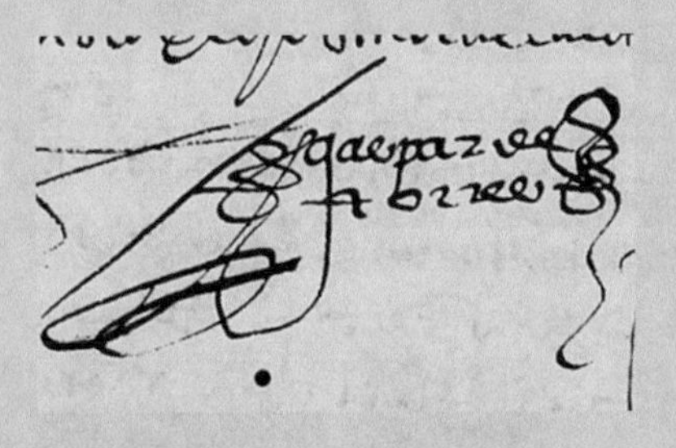

The signature of Gaspar de Torres from the Ramo de Tierras, Vol. 2722–2723, has been located and compared with the signature in the bud of *Ipomoea arborescens*, in folio 1v (Fig. 15.1). The petition signature is in a florid cursive style typical of the sixteenth century with the addition of "de" in the scribal abbreviation format that resembles the Voynichese symbol ff. Furthermore, the terminal letter "s" was written as an "e," typical of Spanish scribal shorthand (Munoz y Rivera 1927). In contrast, the signature in the bud was printed with the word Gaspar, abbreviated to Gasp., and the "de" was omitted. We assume that the print calligraphy and abbreviation were used to fit the name into the small, congested area of the bud.

Biography of Juan Gerson

Juan Gerson was a Hispanicized *indio ladino* (indigenous Nahua) of the sixteenth century known as a *tlacuilo* (indigenous artist) whose baptismal name honored a fifteenth century French theologian, Jean-Charlier Gerson (Azpeitia 1972). Gerson's prominent father, Tomás Tlacochtuectli, later adopted the surname Gerson (Townsend 2016). Juan Gerson attended a school established for native Nahua called La Escuela de San José de los Naturales, founded by Fray Peeter Van Der Moers (also spelled Moere, Moor, or Muer), a Belgian Franciscan monk known as Pedro de Gante (Sanchis Amat 2012). The school included training in the plastic arts. In 1562, Juan Gerson completed a cycle of 28 paintings, plus the Franciscan seal on *amate* (traditional bark paper of the Aztecs) that decorated the ceiling of the Franciscan church, Asuncion de Nuestra Senora, in Tecamachalco, a small city 57 km east of Puebla, in the present-day Mexican state of Puebla. This extraordinary work illustrating the biblical Apocalypse of St. John, a favorite Franciscan theme, was based on European engravings, and has been traced to woodcuts made early in the sixteenth century (Niedermeier 2002).

Juan Gerson stands out from his contemporaries, not only because he was a *tlacuilo* who was honored for his paintings, but also because of his recognition by the Spanish authorities. Juan's father Tomás requested permission to go on horseback in 1555, and in 1592 Juan Gerson was granted the extraordinary rights for an upwardly mobile *indio ladino* (indigenous Nahua) to ride a saddled horse in Spanish clothing and carry a sword and dagger (Carmelo Arredondo et al. 1964; Azpeitia 1972; Landa Albrego 1992; Townsend 2016).

The typographical evidence for Juan Gerson as the illustrator of folio 1v of the *Voynich Codex* is supported by the presence of his initials in the Apocalypse paintings of 1562. In the painting, *The Arresting of the Winds and the Elected* (i.e., John 7: 1–2), Juan Gerson has embedded "JGT" in the lower gown of the anointing angel (Fig. 15.3). The "J" is very obvious, the "G" follows the folds of the gown, and the "T" is fainter and upside down. Juan Gerson was inspired for the paintings at Tecamachalco by the woodcuts of Erhard Schoen (after 1490–1524), Master IF (also called Jacques Lefevre, Jacques Lefèvre, or Jakob Farber, active 1516–1550), and Hans Holbein I (1497–1543) that were copied in Bibles printed in Germany and France (Carmelo Arredondo et al. 1964; Azpeitia 1972). None of these sources had "JGT" embedded in the woodcuts.

Iconographic Comparison of Gerson's Apocalypse Paintings and Voynich Illustrations

To determine if Juan Gerson might have been the illustrator of the *Voynich Codex*, the 28 paintings by Gerson in the church at Tecamachalco have been examined to determine similarities to illustrations in Voynich. Examples of similarities are presented below.

Environmental Images Water and clouds are illustrated in both the *Voynich Codex* and Gerson's Apocalypse images (Fig. 15.4). The techniques of illustrating water and clouds in Voynich (circle 2 in folio 86v) are quite close to the water symbolism in *Noah's Ark and the Deluge*.

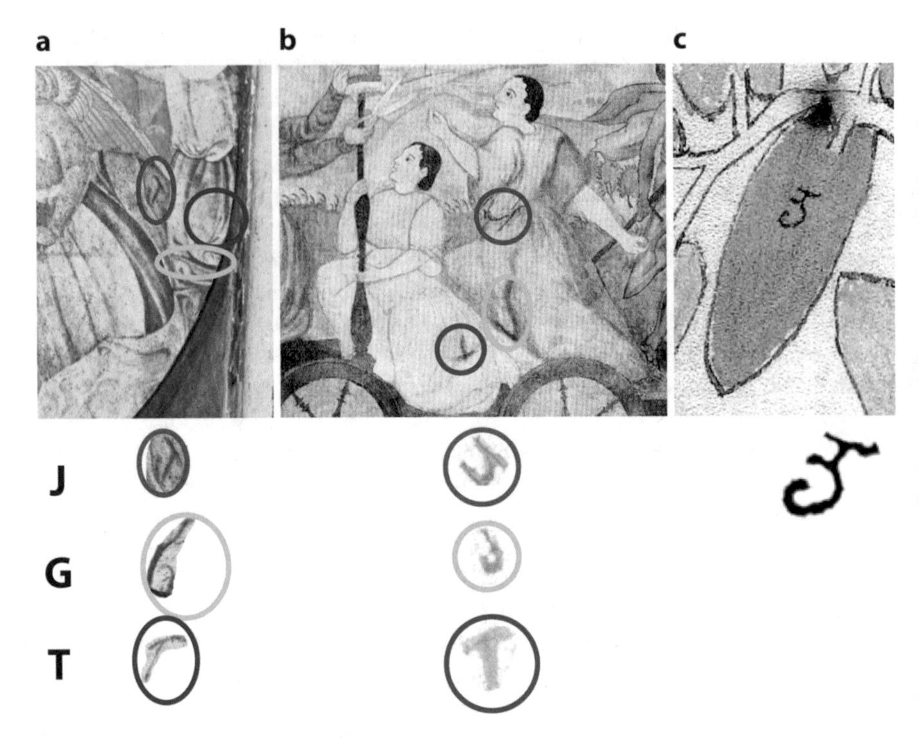

Fig. 15.3 Putative initials of Juan Gerson, *Tlacuilo* ("JGT"), as shown in (**a**) the painting *The Arresting of the Winds and the Elected*, of the Apocalypse ceiling at Tecamachalco; (**b**) Salon of the Triumphs in Casa del Deán; and (**c**) a leaf of *Ipomoea arborescens* on folio 1v of the *Voynich Codex*

Fig. 15.4 Water and clouds in (**a**) the *Voynich Codex* folio 86v, circle 2, and (**b**) Gerson's Apocalypse

Astronomical Images The Cosmological section of the *Voynich Codex* is filled with images of stars (Fig. 15.5a), suns (Fig. 15.5b), moons (Fig 15.5c), and planets (Fig. 15.5d), which are also found in Gerson's ceiling images. The stars are often seven-pointed and a cluster of seven stars known as the Pleiades were found in both the *Voynich Codex* and the Apocalypse images (Fig. 15.5a). Note that both the suns (Fig. 15.5b) and the moons (Fig. 15.5c) have faces and the suns have prominent rays.

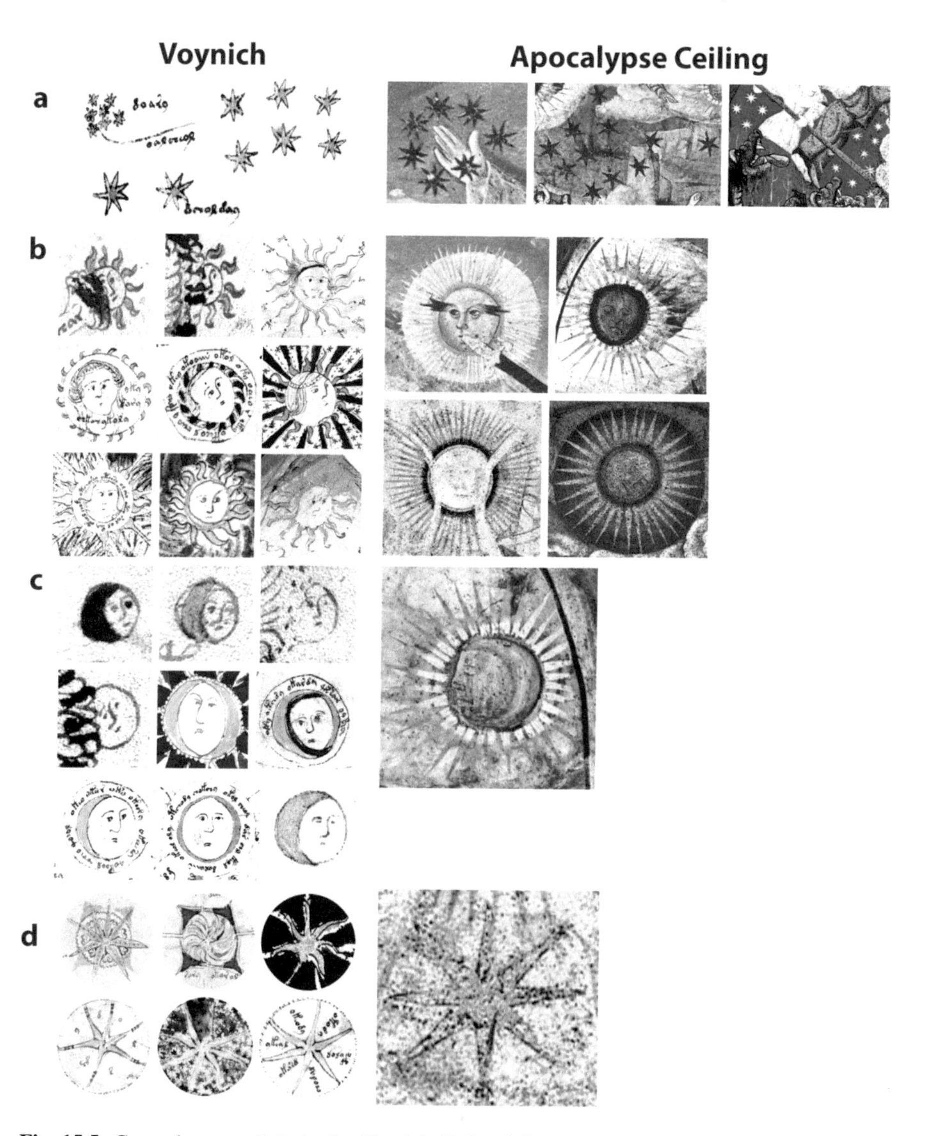

Fig. 15.5 Cosmology symbols in the *Voynich Codex* (left) and Gerson's Apocalypse (right). (**a**) Stars, (**b**) suns, (**c**) moons, and (**d**) planets

We have surmised that a "starfish" symbol (Fig. 15.5d) in the *Voynich Codex* represented planets and this same image was shown in *The Plague of the Locusts with Eight Arms.* The starfish planet symbol of Gerson's painting was derived from a woodcut by the artist named Master IF and could have been the source of the symbol used to represent planets in the Voynich. The close affinity of astronomical images adds plausibility to the assumption that the Voynich images and the Apocalypse images were by the same hand.

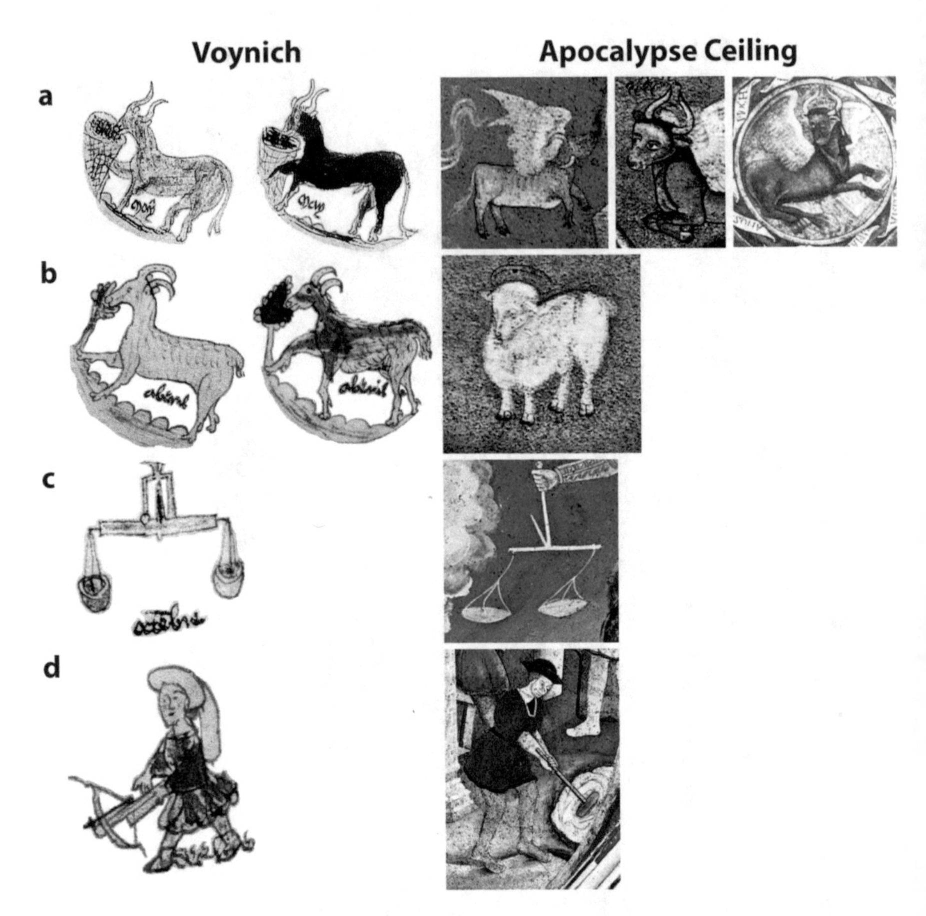

Fig. 15.6 Zodiac signs in the *Voynich Codex* (left) and Gerson's Apocalypse (right). (**a**) Cattle, (**b**) sheep, (**c**) scales, and (**d**) the costumes of the archers with crossbows

Zodiac Images The zodiac images in the *Voynich Codex* (Fig. 15.6) contain signs that are found in Gerson's Apocalypse paintings, including a cattle, cow, and bull representing Taurus (Fig. 15.6a); a sheep, ewe, and ram representing Aries (Fig. 15.6b); scales representing Libra (Fig. 15.6c); and a medieval-dressed figure with a crossbow representing Sagittarius (Fig. 15.6d). There are various bulls in Gerson's paintings, all winged, but the horns in one resemble those in the *Voynich Codex*. The one sheep image in Gerson's painting entitled *Vision of the Lamb on Mt. Zion* is polled (naturally hornless) with a wooly coat and does not resemble the horned sheep in the *Voynich Codex*. The scales in the *Voynich Codex* representing Libra were also included in the scene *The Four Riders of the Apocalypse*. The archer with crossbow in the *Voynich Codex* that represents the constellation Sagittarius was dressed in a very similar manner to a figure in the painting *Erection of the Tower of Babel*. The costume of this figure has been used as evidence for a fifteenth century date for the *Voynich Codex* by Ellie Velinska (2013). The figure in the original woodcut was hatless, but both figures in Voynich and Gerson wear similar hats.

Fig. 15.7 Birds in the *Voynich Codex* (left) and Gerson's Apocalypse (right)

Birds There are two images of caracaras in the *Voynich Codex* and there are a number of birds that appear to be the same bird in the Apocalypse paintings (Fig. 15.7). The third bird image of Gerson is very close to the flying bird image of Voynich.

Structures There are various similar architectural structures in the *Voynich Codex* and Gerson's Apocalypse paintings. Bathing structures in the *Voynich Codex* are similar to a windowed structure in one of the Apocalypse paintings (Fig. 15.8a). There are a number of towers in both (Fig. 15.8b). There appears to be a relationship between Voynich circle A of folio 86v (Chap. 10), which shows a mantle of stars above six qubba (domes), and the *Vision of Jesus Christ and the Seven Candlesticks,* which shows a mystical figure surrounded by large candlesticks (only six shown) pointing to a cluster of seven stars that resembles the Pleiades (Fig. 15.8c). Gerson's the *City of God* (Jerusalem) painting is based on an Albrecht Dürer woodcut (Fig. 15.8d). The painting shows a qubba (dome) in the middle, presumably the Dome of the Rock in Jerusalem, Israel (Fig. 15.8e). The walled city of Jerusalem painting with various domes (or qubba) visible shows a resemblance to circle A of folio 86v, which we assume shows six domes in the new Celestial City of Jerusalem (also called Puebla de los Angeles), constructed in Puebla in 1530 by Motolinía (Chap. 10).

Iconographic Comparison of Casa del Deán Murals and the Voynich Codex

Juan Gerson also appears to be one of the artist painters of mural cycles in the residence, now known as the Casa del Deán, of Don Tomás de la Plaza, the dean of the cathedral in Puebla from 1583 to 1589 (Morrill 2014). The publication of

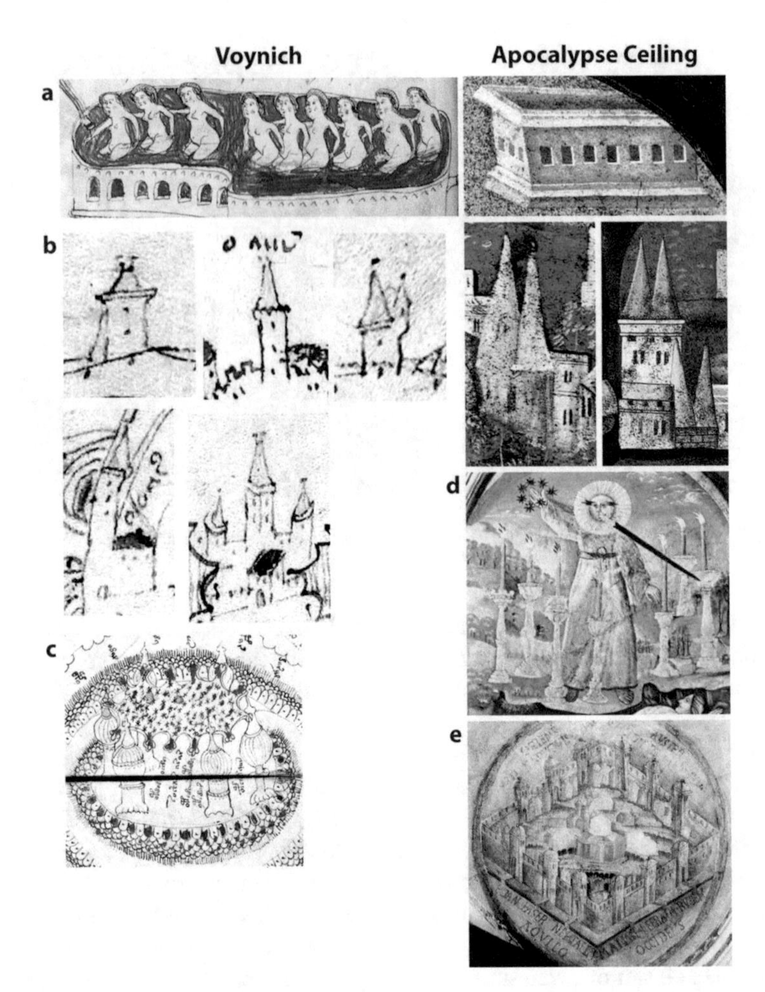

Fig. 15.8 Structures in the *Voynich Codex* (left) and Gerson's Apocalypse (right). (**a**) Baths and (**b**) towers. (**c**) Six qubba (domes) or ciboria in the *Voynich Codex*. (**d**) Seven candlesticks in Gerson's Apocalypse painting. (**e**) A map of Jerusalem

The Casa del Deán: *New World Imagery in a Sixteenth Century Mexican Mural Cycle* (by Morrill 2014) has made available high-resolution images of this unique work by indigenous painters (*tlacuiloque*). The murals painted about 1584 in the residence of Tomás de la Plaza, Cathedral dean at Puebla, include two processions located in rooms known as the Salon of the Sibyls and the Salon of the Triumphs. It was likely that these two processions were painted by different *tlacuiloque*, as it was known that a number of artists were involved in the Casa del Deán murals (Morrill 2014). We have conjectured that the artist of the procession of triumphs could have been Juan Gerson, based on similarities with some of his paintings in the Franciscan church in Tecamachalco (Fig. 15.9).

Art historians Gutiérrez Haces (2002: 77) and Küegelgen (2013: 178) report similar artistic elements between the Casa del Deán murals and the Apocalypse

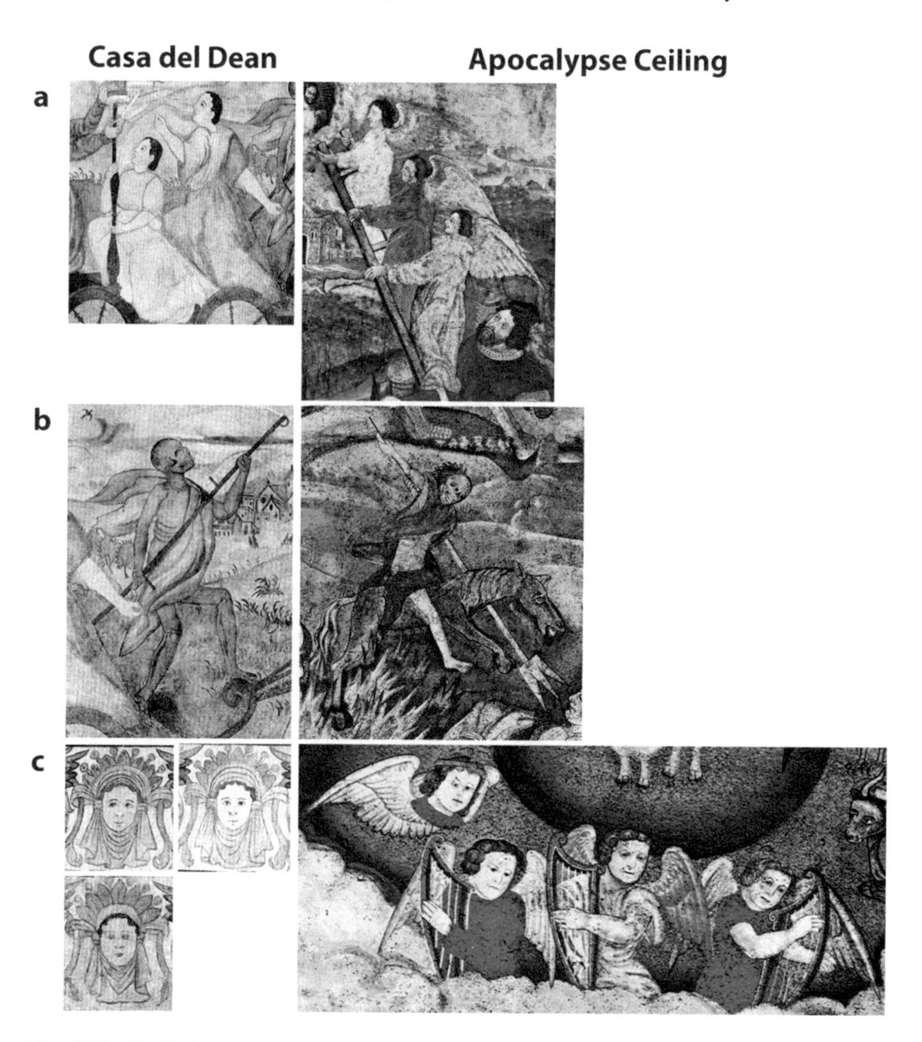

Fig. 15.9 Similarity of Casa del Deán murals (left) with images of Juan Gerson's Apocalypse images in the Franciscan church in Tecamachalco, in the Mexican state of Puebla (right). (**a**) Close-up of women in *Triumph of Death* in Casa del Deán (left) and angels in Gerson's *Jacob's Ladder* (right); (**b**) grim reaper in Casa del Deán (left) and Gerson's figure of Death in the *Four Horsemen of the Apocalypse* (right); (**c**) heads in the frieze of Casa del Deán (left) and angels from Gerson's *Vision of the Lamb on Mount Zion* (right)

paintings of Juan Gerson. The murals dating to 1584 were uncovered from whitewash and wallpaper in the mid-1930s and 1953, and were restored in 2010. This attribution was affirmed by the presence of the initials "JGT" that appear to be embedded in the folds of the gown of the *Triumph of Death*. Although these were faint because of considerable restoration in the twentieth century, the initials were similar to those found in Gerson's Apocalypse paintings.

Iconographic similarities between the Apocalypse paintings and the Casa del Deán murals (Fig. 15.9) support the assumption that Juan Gerson was the artist

Fig. 15.10 Nude figures (**a**) in the *Voynich Codex*, and (**b**, **c**) Casa del Deán murals

involved in both works, in addition to the *Voynich Codex*. For example, nude figures are found in both the *Voynich Codex* (Fig. 15.10a) and the Casa del Deán murals (Fig. 15.10). In the Casa del Deán murals, nudity is present in the friezes: mostly putti, some centaurs (Fig. 15.10b), and there are three nude couples bathing (Fig. 15.10c). There are many nude nymphs in the *Voynich Codex* (Fig. 15.10a): 202 in the Balneological section and 239 in the zodiac folios, all in a frontal view.

Women Images of women from the procession of sibyls in the Casa del Deán murals are compared with bathing nymph images from the *Voynich Codex* (Fig. 15.11). All the sibyls are carefully drawn, European women on horseback, smartly dressed in sixteenth century costumes. However, in the procession of triumphs, there is one woman (the personification of Eternity) with Mexican features (Fig. 15.12). All the women in the procession of sibyls are young and attractive with short dark hair and various head coverings or ornaments. In contrast, the quickly drawn sketches of nymphs (about 500) in the *Voynich Codex*, although clearly European in appearance, appear short and wide-hipped with both short and long blonde hair. Most are nude, but 30 are gowned in two zodiac circles representing Taurus and Aries. Many nymphs in the *Voynich Codex* have head coverings or hair ornaments, but none is identical to those of the women in the procession of sibyls (Fig. 15.13). The quickly drawn images of nymphs in the *Voynich Codex* are stylistically different from the women

Fig. 15.11 Comparison of women in the Casa del Deán murals and the *Voynich Codex*. (**a**) Clothed women riding sidesaddle in the Salon of the Sibyls (note that the first sibyl, Synagoga, blind-folded displaying a broken standard and riding a mule, is the personification of the Old Testament), and (**b**) naked nymphs cavorting in pools in the *Voynich Codex*

of the Apocalypse paintings and the Casa del Deán murals; thus, it cannot be ascertained whether the same hand was involved.

Animals There are many animals portrayed in the Casa del Deán murals and the *Voynich Codex*. The animals in the Casa del Deán murals include various birds (such as swans), plus horses, oxen, unicorns, rabbits, deer, raccoons, coatimundis,

Fig. 15.12 Personification of Eternity as an indigenous Amerindian woman in the Salon of the Triumphs, south wall

jaguars, and monkeys. Animals in the *Voynich Codex* are all either indigenous to Mexico or Spanish imports. Many are an integral part of the zodiac series as substitutions for traditional animals (e.g., alligator gar, perhaps desert bighorn ewe and ram, Andalusian Red cow, Retinta bull, Mexican dwarf crayfish, ocelot, and jaguarundi/jaguar), whereas others appear to be rough sketches or doodles (e.g., alligator gar, Texas horned and earless lizards, *Demorphis* amphibian, frog, iguana, paca, coatimundi, jellyfish, armadillo, and crested caracara), as illustrated in Chap. 6. The animals in common include coatimundi, and possibly jaguar and caracara (Chap. 16, Fig. 16.2).

Plants Plants in the Casa del Deán friezes, including both European and Mexican indigenous species (Tucker and Janick 2017). The European plants appear to be copied from European sources. Some of the Mexican indigenous plants (Fig. 15.14) are also found in the *Florentine Codex* and the *Codex Cruz-Badianus*, but not in the *Voynich Codex*. The plants of the Casa del Deán murals and the *Voynich Codex* are drawn in different styles.

Gaspar de Torres was Doctor of Medicine, Licentiate, Master of Students at Colegio de Santa Cruz, Governor of Cuba

Biography Gaspar de Torres was born in Capitania General de Santo Domingo in the early decades of the sixteenth century. His father, Melchor (also spelled Melchior) de Torres, owned sugar plantations in Hispaniola, and 900–1000 slaves. Melchor de Torres' half-brother, also named Gaspar de Torres, was probably born in Malaga, Spain, and became the leading slaver in Hispaniola from 1535 to 1540. The family included influential and wealthy relatives: Cousin Luis de Villanueva Zapata, *abrogado* (judge) and wealthy hacienda owner in Mexico, and cousin Diego de Villanueva Zapata, *fiscal audiencia* (crown attorney) of Panama (Helps 1857; Chevalier 1982; Cortés López 1995; Kellenbenz and Walter 2001; Schwartz 2004).

Fig. 15.13 (**a**) Heads of nymphs in the *Voynich Codex*. (**b**) Sibyls and (**c**) frieze faces of the Casa del Deán murals

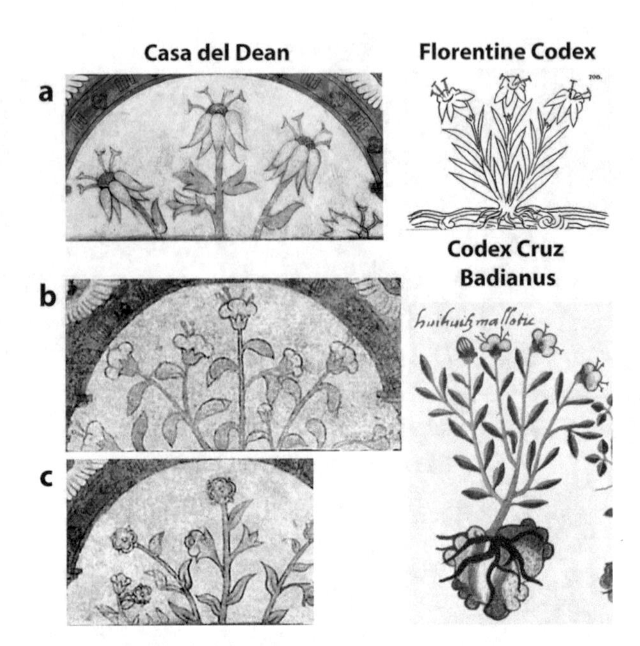

Fig. 15.14 Two indigenous plants of the Casa del Deán murals. (**a**) *Yopixochitl/iopisuchitl*, found in Fig. 708, Book 11 of the *Florentine Codex* (right). (Source: Sahagún 1963). (**b**, **c**) *Acuilloxochitl/acuillosuchitl/acujllosuchitl/huihuitz mallotic* (full of needles) = *Mentzelia hispida*, found in plate 59 (left) of the *Codex Cruz-Badianus* (right). (Source: Emmart 1940)

The younger Gaspar de Torres was well educated and attended La Real y Pontificia Universidad de México (Royal and Pontifical University of Mexico), the first university of Mexico established on 21 September 1551, for the education of the Spanish (Carreño 1961). Gaspar de Torres was awarded the doctorate degree in medicine on 31 August 1553. Later, in a career change, he was awarded the degree of licentiate in 1569. The licentiate degree was influential, as explained by Herrera (2003): "*On a much higher plane stood the small number of licentiates working the Audiencia* [court established to hear royal justice]. *They mainly adjudicated disputes over land and inheritance. The majority of those who held licentiates in law arrived as officials attached to the Audiencia. As a result of their university training and association with royal institutions, licentiates had a ready-made high-status niche that outstripped almost any other position that notaries and other similar professions could expect to attain.*" He was also employed at La Real y Pontificia Universidad from 14 December 1569 to 26 August 1571, notably in the company of his older friend, Damián Sedeño (Vazquez 2002). Gaspar de Torres then became a staff member as *maestro de niños* (master of students) of the Colegio de Santa Cruz, based on records from 1568 to 1572. This institution was originally created to educate the sons of the Aztec nobility for possible ordination as priests. The Colegio had been formally established on 6 January 1536, but had, in fact, existed since 8 August 1533. On 21 January 1568, Torres was paid to teach students to read and write, and there are records of other payments in 1572 (Anon 1892; Canedo 1982; Kobayashi 2007; SilverMoon 2007). Most importantly, the Colegio had the only scriptorium in New Spain (Gravier 2011), which would have had vital supplies of inks, pigments, paper, parchment, and vellum not ordinarily available elsewhere. Juan Gerson may have been associated with the Colegio during this time, although no period citation has yet been uncovered.

Gaspar de Torres also signed two documents as an interpreter and scribe for land rights of natives in Tulancingo, Hidalgo, in 1569 (Carrasco 1963). This would indicate that he was fluent in Nahuatl, the language of the Aztecs, and familiar with other languages in the region.

There is evidence that Gaspar de Torres moved to Santo Domingo and records place him there on 9 July 1572, and in 1578 and 1579 (Autos entre partes Santo Domingo 1575–1580; Caudevilla y Escudero et al. 1788; Cipriano de Utera 2014; Poder Judicial de la República Dominicana 2016). In 1579, Gaspar de Torres was appointed by the *Audiencia* to Cuba as governor *pro tempore* on half salary. He arrived on 3 October 1579, in Cuba and assumed office in January 1580. Yet, he fled 8 months later before the arrival of his successor, D. Gabriel de Lujan, as a result of accusations by the royal treasurer, Juan Bautista de Rojas, that he had looted the treasury (Blanchet 1866; Wright 1916; Johnson 1920).

Gaspar de Torres was reported to be in Guatemala in 1584, and eventually worked in San Isidoro (now part of Panama City, Panama) from 1595 to 1598 as an estate lawyer. He was probably aided by his influential cousin of the *Audiencia* in Panama, Diego de Villanueva Zapata (Herrera 2003; Viforcos Marinas 2005).

Association with the Voynich Codex

The brief biography of Gaspar de Torres above has many gaps, but a number of facts can be gleaned from it that relate to the conjecture that he was associated with the *Voynich Codex*.

1. The key evidence is that his name (Gasp.Torres) along with the ligated initials of Juan Gerson, Tlacuilo, appear on the first botanical image on folio 1v. This suggests a collaboration between him and the painter Juan Gerson.
2. As a master of students at the Colegio de Santa Cruz and scribe and interpreter for natives of Tulancingo, Hidalgo, we can assume that Gaspar de Torres was sympathetic to the Aztec culture and familiar with the Aztec intellectual community at the Colegio, which included Antonio Valeriano, Martin Jacobita, Martin de la Cruz, Juan Badiano, Pedro de San Buenaventura, and Juan Bautista

de Pomar, in addition to Fray Bernardino de Sahagún (Klor de Alva et al. 1988), at one time dean of the Colegio and pioneering ethnographer of the Aztecs.

3. His biography suggests that Torres was fluent in many languages, including sixteenth century Latin, American Spanish, Nahuatl, Taino, and perhaps other languages of central Mexico. This extraordinary linguistic ability, along with his doctorate in medicine, could explain the Spanish- and Nahuatl-derived names of the plants deciphered in the *Voynich Codex* (Tucker and Talbert 2013; Tucker and Janick 2016) (see Chap. 4). Taino was the language of the indigenous Indians in Hispaniola, and some of the deciphered plant names in the *Voynich Codex* are phonetic derivatives from that language, e.g., *māguoey* or *maguey* (*Agave* sp. on folio 100r#4) and *macanol*, after the macana-like leaf (*Philodendron* sp. on folio 100r#7).
4. Another clue to the approximate date of the *Voynich Codex* is indicated by the *Codex Osuna*. The *Codex Osuna* is a set of seven documents created to present evidence against the government of Viceroy Luis de Velasco during the inquiry of Jerónimo de Valderrama in 1563–1566. It was an administrative document, not comparable with other codices from the era, and because it was created on order of Phillip II, it would have been sent to Seville, Spain, shortly after completion. Not only are many symbols of the *Voynich Codex* similar to the handwriting in the *Codex Osuna*, but the *Codex Osuna* included claims of nonpayment for various goods and services, including building construction and domestic help, that is, the estate law of interest to Gaspar de Torres. Most significantly, Torres' close friend, Damián Sedeño, gave testimony in the *Codex Osuna* in 1565 (Chávez Orozco 1947).
5. His influential family and rise to governor of Cuba would indicate political skills and a wide circle of wealthy and influential acquaintances.
6. His abrupt leave from Mexico to Santo Domingo in 1572 suggests that Torres might have had something to hide. It is clear that the text of the *Voynich Codex* was written after the illustrations were drawn; thus, it is likely that the author, not the illustrator, would have been in possession of the manuscript. One reason for his departure to Santo Domingo may have been fear of the Inquisition. "*of book of spells or other superstition or any other type of prohibited book, they should be denounced. And we order that anyone who should possess these books or have them read*". Although the Spanish Inquisition had been spottily applied in New Spain since 1536, the Spanish Inquisition was formally and vigorously introduced into New Spain in 1571. In 1571 and 1572, a new inquisitor, Dr. Pedro Moya de Contreras, began a widespread purge of prohibited books. An Edict of Faith issued by Dr. Contreras in Mexico City on 10 October 1576 stated: "*Also, if you should know about any person who uses spells, incantations, charms, or conjures up spirits, or commits any other type of superstitious enchantments, or uses any other type of witchcraft, even if they are medicinal curers, or if anyone should have a copy of any type them should denounce themselves within the space of fifteen days before Us or Our other Inquisitors or their commissaries so that these writings can be examined and reviewed*" (Chuchiak 2012). The contents of the *Voynich Codex* would have been particularly offensive

to the inquisitor of the Spanish Inquisition, in view of the presence of the kabbalah imagery and the fact that the text was written in a symbolic cipher. Not only would it have been burned if found, but the owner would have been punished appropriately. Further examples are chronicled by Francisco Lossa (1642), Carmichael (1959), Greenleaf (1961), and Starr (1987). Santo Domingo offered sanctuary, beyond the pale of the Spanish Inquisition, because its economy had severely declined since the conquest of Mexico in 1521, later serving only as a way station for slavers and dealers in contraband (Vizuete Picon 1890).

The biography of Gaspar de Torres suggests possible paths of the *Voynich Codex* from the New World to Rudolf II, emperor of the Holy Roman Empire. We conjecture that Gaspar de Torres, as the author of the manuscript now known as the *Voynich Codex*, was in possession of it when he left Tlatelolco to go to his home in Santo Domingo. He either left it in Santo Domingo when he became governor of Cuba in 1580 or brought it back when he fled Cuba the same year. If so, the manuscript could have been part of the booty captured by the expedition of Sir Francis Drake during his military campaign in the Caribbean, from 1585 to 1586 (Sugden 2006; Dean 2010; Kelsey 1998). From 1–31 January 1586, Drake's forces destroyed more than a third of Santo Domingo, then pillaged Cartagena, raided Saint Augustine, and rescued the English colonists at Roanoke, returning to Plymouth, England on 26 July 1586. The soldiers who plundered Santo Domingo were under the command of Christopher Carleill on the ship, Tiger. Carleill was the stepson of Sir Francis Walsingham, spymaster of Queen Elizabeth I, and friends with John Dee. John Dee (1527–1608) occult philosopher, mathematician, and kabbalist, was a secret agent (the original Agent 007) to Walsingham.

Dee has long been considered either an author or conveyer of the *Voynich Codex.* If part of the booty of Drake's fleet, the codex would have belonged to the government (Alford 2012), but may somehow have reached Dee, who was in Central Europe from 1583 to 1589, including Poland, Prague, and Třeboň. John Dee had a private audience with Emperor Rudolf II in 1584, and Roberts and Watson (1990: 172) state that Dee in frequent contact with the emperor from 1584 to 1586. Dee was also in correspondence with Walsingham between 1586 and 1587, and Francis Garland acted as a courier between Dee and Walsingham (Parry 2011). A 1665 letter by Joannes Marcus Marci to Athanasius Kircher contains hearsay evidence that the book (*Voynich Codex*) belonged to Emperor Rudolf II who gave 600 ducats to the bearer and thought it was the work of Roger Bacon (Kennedy and Churchill 2008: 20–21). In Dee's spiritual diary of 17 October 1586, there is evidence that he had 630 ducats on that day. At the time, Dee was desperate for funds to pay off Francesco Pucci, who would have dragged him off to Rome under orders of the Pope. The rumor of Rudolf II paying 600 ducats for the *Voynich Manuscript* and Dee having 630 ducats in 1586 seems more than a coincidence, suggesting that Dee was "the bearer of the book," but there are many skeptics, including Zandbergen (2016: 4). John Dee's occultist friend, Edward Kelley, was also retained as alchemist by Rudolf II after 1589. Finally, the book could have been purchased in London by agents of Rudolf II dispersed throughout Europe to collect for his *Kunstkammer*, or cabinet of wonders (Fučiková et al. 1997).

Conclusion

Based on plant, animal, and mineral identifications, we are convinced that the *Voynich Codex* is a sixteenth century New World document and thus dismiss all European claimants as authors. The 1425 dating of the vellum by mass spectrometry (Stolte 2011) does not date the actual manuscript, as vellums were reused (i.e., a palimpsest). We see in the *Voynich Codex* the fusion of the submerged, dying, indigenous culture of the Aztecs and the medieval culture of Spain influenced by conquistadors, the Catholic Church, Jews, and Arabs.

Iconographic analyses of the ligated "JGT" initials on folio 1v of the *Voynich Codex*, the Apocalypse paintings in Tecamachalco (1562), and the Casa del Deán murals in Pueblo (1584), suggest that Juan Gerson, *Tlacuilo,* might be associated with all three works. The artistic similarities among the paintings at Tecamachalco of 1562, some of the murals of Casa del Deán of 1584, and the *Voynich Codex* underscore the extraordinary accomplishments of the artist Juan Gerson, an *indio ladino* (indigenous Nahua), a native Amerindian of two worlds – the remnants of the Aztec empire and the emerging colonial power of New Spain.

The presence of the name Gasp.Torres on folio 1v indicates that Gaspar de Torres was also associated with this manuscript, probably as the author or co-author of the text. Gaspar de Torres – well educated, multilingual, and an advocate of Indian land rights – was master of students between 1568 and 1572 of the Colegio de Santa Cruz in Tlatelolco, Mexico, established in 1536 for the education of sons of Aztec nobility. We conjecture that Juan Gerson, as an indigenous, well-connected, talented Nahua, might have been associated with the Colegio in some manner and could have used its Scriptorium as a source of vellum and pigments. This would explain his collaboration with Gaspar de Torres. The flight of Gaspar de Torres to Santo Domingo in 1572 suggests fear of severe punishment by the Inquisition for possession and authorship of what would have been considered a heretical manuscript.

Based on the presence of the ligated initials of Juan Gerson and the name Gasp. Torres on the first herbal folio, in addition to the iconographic and the biographical evidence, we propose Juan Gerson and Gaspar de Torres as plausible claimants for illustrator and author of the *Voynich Codex*. Our assumption is that the Voynich was composed in Tlatelolco between 1565 and 1572, probably during Torres' tenure at Colegio de Santa Cruz, a few years after the Apocalypse murals by Juan Gerson were completed, and soon after the *Codex Osuna* was created, from which some of the symbols of the text are derived.

Strikingly, when the careers of these two men are amalgamated, the composite fits in many ways with the personality profile created for the author/artist before this new information was discovered, as discussed in Chap. 14. It was predicted that the author and artist were sons of the Aztec nobility and trained at a school established by the Spanish to educate sons of the Aztec nobility. We were half right. Juan Gerson, an indigenous Nahua, fitted this description and attended La Escuela de San José de los Naturales. Gaspar de Torres, born in Santo Domingo from Spanish parents (a *criollo*), was master of students at the Colegio de Santa Cruz, and would have been in contact with Bernardino de Sahagún, who was involved with extraordinary pioneering studies on Aztec ethnology and culture. We assumed that the author and artist were polymaths and it is clear that Torres was highly educated, with degrees in medicine and law. Both Gerson and Torres must have been multilingual, with abilities in Nahuatl, Spanish, Latin, and perhaps Taino for Torres, who was born in Hispaniola. Gerson, the son of a prominent Nahua, was esteemed as a painter of the Apocalypse, based on European etchings and given unique recognition by the Spanish authorities. Both were trained in Spanish schools and brought up as Christians. Gerson was baptized and painted religious paintings at Tecamachalco, but would also have been familiar with Aztec culture. Torres was a supporter of the rights of Amerindians and worked on their behalf, as indicated by his signatures in lawsuits to reclaim Indian land. Torres was likely to have been aware of the herbals being created by Martin de la Cruz and Juan Badiano at the Colegio de Santa Cruz and, based on his medical training, would have been familiar with medicinal plants. Both men trained in Franciscan institutions with extensive libraries and an environment of liberalism. They could be expected to be aware of Jewish influences, such as kabbalah and Moorish architecture that were of interest to Franciscans and recreated in pageants. It is intriguing that at one time, Gerson was suspected of being Jewish (Pedro Rojas 1963, cited in Azpeitia 1972). Furthermore, the surname Torres was associated with Jews, including the *converso* (convert) Luis de Torres, the translator of Columbus, and many with that name were condemned by the Inquisition as crypto-Jews. We can only speculate on the unique and close collaboration between these two talented men.

Literature Cited

Alford, S. 2012. *The watchers: A secret history of the reign of Elizabeth I*. New York: Bloomsbury Press.

Anonymous. 1892. *Codice Mendieta: Documentos Franciscanos, Siglos XVI y XVII*. Vol. 2. México: Imprenta de Francisco Diaz de Leon.

Autos entre partes Santo Domingo. 1575–1580. Retrieved 2016 July 2 from http://censoarchivos.mcu.es/CensoGuia/fondoDetail.htm?id=930360.

Azpeitia, R.C. 1972. *Juan Gerson, Pintor indigena del siglo XVI – simbolo del mestizage Tecamachalco Puebla*. Mexico: Fondo Editorial de la Plastica Mexicana.

Blanchet, E. 1866. *Compedia de la historia de Cuba*. Matanzas: La del Yumuri, de Jose Curbelo y Herm.

Du Bron, M. 2010. *Le cheval Mexicain en Nouvelle-Espagne entre 1519 et 1639*. Ph.D. thesis, École des Hautes Études en Sciences Sociales, Paris.

Brumbaugh, R.S. 1978. *The most mysterious manuscript*. Carbondale: Southern Illinois Press.

Buonafalce, A. 2007. Review of the curse of the Voynich. The secret history of the World's most mysterious manuscript by Nicholas Pelling. *Cryptologia* 31 (4): 361–362.

Canedo, L.G. 1982. *La educación de los marginados durante la Época Colonial: Escuelas y colegios para Indios y Mestizos en la Nueva España*. México: Editorial Porrúa.

Carmelo Arredondo, R., J. Gurría Lacroix, and C. Reyes Valerio. 1964. *Juan Gerson, Tlacuilo de Tecamachalco*. México: Departamento de Monomuntos Coloniales, Instituto Nacional de Anthropolgia e Historia.

Carmichael, J.H. 1959. Balsalobre on idolatry in Oaxaca. *Boletín de Estudios Oaxaqueños* 13: 1–9.

Carrasco, P. 1963. Dos cacques chichimecas de Tulancingo. *Estudios de Cultura Náhuatl* 4: 85–91.

Carreño, A.M. 1961. *La Real y Pontificia Universide de Mexico 1536–1865*. México: Universidad Nacional Autónoma de México.

Carthago, R. 2006, 2015. Das Voynich Manuscript – Übersetzubg der ersten 13 Seiten. GMA mbH, Germany.

Caudevilla y Escudero, J., B. Mendel, Consejo de Indias (Spain). Consejo de Indias (Spain). 1788. Memorial ajustado, hecho de orden del real y supremo Consejo de Indias, con citatcion y asistencia de las partes, en el pleyto, que en grado de revista se sigue en el por el Senor Don Mariano Colón de Larreategui num. 64, del Consejo de Castilla, Superintendente General de Policía, Madrid.

Chávez Orozco, L. 1947. *Codice Osuna*. Mexico: Ediciones del Instituto Indigenista Interamericano.

Chevalier, F. 1982. *Land and society in Colonial Mexico: The great haciendas*. Trans. A. Eustis. Berkeley: University of California Press.

Chuchiak, J.F. 2012. *The inquisition in New Spain, 1536–1820: A documentary history*. Baltimore: Johns Hopkins Press.

Cipriano de Utera, Fray. 2014. Historia militar de Santo Domingo (Documentos y noticias). Tomo II. D. N. República Cominicana, Santo Domingo.

Comegys, James C. 2001. Keys for the Voynich scholar. Necessary clues for the decipherment and reading of the world's most mysterious manuscript which is a medical text in Nahuatl attributable to Francisco Hernández and his Aztec Ticiti collaborators.

Cortés López, J.L. 1995. 1544–1550: el período más prolífico en la exportación de esclavos durante el s. XVI. Análisis de un interesante document extrído del Archivo de Simancas. Espacio, Tiempo y Forma Ser. IV. 8: 63–86.

Dean, J.S. 2010. *Tropics bound: Elizabeth's seadogs on the Spanish main*. Stroud: History Press.

Dos Santos, M. 2005. *El manuscrito Voynich: El libro más enigmático De todos los tiempos*. Madrid: Aguilar.

Dragoni. 2012. *Il manoscritto Voynich: Un po' di chiarezza sul libro più misterioso del mondo*. Unpublished pdf. http://www.cropfiles.it/altrimisteri/Voynich.pdf. 1 Sept 2014.

Emmart, E.W. 1940. *The Badianus manuscript (Codex Barberini, Latin 141) Vatican Library. An Aztec herbal of 1552*. Baltimore: Johns Hopkins Press.

Fučiková, E., J.M. Bradburne, B. Bukovinská, J. Hausenblasová, J.L. Konečný, I. Muchka, and M. Šronék, eds. 1997. *Rudolf II and Prague: The imperial court and residential city as the cultural and spiritual heart of Central Europe*. Prague: Prague Castle Administration.

Gravier, M.G. 2011. Sahagún's codex and book design in the indigenous context. In *Colors between two worlds*, ed. G. Wolf and J. Connors, 156–119. Milan: The Florentine Codex of Bernardino de Sahagún. Officina Libraria.

Greenleaf, R.E. 1961. *Zumárraga and the Mexican inquisition 1536–1543*. Washington, DC: Academy of American Franciscan History.

Gutiérrez, Haces J. 2002. ¿La pintura novohispanica como una *koiné* pictórica Americana? *Anales del Instituto de Investigacioines Estéticas* 24 (80): 47–99.

Helps, A. 1857. *The Spanish conquest in America, and its relation to the history of slavery and to the government of colonies*. New York: Harper & Brothers.

Herrera, R. 2003. *Natives, Europeans, and Africans in sixteenth-century Santiago de Guatemala*. Austin: University of Texas Press.

Janick, J., and A.O. Tucker. 2017. Were Juan Gerson the illustrator and Gaspar de Torres the author of the Voynich codex? *Notulae Botanicae Horti Agrobotanici Cluj-Napoca* 45 (2): 343–352. https://doi.org/10.15835/nbha45210693.

Johnson, W.F. 1920. *The history of Cuba*. Vol. 1. New York: B.F. Buck & Co.

Kellenben, H.W.R., and S. Walter. 2001. Oberdeutsche Kaufleute in Sevilla und Cadiz (1525–1560): Eine Edition von Notariatsakten aus den dortigen Archiven. Deutsche Handelsakten des Mittelalters und der Neuzeit. Franz Steiner Verlag vol. 21.

Kelsey, H. 1998. *Sir Francis Drake, the Queen's pirate*. New Haven, CT: Yale University Press.
Kennedy, G., and R. Churchill. 2008. *The Voynich manuscript*. Rochester: Inner Traditions.
Klor de Alva, J.J., H.B. Nicholson, and E.Q. Keber, eds. 1988. *The work of Bernardino de Sahagún: Pioneer ethnographer of sixteenth-century Aztec Mexico*. Institute for Mesoamerican Studies, University at Albany, NY.
Kobayashi, J.M. 2007. *La educación como conquista (empresa franciscana en México)*. México: El Colegio de México.
Küegelgen, H., ed. 2013. *Profecía y triunfo: la Casa del Deán Tomás de la Plaza: facetas plurivalentes*. Madrid: Iberoamericana.
Landa Abrego, M.E. 1992. Juan Gerson, tlacuilo. Gobierno del Estado de Puebla, Comisión Puebla V Centenario, Mexico.
Lossa, F. 1642. *Vida que el siervo de Dios Gregorio Lopez hizo en algunos lugares de la Nueva España*. Madrid: Imprenta Real.
Manly, J.J. 1931. Roger Bacon and the Voynich MS. *Speculum* 6: 345–391. https://doi.org/10.2307/2848508.
Morrill, P.C. 2014. *The casa del Deán: New World imagery in a sixteenth century Mexicmural cycle*. Austin: University of Texas Press.
Niedermeier, M. 2002. Finalidad y función de modelos gráficos europeos. El ejemplo del ciclo de Juan Gerson en el convento de Tecamachalco, Puebla. In *Herencias indígenas, tradiciones europeas y la mirada europea. Indigenes Erbe, europäische Traditionen und der europäische Blick. Actas del Coloquio de la Asociación Carl Justi y del Instituto Cervantes de Bremen, Bremen, del 6 al 9 de abril de 2000*, ed. H. von Kügelgen, 95–121. Bremen: Akten des Kolloquiums der Carl Justi-Vereinigung und des Instituto Cervantes Bremen/Iberoamericana, Vervuert 6–9. April 2000.
Parry, G. 2011. *The arch-conjuror of England*. John Dee: Yale University Press.
Pelling, N. 2006. *The curse of the Voynich: The secret history of the world's most mysterious manuscript*. Surbiton: Compelling Press.
Poder Judicial de la República Dominicana. 2016. *Historia del Poder Judicial*. http://www.poderjudicial.gob.do/poder_judicial/info_gral/historia_poder_judicial.aspx. Retrieved 10 Oct 2016.
Prinke. R.T. n.d. *Did John Dee really sell the Voynich MS to Rudolf II?*. http://main2.amu.edu.pl/~rafalp/WWW/HERM/VMS/dee.htm. Accessed 30 June 2017.
Resendez, A. 2016. *The other slavery: The uncovered story of Indian enslavement in America*. Boston: Houghton Mifflin Harcourt.
Roberts, R.J., and A.J. Watson, eds. 1990. *John Dee's library catalogue*. London: The Biographical Society.
Rugg, G. 2004a. The mystery of the Voynich manuscript: New analysis of a famously cryptic medieval document suggests that it contains nothing but gibberish. *Scientific American* 291 (1): 104–109.
———. 2004b. An elegant hoax? A possible solution to the Voynich manuscript. *Cryptologia* 28: 31–46.
Sahagún, B. de. 1963. *Florentine Codex. General history of the things of New Spain. Book 11 – Earthly things*. Trans. C.E. Dibble, and A.J.O. Anderson. Salt Lake City: University Utah Press.
Sanchis Amat V.M. 2012. *Francisco Cervantes de Salazar (1518–1575) y la patria del conocimiento: la soledad del humanista en la ciudad de Mexico*. PhD Thesis, Alicante, Spain.
Schwartz, S.B., ed. 2004. *Tropical Babylons: Sugar and the making of the Atlantic world, 1450–1680*. Chapel Hill: University of North Carolina Press.
Sherwood, E. 2002. *Leonardo and the Voynich manuscript*. www.edithsherwood.com/voynich_author_da_vinci/. 5 June 2017.
SilverMoon. 2007. *The imperial college of Tlatelolco and the emergence of a new Nahua intellectual elite in New Spain (1500–1760)*. Ph.D. thesis, Duke University Durham.
Starr, J. 1987. Zapotec religious practices in the valley of Oaxaca: An analysis of the 1580 "Relaciones Geograficas" of Philip II. *Canadian Journal of Native Studies* 7: 367–384.

Stolte, D. 2011 *UA experts determine age of book 'nobody can read'*. Retrieved 2012 July 25 from http://uanews.org/node/37825

Sugden, J. 2006. *Sir Francis Drake*. London: Pimlico.

Townsend, C. 2016. *Annals of North America: How the Nahuas of Colonial Mexico kept their history alive*. New York: Oxford University Press.

Tucker, A.O., and R.H. Talbert. 2013. A preliminary analysis of the botany, zoology, and mineralogy of the Voynich manuscript. *HerbalGram* 100: 70–85.

Tucker, A.O., and J. Janick. 2016. Identification of phytomorphs in the Voynich Codex. *Horticultural Reviews* 44: 1–64.

Varey, S., ed. 2000. *The Mexican treasury: The writings of Dr. Francisco Hernández*. Stanford: Stanford University Press.

Vázquez, R. 2002. Síntesis sobre la real y pontificia Universidad de México. *Anuario Mexicano de Historia del Derecho* 15: 265–342.

Velinska, E. 2013. *The Voynich manuscript: Hunting Tapestries*. ellievelinska.blogspot.com/2013/09/the-voynich-manuscript-hunting.html. 5 June 2017.

Viforcos Marinas, I. 2005. La volatilidad de los legados Indinos. El caso de Ruy Díaz Ramírez de Quiñones y sus disposiciones testamentarias. *Estudios Humanísticos Historia* 2005 (4): 263–293.

Vizuete Picon, P. 1890. *Diccionaro enciclopedico Hispanico-Americano de literature, ciencias y artes,* Vol. 5(2). Barcelona: Montaner y Simón.

Wright, I.A. 1916. *The early history of Cuba, 1492–1586*. New York: Macmillan Co..

Zandbergen, R. 2016. Earliest owners. In *The Voynich manuscript*, ed. R. Clemens, 1–9. New Haven: Yale University Press.

Chapter 16
Conclusions, Conjectures, and Future Studies

Jules Janick and Arthur O. Tucker

Summing Up

Our attempts to unravel the *Voynich Codex* involve a combination of hard evidence based on plant and animal identifications, plus conjecture based on iconographic and historical analysis. The present work covers a number of rolling objectives, as follows:

1. Confirm the conjecture that the *Voynich Codex* is a sixteenth century Mexican manuscript.
2. Decode the Voynichese symbols.
3. Decipher the text.
4. Determine the author and illustrator.
5. Place the *Voynich Codex* in context in time and place.

In this final chapter, we attempt to summarize our answers to these questions and to suggest the direction of future studies.

Voynich Codex as a Sixteenth Century Work of Colonial Mexico

We are convinced, in fact certain, that the *Voynich Codex* is a New World document. The botanical and zoological identifications of plants and animals in the codex are hard evidence that it is a post-Columbian manuscript. All the plants identified are indigenous to New Spain, confirming the pioneering work of Rev. Dr. Hugh O'Neill (1940), who identified the sunflower and capsicum pepper (Tucker and Talbert 2013; Tucker and Janick 2016). Charles Singer, the renowned historian of science, also considered the *Voynich Codex* to be a late sixteenth century manuscript (Tiltman 1967). In the present work, the identifications made by Tucker and Talbert (2013)

J. Janick, A. O. Tucker, *Unraveling the Voynich Codex*, Fascinating Life Sciences, https://doi.org/10.1007/978-3-319-77294-3_16

and Tucker and Janick (2016) have been expanded to 60 phytomorphs. Furthermore, 18 of the 20 animals found in the *Voynich Codex* are native to the New World, as discussed in Chap. 6. (Cattle, and perhaps sheep, are Spanish imports). Simply put, there is no way a manuscript written on vellum that contains a sunflower and an armadillo could have been written before 1492. There is no way of getting around all conjectures that the manuscript is a fifteenth century European manuscript being nonsense.

We find it difficult to explain the inability of many Voynich bloggers to equivocate on the identification of the sunflower and the armadillo, in spite of their obvious resemblance and despite any clear evidence to the contrary. The discovery of the manuscript in Italy and the carbon dating of its vellum (not the text) have led to the general hypothesis that the manuscript might be fifteenth century European, but the plant, animal, and mineral identifications by professional biologists belie this assertion. Defending theories despite contrary evidence is the purview of fanatics rather than scientists. Rather than revising their assertions, the deniers resort to far-fetched, incredulous explanations to contradict the obvious. One hypothesis of bloggers is that the *Voynich Codex* might have been an elaborate hoax created before 1912. However, adherents to this hypothesis would be wise to consider what we have stated under *Viola bicolor* in Chap. 4: "Thus, because *V. bicolor* has only been known as a North American endemic since 1961, any attempt to propose a forgery before 1912 (Barlow 1986) would have to explain this anomaly."

Our iconographic analysis of the manuscript confirms the origin and date of the manuscript, as summarized below.

1. Sixty phytomorphs (plant illustrations) in the *Voynich Codex*, including 57 botanical species, are indigenous to the New World. With the exception of circumboreal species, no Old World plants have been identified. The wide variety of plants is consistent with the botanical gardens of the Aztecs and their trading networks of healers (*curanderos* or *curanderas*).
2. The geomorph (geological illustration) boleite, which is abundant in Baja California Sur and associated with atacamite (a possible source of the green pigment used in the codex), is included.
3. Some zoomorphs (animal illustrations) are common Mesoamerican animal species, such as alligator gar, jaguarundi, coatimundi, ocelot, horned lizard, and armadillo, whereas others, such as cattle and sheep, represent common domestic introductions to New Spain by the Spanish in the sixteenth century.
4. A map (folio 86v) with kabbalistic imagery suggests four cities identified as Huejotzingo, Vera Cruz, Tecamachalco, and Tlaxcala, and three volcanoes identified as Popocatepetl, La Malinche, and Pico de Orizaba, all surrounding Puebla de los Angeles, a city founded by the Franciscan Fray Toribio de Benavente (Motolinía) in 1530. This places the work in central Mexico after 1530.
5. Voynichese symbols used in the text are similar to many common scribal abbreviations and are also found in many works from sixteenth century New Spain. Other symbols include the so-called "bird glyphs" found on the opening page (folio 1r) of the *Voynich Codex*, which are common paragraph markers known as *calderón* (Mackenzie 1997). T-O maps are found at least four times in the

Voynich Codex, and are also observed in the *Florentine Codex*. Symbols for celestial bodies (sun, moon, planets, and stars) in the *Voynich Codex* are also found in the Apocalypse paintings of Juan Gerson of 1562.
6. The ligated initials ("JGT"), identified as those of a sixteenth century indigenous painter of New Spain (Juan Gerson, *Tlacuilo*), and the name "Gasp.Torres" (Gaspar de Torres), master of students of El Colegio de Santa Cruz de Tlatelolco (also called Colegio de Santa Cruz), an institution formed to train sons of the Aztec nobility for the priesthood (SilverMoon 2017), are found embedded in the first phytomorph (folio 1v) of the *Voynich Codex*. We have conjectured that indigenous artist Gerson is likely to be the illustrator, and that Torres, a Spaniard born in Santo Domingo, could be the author of the *Voynich Codex*. There are iconographic similarities among Gerson's 1562 paintings of the Apocalypse at Tecamachalco, the mural at Casa del Deán, and the *Voynich Codex*.
7. Dating of the manuscript can be inferred to be between 1565 and 1572. Jacobus Horčicky, whose ennobled name Jacobi à Tepenecz has been inscribed on the manuscript, died in 1622 (Zandbergen 2017). Thus, the manuscript must have been written later than 1492 (based on images of New World plants and animals on its vellum) and before the death of Horčicky in 1622. This date was narrowed by circumstantial evidence to between 1565 and 1572 based on the founding date of Puebla de los Angeles in 1530, the date of the Tecamachalco murals in 1562, and the date of the *Codex Osuna* in 1565. The establishment of the Inquisition Tribunals in 1571 makes it unlikely that the manuscript was written after that date, because the presence of kabbalah symbolism would have been considered heretical.

Decoding and Decipherment

The most vexing problem with Voynich is decipherment. We have separated this into two issues. The first is the decoding of the Voynichese symbols and the second is the decipherment of the text. The symbol decoding presented in Chap. 5 was an independent discovery by Tucker and Talbert (2013), but a subsequent search of the literature has shown other attempts at symbol decipherment, including those by John Tiltman, the William Friedman study groups, Prescott Currier, Mary E. D'Imperio, Edith Sherwood, James C. Comegys, and Stephen Bax. These competing decipherments were analyzed in the cryptology chapter (see Chap. 11) and found wanting.

The Tucker and Talbert (2013) decoding of the alphabet (see Chap. 5) was based initially on two plants, identified by phonetic names as *nāshtl* (prickly pear) and *māguoey* (agave), and then expanded to six others, which completed the alphabet. The letters were internally consistent, but the deciphered names did not fit known plant names in Nahuatl, although many of the names were plausible based on their meanings in Nahuatl or Latin American Spanish. However, we were buoyed by the ability of the symbol translation to identify Huejotzingo (circle 2 in folio 86v) and the nickname "water seller" for Vera Cruz (see Chap. 10). The decoded alphabet suggested names of some other cities. The decoded alphabet was then checked for star names and many appeared to be redolent of Arabic, which was used in Spanish astronomy (see Chap. 13, Fig. 13.1). For example, a number of strange deciphered names, such as *ātlāaaā,* appeared to be close in pronunciation to the Arabic *al'Awwa* ("barking dog") assuming the "atl" could also be pronounced as the common *al* in Arabic. This provided a fit for stars in the constellation Virgo and confirmed that the Voynichese alphabet was phonetic and could be used to translate names in a number of languages, such as Nahuatl, Spanish, Taino, or Arabic. However, to our chagrin, the alphabet decoding failed at our attempts to decipher the Voynichese text. There were two explanations: the alphabet decoding was not completely correct or, more likely, the Voynichese words were not Classical Nahuatl but rather another Mesoamerican language. However, we have failed to discover which language, despite our review of 50 Mesoamerican dictionaries. Perhaps the language is lost, such as Chicimecan or Acolhuacatlatolli. We now suggest that the language of Voynich might be a mixed synthetic language, as suggested by William Friedman, and perhaps the lingua franca used by Aztec traders (Dakin 1981). We acknowledge that we need help from linguists with expertise in synthetic languages.

Noun Frequency

It is clear to us that future studies require linguistic analysis. Rather than more computerized cryptological analysis, we suggest that the path to follow might be to analyze elements in the illustrations that are labeled. We propose the most likely method to pursue is the analysis of nouns. There are various lists of named objects in the *Voynich Codex* that appear to be nouns. These include the following:

1. 26 words in circles 5, 7, and 9 of folio 86v that are presumed to be cities (see Appendix A).
2. 288 nymphs, each carrying a star on a string, in ten zodiac images (see Appendix B).
3. 63 stars in folios 68r(1), 68r(2), and 68r(3) (see Appendix C).
4. 31 jar names in the Pharmaceutical section (see Appendix D).
5. 188 plants in the Pharmaceutical section (see Appendix E).

In two of the lists (cities and nymphs), there were associated words in rings around the circles. It was assumed that the words in the lists were nouns, whereas the words in the two rings included other parts of speech. All words in each list were decoded based on the syllabary of Tucker and Talbert (2013) and then alphabetized. A master spreadsheet was made of all the names. This allowed a number of comparisons:

1. The comparison of word occurrence in five lists (cities, stars, nymphs, jars, and plants) to the 48 most frequent words in the *Voynich Codex* compiled by Kevin Knight (2009), the list of 30 optimal and 30 thematic words of Montemurro and Zanette (2013), and those words found in more than one list (Table 16.1).
2. An analysis of the commonality of words within and among lists (Table 16.2).
3. A comparison among city and nymph names with words in the rings around the lists (Table 16.3).

Table 16.1 Lists of objects in the *Voynich Codex*

Variable	Words in circles 5, 7, and 9 of folio 86v (cities)	Star names in folios 68r(1), 68r(2), and 68r(3)	Nymph names 10 zodiac circles	Jar names Pharmaceutical section	Plant names Pharmaceutical section
No. names	26	66	288	31	206
No. unique names	26	66	260	31	199
Repeats			āaaācholl 2×		ātlācâ 2×
			āaachi 2×		ātlācâchi 3×
			ātlaāchi 3×		ātli 2×
					ātlocâ 2×
			ātlaāchi 3×		ātloei 3×
			ātlaai 2×		
			ātlaān 3×		
			ātlacâi 2×		
			ātlachi 3×		
			ātlaocâ 2×		
			ātli 3×		
			ātlichi 2×		
			ātlocâ 4×		
			ātlocâ 2×		
			ātlocâi 3×		
			ātlocâi 3×		
			ātlocâoe 2×		
			ātlocâoll 2×		
			itlaāch 2×		
			mcuai 2×		

(continued)

Table 16.1 (continued)

Variable	Words in circles 5, 7, and 9 of folio 86v (cities)	Star names in folios 68r(1), 68r(2), and 68r(3)	Nymph names 10 zodiac circles	Jar names Pharmaceutical section	Plant names Pharmaceutical section
No. frequent names based on Knight 2009	2 (7.69%) ātlachi ātlachi	0	7 (2.68%) oe ātlaai ātlachi ātloll mchi ātlocâ ātlocâ	0	5 (2.51%) māe ātloll ātlocâ ātloe ātloe
No. names in optimal lists Montemurro and Zanette	2 (7.69%) ātlachi ātlachi	0	4 (1.54%) oe ātlaai ātlachi māe	0	0
Thematic	0	0	1 (0.38%) oe	0	1 (0.50%) māe
Names found in other lists	ātlācâoe (nymph) ātlachi (nymph) ātlāei (star, plant) ātlichi (nymph)	ācui (plant) ātlmāchi (nymph) ātlān (plant) ātlācâ (plant) ātlāei (plant, city) ātlichi (nymph) itlmchi (plant)	ātlaācâi (plant) ātlaāchi (plant) ātlācâi (plant) ātlācâoe (city) ātlachi (city) ātlāchi (plant) ātli (plant) ātlichi (city) ātlmāchi (star) ātlocâ (plant) ātlocâchi (jar) ātlocâi (plant) ātloei (plant) itlaāchi (plant) tsaācâ (plant)	ātlocâchi (nymph) ātlāeoha (plant)	ātlaācâi (nymph) ātlaāchi (nymph) ātlācâi (nymph) ātlāchi (nymph) ātlāeoha (jar) ātli (nymph) ātlocâ (nymph) ātlocâi (nymph) ātloei (nymph) itlaāchi (nymph) tsaācâ (nymph)

Table 16.2 Number of city and nymph names compared with words in their respective rings and compared with frequent words in Knight (2009) and optimal and thematic words in Montemurro and Zanette (2013)

			Number of words (% of unique)		
Source of words	Total	Unique	Knight	Montemurro and Zanette	
				Optimal	Thematic
City ring (folio 86v)	93	84	11 (13.10%)	6 (7.14%)	4 (4.76%)
			aca, āe, atlachi, atloca, atloca, atloe, atloe, machi, oe, oll, tsachi	aca, atlachi, machi, oe, oll, tsachi	machi, oe, oll, tsachi
Cities (folio 86v)	26	26	2 (7.69%)	2 (7.69%)	0 (0.00%)
			ātlachi, ātlachi	ātlachi, ātlachi	
Zodiac ring	848	607	35 (5.77%)	22 (3.62%)	16 (2.64%)
			ācâ, ātlaai, ātlaai, ātlachi, ātlachi, ātlocâ, ātloe, ātloll, ātloll, chi, chocâ, choe, chol, choll, maācâ, maai, mācâ, machi, macui, māe, mai, mai, mchi, mi, noll, ocâ, oe, oll, quātlaai, tsa, tsā, tsācâ, tsachi, tsai, tsāi	ācâ, ātlaai, ātlachi, ātlachi, chi, chocâ, chol, choll, cui, mācâ, machi, māe, mi, ocâ, oe, oll, quātlaai, tsa, tsā, tsachi, tsai, tsāi	chi, chol, choll, cui, mācâ, machi, māe, mi, ocâ, oe, oll, quātlaai, tsa, tsā, tsācâ, tsachi
Nymphs	288	259	7 (2.70%)	3 (1.16%)	1 (0.39%)
			ātlaai, ātlachi, ātlocâ, ātlocâ, ātloll, mchi, oe	oe, ātlaai, ātlachi	oe

Table 16.3 Words from Voynich folio 1v and folio 103r, lines 1–4, compared with frequent words in Knight (2009) and optimal and thematic words in Montemurro and Zanette (2013)

			Number of words (% of unique)		
Source of words	Total	Unique	Knight	Montemurro and Zanette	
				Optimal	Thematic
folio 1v Herbal section	87	68	10 (14.7%)	6 (8.8%)	7 (10.3%)
			ācâ, āe, ātlocâ, chocâ, choe, choll, maācâ, mācâ, mi, tzācâ,	ācâ, chocâ, choll, cui, mācâ, mi	choll, cuācâ, cui, mācâ, mi, tzācâ, tzāe
folio 103r lines 1–4 Recipe section	48	46	7 (15.2%)	6 (13.0%)	4 (8.7%)
			ācâ, ātlachi, ātlocâ, câmachi, chol, machi, oe	ācâ, ātlachi, câmachi, chol, machi, oe	câmachi, chol, machi, oe

All the cities, stars, and jars had unique words, but there were repeated words in nymphs (19 words, 7.31%) and plants (five words, 2.5%; Table 16.1). None of the 66 star and 31 jar names were included in the most frequent words of Knight (2009), but two of the 26 city names (7.69%), seven of the 260 unique nymph names (1.54%), and five (2.61%) of the 199 unique plant names were included (Table 16.2). Two city names (7.96%) and four nymph names (1.54%) were included in the optimal list of Montemurro and Zanette, and one nymph (0.38%) name and one plant name (0.50%) were in the thematic list. There were a number of names that were in more than one list.

In a comparison of the frequency of city and nymph names with the frequency of words in the associated rings in the lists of Knight (2009) and Montemurro and Zanette (2013), it was apparent that the ring words had a much higher occurrence. The exception was optimal names for city and ring words, which were similar (7.14% and 7.69%). This was evidence that the ring words include various parts of speech (Table 16.2). We were surprised to find that 19 nymph names and five plant names were repeated. Furthermore, three words – ātlācâchi, ātli, and ātlocâ – were found in all lists.

Word Frequency in Textual Passages

Two different sections of text were analyzed: the text in folio 1v describing the soap plant (*Ipomoea arborescens*; 68 unique words) and the text in folio 103r (lines 1–4, the first paragraph in the Recipe section; 46 unique words). They were examined to find out how many of these words were also found in the list of Knight and that of Montemurro and Zanette. The results in Table 16.3 show that these two lists had between 8.70% and 15.22% commonality, compared with 2.64% and 13.10% for the ring words of cities and the zodiac. We conclude that the noun names in the lists were more restrictive than the names in the textual passages.

Authorship

The authorship of the *Voynich Codex* was serendipitously discovered by identifying their autographs on the first botanical illustration (Janick and Tucker 2017). The ligated initials of Juan Gerson, *Tlacuilo,* and the name Gasp.Torres were found embedded in the phytomorph *Ipomoea arborescens* in folio 1v. As a result, we have conjectured that Gerson, an indigenous artist, was the illustrator, and that Torres was the author of the *Voynich Codex.* Our claim regarding Gerson was supported by his initials in the Codex, the Apocalypse paintings, and the Casa del Deán murals, and iconographic similarities among the three works. The biography of Torres supports our claim that he was the possible author. He was highly educated as a medical doctor and physician, master of students at the Colegio de Santa Cruz from 1569 to 1572, a supporter of Indian rights, and governor of Cuba in 1580. The amalgamated biography of the two men fits the predictive profile of the author/artist based on the imagery and content of the *Voynich Codex* in Chap. 14. This is discussed more fully in the Afterword that follows.

Spanish-*Tlacuilo* Art in New Spain

Although the *Voynich Codex* at first glance appears unique, we present iconographic evidence that it fits into the genre of Spanish-*Tlacuilo* art, i.e., art made by Nahua-trained artists. Our evidence is based on comparisons of illustrations in the *Voynich Codex* with various sixteenth century works that involve *tlacuiloque*. These include the *Codex Cruz-Badianus* of 1522 (Emmart 1940), the *Florentine Codex* of Bernardino de Sahagún, written between the 1560s and 1592 (Sahagún 1951–1982), the *Codex Mexicanus*, 1570 (*Codex Mexicanus* 1571–1590; Brotherston 1998), and some of the botanical works of Francisco Hernández, illustrated between 1570 and 1577, as the original artwork was made by indigenous artists (Varey 2000). In addition, we include artwork in New Spain, such as Gerson's ceiling paintings (completed in 1562) in the Franciscan church, Asuncion de Nuestra Senora, in Tecamachalco (Azpeitia 1972), in addition to the murals in the Casa del Deán in Puebla, completed in 1584 (Morrill 2014), which we believe also include his work (see Chap. 15, Fig. 15.2). The iconographic similarities among the Voynich and other sixteenth century codices and paintings by *tlacuiloque* (indigenous artists) in New Spain indicate that the *Voynich Codex* fits into the genre of Spanish-*Tlacuilo* art. Thus, we conclude that the *Voynich Codex*, bizarre and strange as it is, must be considered an integral part of this body of work. Examples are as follows.

Botanical and Zoological

The commonality of plants (Fig. 16.1) and animals (Fig. 16.2) illustrated in the *Voynich Codex* with various sixteenth century codices indicates that Voynich artwork falls within the genre of sixteenth century illustrations by *tlacuiloque* (indigenous artists).

Judaic Allusions

There are Judaic allusions in the *Voynich Codex*, such as Juan Gerson's 1562 paintings of the Apocalypse and the 1582 Casa del Deán murals in Puebla. The kabbalah image in folio 86v of the *Voynich Codex* (Fig. 16.3a) doubles as a map centered around the Celestial City of Jerusalem, and is consistent with the Franciscan incorporation of

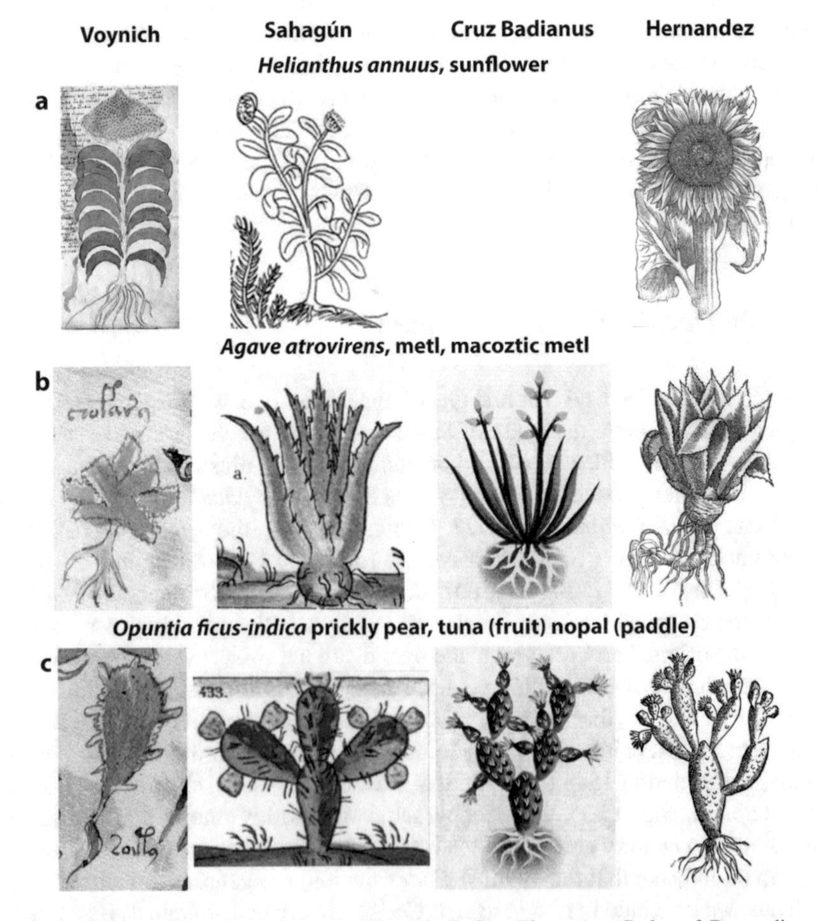

Fig. 16.1 Commonality of plants in the *Voynich Codex*, *Florentine Codex* of Bernardino de Sahagún, *Codex Cruz-Badianus*, and Hernández. (**a**) Sunflower; (**b**) agave, and (**c**) prickly pear

Fig. 16.2 Commonality of animals in the *Voynich Codex*, *Florentine Codex* (Sahagún), *Codex Mexicanus*, Casa del Deán mural, and the Apocalypse ceiling: armadillo, caracara, cattle, coatimundi, crayfish, frog, iguana, ocelot, and sheep

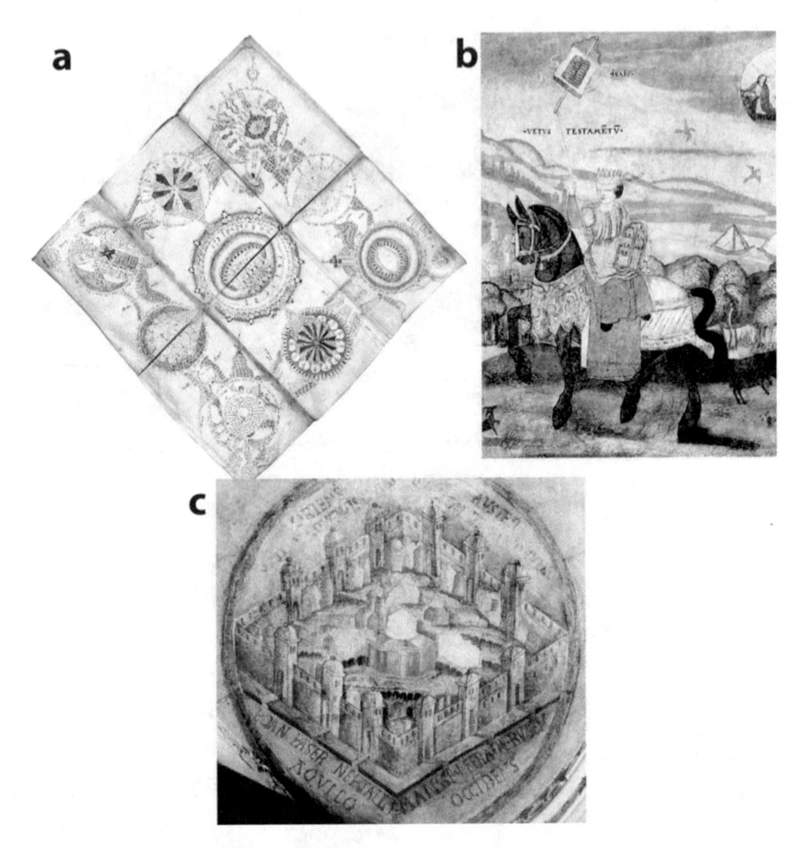

Fig. 16.3 Judaic allusions. (**a**) The kabbalah Sephirothic Tree of Life symbol in the *Voynich Codex*, (**b**) the Synagoga sibyl in the Casa del Deán procession, and (**c**) Jerusalem image in Gerson's Apocalypse ceiling painting

kabbalah (or *cabala* in its Christian formulation). The image of Jerusalem (Fig. 16.3c) in Gerson's 1562 murals is consistent with the Franciscan interest in religious reform and renewal and the development of a New Jerusalem in America (Lara 2004, 2008, 2013). Furthermore, many of the Apocalypse paintings have Old Testament allusions, including Noah's ark, the Tower of Babel, the sacrifice of Isaac, Jacob's ladder, the plague of locusts, and the fall of Babylon. In the Casa del Deán murals, the first sibyl in the Salon of the Sibyls represents the blindfolded sibyl Synagoga, a personification of the Jewish religion (Fig. 16.3b), and is consistent with Roman Catholic theology of the unwillingness of the Jews to accept the New Testament (Morrill 2014).

Zodiac

The zodiac (circle of animals), an ancient occult construct of the West, took hold in medieval Europe and was introduced to New Spain (see Chap. 8, Table 8.1 and Fig. 8.13). The zodiac images are found in the *Voynich Codex* with New World

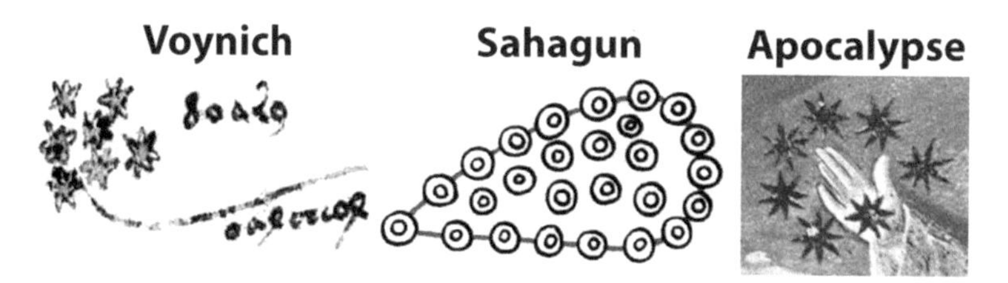

Fig. 16.4 The Pleiades: *Voynich Codex*, *Florentine Codex* (Sahagún), and Gerson's Apocalypse ceiling

animals replacing many of the traditional signs (see Chap. 8, Table 8.1). The zodiac is also found in the *Codex Mexicanus* (see Chap. 8, Fig. 8.13). It incorporates the Aristotelian elements of air, water, fire, and Earth that are found in kabbalah and substitutes a crayfish for the crab in Cancer. The sign for Gemini, usually portrayed as male twins in medieval zodiacs, is represented by a heterosexual couple in the *Voynich Codex*. The couple appears in coitus in the *Codex Mexicanus*.

Astronomical Images

A comparison of astrological images in Voynich with those of Gerson's Apocalypse paintings is found in Fig. 15.4 in Chap. 15. Three illustrations of the star cluster known as the Pleiades, which was so important to Aztec cosmology and religion, are illustrated in the *Voynich Codex*, the *Florentine Codex*, and Gerson's Apocalypse paintings (Fig. 16.4).

Nude Bathing

The plethora of nude nymphs – all appearing to be the same blonde, curvaceous woman – cavorting in pools (Fig. 16.5a) is one of the fascinating things about the *Voynich Codex*. A nude woman bather is portrayed outside a steam bath in the *Florentine Codex* (Fig. 16.5b), and three pairs of nude bathers are found in the background of the west wall in the Salon of the Sibyls in the Casa del Deán mural (Fig. 16.5c). These bathers (two couples and two children) have been identified as Amerindians by Penny C. Morrill (2014: 93).

Wit and Whimsy

A number of drawings in the *Voynich Codex* indicate that the artist had a sense of wit and whimsy, which was not unknown in the art of the *tlaculoque* (Fig. 16.6). Whimsical images in the *Voynich Codex* include fantastical root elaborations in the

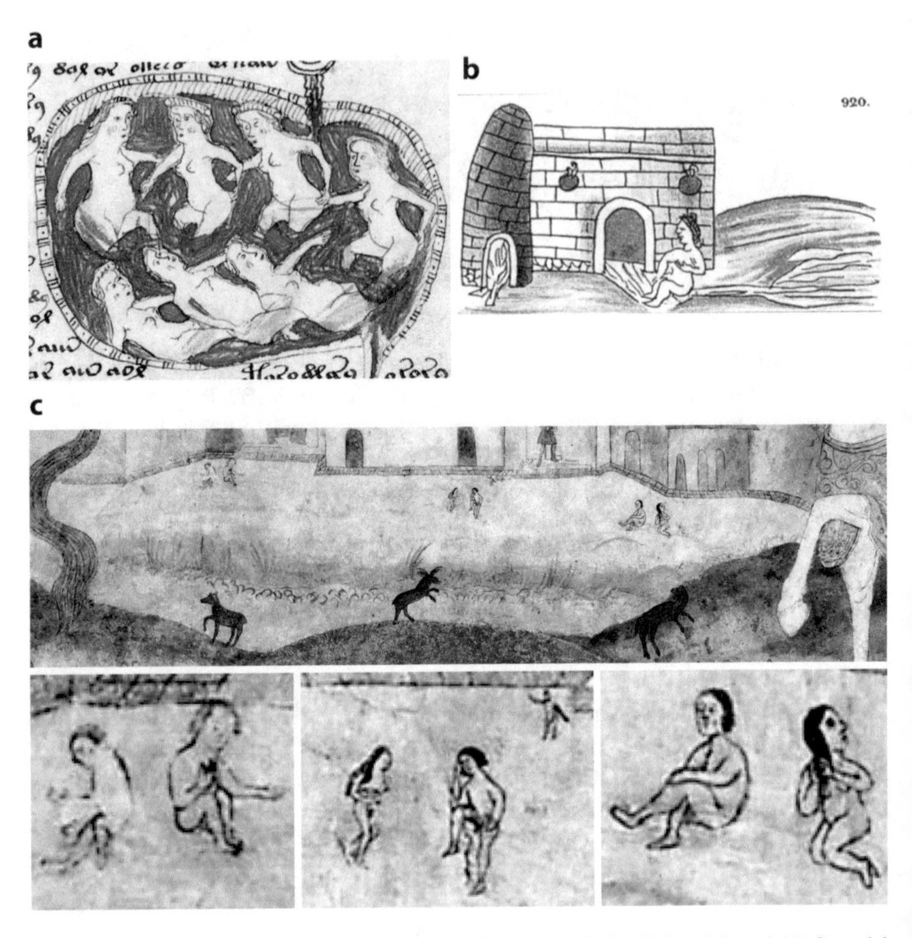

Fig. 16.5 Nude bathing. (**a**) *Voynich Codex*, (**b**) *Florentine Codex* (Sahagún), and (**c**) Casa del Deán murals

shape of animals (Fig. 16.6a); the inclusion of faces in roots (Fig. 16.6b), the sun (Fig. 16.6c), the moon (Fig. 16.6d), and T-O maps representing the Earth (Fig. 16.6e); doodles of an alligator gar swallowing a woman (see Chap. 6, Fig. 6.1); fantastical elaborations of plumbing or vascular systems in the Balneological section (see Chap. 7, Fig. 7.1); and detailed drawings of apothecary jars (see Chap. 11, Table 11.4). Humorous depictions of native, anthropomorphized animals, such as coyote, monkey, rabbit, deer, and coatimundi, can be found in the Casa del Deán murals (Fig. 16.6f). We conclude that the artist had an imagination not unlike that of surrealist artists, such as Salvador Dali. The eagle image in folio 46v resembles the body of the Hapsburg double-headed eagle, the symbol of Carlos V, king of Spain from 1516 to 1556.

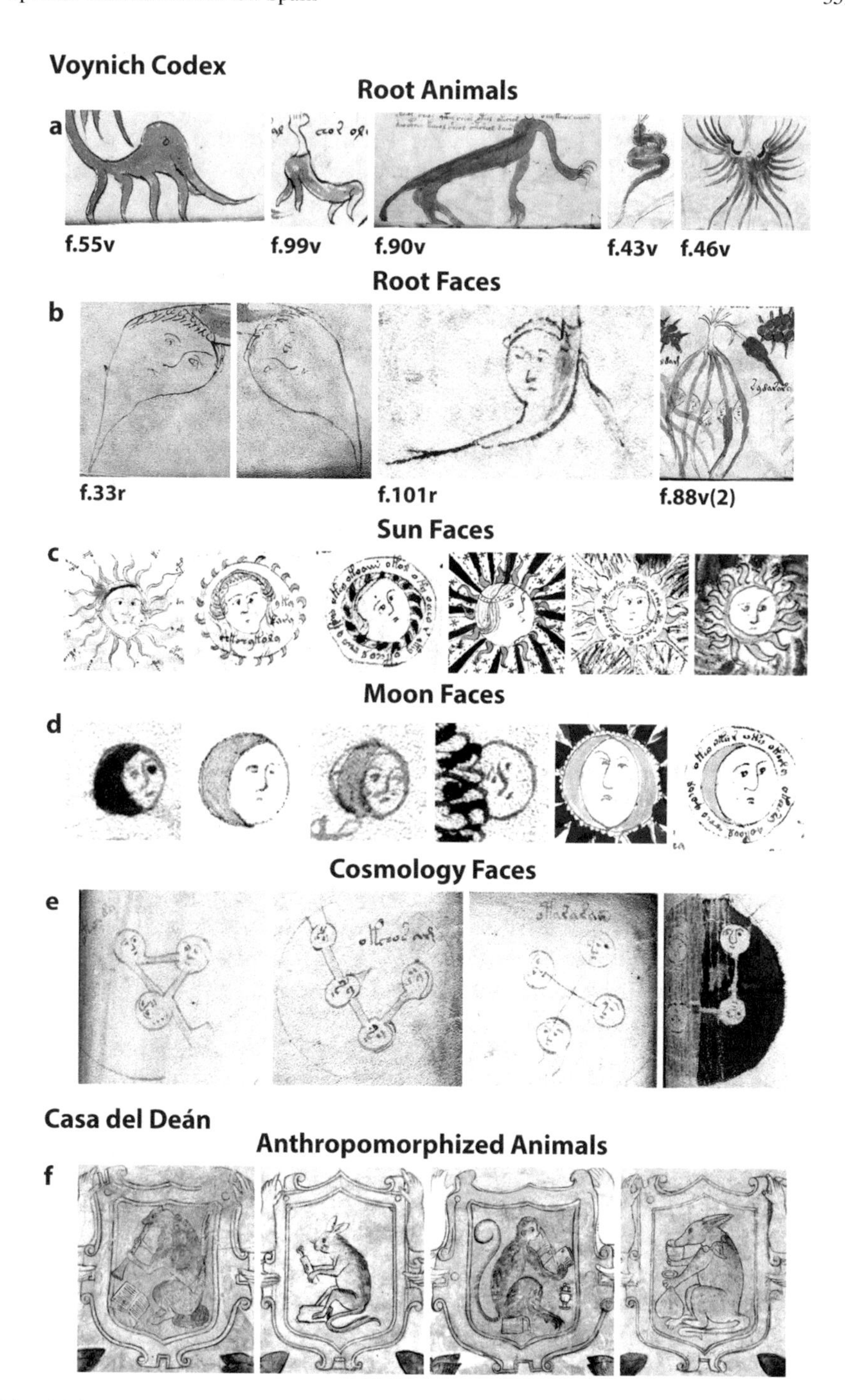

Fig. 16.6 (**a**–**e**) Whimsical images in the *Voynich Codex* and (**f**) the Casa del Deán mural. (**a**) Roots in the shape of animals, (**b**) faces associated with roots, (**c**) sun faces, (**d**) moon faces, (**e**) faces in T-O maps, and (**f**) anthropomorphized animals (coyote, rabbit, monkey, and coatimundi)

Contrary Evidence

Is there substantial, solid evidence in the *Voynich Codex* that could negate our hypothesis? We offer a few examples that would be expected from alternative opinions in the literature and provide our preemptive explanations.

Plant and Animal Identifications

We have avoided participating in Voynich wars, but in this case, we cannot resist. Examples include the online essays by Nicholas (Nick) Pelling, one of the most expansive and intemperate of bloggers and author of a self-published book, *The Curse of the Voynich: The Secret History of the World's Most Mysterious Manuscript,* in 2006. In an essay published 29 May 2009 in Cipher Mysteries entitled "Is This the Way to Armadillo?" (http://ciphermysteries.com/2009/05/29/is-this-the-way-to-armadillo), Pelling discusses the "claimed" armadillo zoomorph on folio 80v. He makes the prescient statement: "*if this can be proven to be intentionally depicting an animal from the New World, then a lot of other dating evidence becomes secondary*." ("Secondary" is a strange word. The better word would be "incorrect.") Pelling goes on: "*But of course this kind of controversy is nothing new: you only have to think of the decades-long hoo-ha [sic] over the (claimed) New World sunflowers.*" Pelling is absolutely correct. The sunflower (discussed below) and armadillo identifications would demolish the theory of a fifteenth century date for the *Voynich Codex*. In the case of the armadillo (see Chap. 6, Fig. 6.11), Pelling simply rejects the obvious identification and offers two explanations for the pesky creature: "*[The image] is probably a depiction of catoblepas – a fearsome creature Leonardo da Vinci (and doubtless many of his contemporaries) believed lived at the source of the Niger river, and whose bull-like head was so heavy that it permanently hung down near the ground.*" The problem is the armadillo depiction does not have large horns. What he identifies as horns are ears. Furthermore, medieval pictures of catoblepas show a bull-like creature without scales. His second suggestion is that the drawing was "*emended...to strengthen that resemblance?*" Thus, Pelling inadvertently concedes that there is a resemblance. In contrast, well-known blogger Richard SantaColoma (2015) has accepted the armadillo (and sunflower) identification, and uses it to support the theory of a 1610–1620 date for the manuscript, which he considers to be a faux document. This is based on his identification of one of the apothecary jars that closely resembles one of Galileo's microscopes of 1610 or 1614.

We would also be remiss if we did not mention alternative plant identifications for the Herbal section of the *Voynich Codex* made by non-botanists, with the assumption that the plants are European. Table 16.4 includes tentative identifications made by Theodore C. Petersen, Coptic linguist and historian; identifications made by Sherwood and Sherwood (2008), organic chemist and daughter; Velinska (2013a);

Table 16.4 Species identification of *Voynich Codex* phytomorphs in the Herbal section by Tucker and Janick (Chap. 4), pre-1960s tentative notations after Zandbergen (2017), Sherwood and Sherwood (2008), Velinska (2013a, b), and Bax (2014)

Voynich Codex folio	Tucker and Janick (Chap. 4)	Pre-1960s unedited tentative notations after Zandbergen (2017)	Sherwood and Sherwood (2008)	Velinska (2013a, b)	Bax (2014)
1v	*Ipomoea arborescens* (Convolvulaceae)	Solanum. Solatrium, Belladonna (Petersen) *Atropa belladonna* (ELV)	*Atropa belladonna*	Hypericum (*Atropa belladonna*)	
2r	*Lactuca graminifolia* (Asteraceae)	*Cyanus silvestris* (O'Neill) Centauria (Holm)	*Centaurea diffusa*	*Centaurea*	*Centaurea*
2v	*Nymphoides aquatica* (Menyanthaceae)	coliocasia, Egyptian lotus (Petersen) Colosia? Villarsia, Limnantherum (ELV)	*Nymphoides*	*Nymphoides*	
3r		Crassulatea (Cretan dittany), [illeg.] (Petersen) Dictamnus? (ELV)	*Celosia argentea*	*Amaranthus tricolor*	
3v			*Helleborus foetidus*		Hellebore
4r		Hypericum, Centarium Erythaea (O'Neill) Hypericum? (ELV)	*Saxifraga cespitosa*	*Linum usitatissimum*	
4v	*Cobaea* sp., cf. *C. biaurita* (Polemoniaceae)	Conbolvula ipomea? (Petersen) Ipomoea? Scammonia? (ELV)	*Campanula rapunculus*	*Clematis integrifolia*	
5r		Paris, Indian cucumber? (Petersen) Paris? (ELV)	*Arnica montana*	Herb-paris	
5v	*Jatropha cathartica* (Euphorbiaceae)	Parietaria (O'Neill) Geranium (ELV)	*Malva sylvestris*	*Malva sylvestris*	
6r		Asclepiades (Petersen) Aristolochia? (ELV)	*Acanthus mollis*	*Philodendron*, *Reseda lutea*	
6v	*Cnidoscolus texanus* (Euphorbiaceae)	Black gum tree? Arctium? Bidenus hypnis (O'Neill) castor oil, ricinus (Petersen) Ricinus, Castor oil plant (ELV)	*Eryngium maritimum*	*Ricinus communis*	
7r		Nymphaea alba, Nenufar (Petersen) Paeonia (ELV)	*Trientalis europaea*	*Geropogon hybridus*	
7v		Polygonum persicarum (ELV) Bisangia? Potentilla selvaticus? (Petersen)	*Myrica gale*	*Rhododendron hirsutum*	

(continued)

Table 16.4 (continued)

Voynich Codex folio	Tucker and Janick (Chap. 4)	Pre-1960s unedited tentative notations after Zandbergen (2017)	Sherwood and Sherwood (2008)	Velinska (2013a, b)	Bax (2014)
8r		Praenanthes (O'Neill), Atriplex hastata, Atriplex, Orache (ELV)	*Pisum sativum*	*Cucumis sativus*	
8v		Silene (Petersen) Oleander? Verbena? (ELV)	*Symphytum officinale*	*Silene dioica*	
9r		Chalidonium magna (Petersen) Delphinium (ELV)	*Ricinus communis*	*Vitis vinifera*, *Morus alba*	
9v	*Viola bicolor* (*V. rafinesquei*) (Violaceae)	Viola tricoloris, herba trinitatis (O'Neill) Viola, pansy (ELV)	*Viola*	*Viola tricolor*	
10r		Scabiosa (Petersen)	*Cichorium pumilum*	*Centaurea cyanus*	
10v		Hellebora? Orientalis (Petersen) Helleborus (ELV)	*Linnaea borealis*	*Linnaea borealis*	
11r		Silene scambis (Petersen)	*Rosmarinus officinalis*		
11v		Leonotus, Leonarus (Petersen)	*Curcuma longa*	*Curcuma longa*	
13r	*Petasites* sp., cf. *P. frigidus* var. *palmatus* (Asteraceae)	Tussilago petronites (Petersen) Tussilago (ELV)	Banana	*Musa*	
13v		Crassolatea Fetthenne (Petersen)	*Lonicera periclymenum*	*Lonicera*	
14r		Pfeilkraut – sagittaria (Petersen)	*Scorzonera*		
14v		Osmodei (Petersen) Botrychium (ELV)	*Stachys monnieri*	*Stachys*	
15r		[Some thistle] (Petersen)	*Sonchus oleraceus*	*Carlina vulgaris*	
15v		Herba Paris (Petersen) Lilium? (ELV)	*Paris quadrifolia*	Arapabaca, *Spigelia anthelmia*	
16r		Canabis (hemp) (Petersen) Aconitum? (ELV)	Cannabis	*Cannabis sativa*	*Juniperus oxycedrus*

16v	*Eryngium* sp., cf. *E. heterophyllum* (Apiaceae)	Eryngium? (ELV)	Chrysanthemum	*Eryngium jacea nigra*	
17r			*Catananche caerulea*	*Artemisia dracunculus*	
17v	*Dioscorea composita* (Dioscoreaceae)	Smilax, Tamus communis (Petersen) Tamus? (ELV)	*Dioscorea*	*Polygonum convolvulus*	
18r		Ringelblume (O'Neill), Calendula officinaris (Petersen) Aster? Calendula (ELV)	*Aster alpinus*	*Calendula officinalis*	
18v				*Leontopodium*	
19r			*Polemonium caeruleum*	*Polemonium caeruleum*	
19v			*Draba nivalis*	*Galium aparine*	
20r		Moss polytrichnum (O'Neill), cf. cranberry (Petersen) Arctostaphylis, Poterium? (ELV)	*Astragalus hypoglottis*	*Vaccinium myrtillus*	
20v		Fam. compositae (O'Neill), chicory? salsify? (Petersen) Tragopogon, salsifica (ELV)	*Cynara cardunculus*		
21r	*Euphorbia thymifolia* (Euphorbiaceae)	Polygonum, Eisenkraut, Tymium (Petersen) Herniaria (ELV)	*Anagallis arvensis*	*Polygonum aviculare*	
21v		Salvia?, Steinbrech (Petersen) Crassula? Sedum? (ELV)	*Dictamnus albus*	*Sedum dasyphyllum*, *Sedum spurium*	
22r			*Verbena officinalis*	*Sedum maximum*	
22v		Sedum? (ELV)	Tulip	*Pulsatilla alpina*	
23r	*Ribes malvaceum* (Grossulariaceae)	Water Veronia, Helleborum, stinkende Mierwurz (Petersen) Veronica (ELV)	*Pulsatilla vulgaris*	*Dasiphora fruticosa*	

(continued)

Table 16.4 (continued)

Voynich Codex folio	Tucker and Janick (Chap. 4)	Pre-1960s unedited tentative notations after Zandbergen (2017)	Sherwood and Sherwood (2008)	Velinska (2013a, b)	Bax (2014)
23v	*Passiflora* Subgenus *Decaloba*, cf. *P. morifolia* (Passifloraceae)		*Borago officinalis*	*Borago officinalis*	
24r	*Silene* cf. *S. menziesii* infected with *Microbotryum violaceum* (Caryophyllaceae)	Libnis alba Behen (Petersen) Lychnus? (ELV)	*Cucumis sativus*	*Silene maritima*	
24v		Origophyllus (???) [illeg.] (Petersen) Lychnis? (ELV)	*Ficus religiosa*		
25r	*Urtica* sp., cf. *U. chamaedryoides* (Urticaceae)	Menta hortensis, Herzkraut Marrube, Mercurialis, Mentha arvensis, urtica, Mentastrum (O'Neill) Mentha piperata (Petersen) Parietaria, Urtica? (ELV)	Wild thyme	*Commiphora wightii*	
25v		Leontopodium, Edelweiss, Plantago (Petersen) Plantago (ELV)	*Isatis tinctoria*		
26r	*Wigandia urens* (Boraginaceae)	gnaphalium (Petersen)	*Prunella vulgaris*	*Veronica austriaca*	
26v		Verbena foenica, Creutzwurz (Petersen)	*Lens culinaris*		
27r		Asarium, Haselwurz [the smaller one] (Petersen) Datura (ELV)	*Spinacia oleracea*	*Laurus nobilis* + *Centella asiatica*, *Alliaria petiolata*	*Colchicum autumnale*
27v			*Dianthus superbus*	*Clematis vitalba*	
28r		Arum, Anisarum (Petersen) Rumex (ELV)	*Aristolochia* sp.	*Arisaema triphyllum*	
28v		Arnica (ELV)	Rhododendron	*Jasminum angustifolium*	

29r		lappa bur [… /illeg] (Petersen) Arctium? Xanthium? Burdoch or the nearly allied Burweed (ELV)	*Lactuca sativa* var. *longifolia*	*Ptilostemon casabonae*	
29v	*Anemone tuberosa* (Ranunculaceae)		Nigella sativa	Nigella sativa	*Nigella sativa*
30r		Schuppenwurz?, Menanthes trifolia (Petersen) Menyanthes (ELV)	*Prunella vulgaris*	*Urtica membranacea*	
30v	*Penthorum sedoides* (Penthoraceae)	Boragine (Petersen) Lithspermum (ELV)	*Cuscuta europaea*		
31r		Hyrciacum Aratum, henbane (Petersen) Hyasciamus, Henbane (ELV)	*Erigeron acris*		*Gossypium*
31v		Heracleum cervafolium (Petersen)	Valerian	*Hyoscyamus niger*	
32r	*Ocimum campechianum* (*O. micranthum*) (Lamiaceae)	Mentastrum (O'Neill), Rosenmüntz, brunella vulgaris (Petersen) Mentha, some kind of mint (ELV)	*Veronica triphyllos*	*Lantana camara*	
32v	*Ipomoea pubescens* (Convolvulaceae)	Campanula, Rapunculus, Convolvulus orectus? (Petersen) Malva? (ELV)	*Campanula rotundifolia*	*Wahlenbergia hederacea*	
33r	*Allionia incarnata* (Nyctaginaceae)	Mandragora (ELV)	*Silene vulgaris*	*Silene fimbriata*	
33v	*Psacalium* sp.? *Pippenalia* sp.? (Asteraceae)	Scabiosa (Petersen) Knautia, Scabiosa (ELV)	*Tanacetum parthenium*	*Scabiosa*	
34r		[illeg.], scabiosa (Petersen) Scabiosa (ELV)	*Anemone hortensis*	Astrantia major/ minor	
34v		libanotis umbilifera (Petersen) Eryngium? (ELV)	*Lunaria annua*	*Lunaria annua*	
35r		Leonuris? orchid root? (Petersen)	*Cichorium intybus*	*Caryota urens*	
35v		uva quercina, oak quercus gall apple? No! grapevine? (Petersen) Uva quercina, The fabulous "grapes of oak" (ELV)	*Ribes nigrum*	Oak/ivy combination	

(continued)

Table 16.4 (continued)

Voynich Codex folio	Tucker and Janick (Chap. 4)	Pre-1960s unedited tentative notations after Zandbergen (2017)	Sherwood and Sherwood (2008)	Velinska (2013a, b)	Bax (2014)
36r		Geranium (ELV)	*Delphinium staphisagria*	*Geranium macrorrhizum*	
36v	*Dorstenia contrajerva* (Moraceae)	[illeg] (Petersen) Cannabis (ELV)	*Lamium amplexicaule*	*Alchemilla alpina*	
37r		bladrian, valerian (Petersen) Sambucus? Dwarf Elder (Danewort) (ELV)	*Mentha longifolia*	*Chenopodium capitatum*	
37v			*Emilia fosbergii*		
38r		Aster luzei, asarum? (Petersen)		*Etlingera elatior*	
38v		(Cichrium?), Lactuca (Petersen) Cynarum? (ELV)	*Euphorbia myrsinites*		
39r		Colchicum (or Crocus) (ELV)		*Erythronium dens-canis*	
39v	*Phacelia crenulata* (Boraginaceae)	[illeg.] (Petersen) Primula? (ELV)			
40r		[illeg.] (Petersen) Primula (ELV)	*Erodium malacoides*		
40v	*Smallanthus* sp. (Asteraceae)	Thistle, Jerusalem artichoke, Helianthus tuberosus (Petersen)	*Crocus vernus*	*Celosia cristata*	
41r		Thistle (Petersen) Carduas? (Cnicus?) (ELV)	*Origanum vulgare*	*Anethum graveolens*, *Carduus* thistle, Common agrimony	
41v	*Lomatium dissectum* (Apiaceae)	[illeg.], Fern, Maidenhair fern???, Tansy? (Petersen) Daucus (ELV)	*Coriandrum sativum*	*Petasites frigidus* var. *palmatus*	Coriander
42r		Arum sagittarium (Petersen) Ranuncus? (ELV)		*Arum italicum*, *Fragaria vesca*	
42v	*Calathea* sp., cf. *C. loeseneri* (Marantaceae)	(ELV) Ranuncularia [illeg.] (Petersen) Aquilegia (ELV)	*Aquilegia vulgaris*	*Centaurea solstitialis*	

43r		Antemum Gnaphalium (Petersen) Belledactrum (ELV)	Stellaria media	Cypress or Ebony	
43v		polygonum?, Bistorta? (Petersen) Polygonum (ELV)	*Elytrigia repens*	*Dipsacus fullonum*	
44r		Veratrum (Petersen) Veratrum? (ELV)	*Mandragora officinarum*		
44v			*Apium graveolens*		
45r	*Salvia cacaliifolia* (Lamiaceae)	Glycyrrhiza (ELV)	*Atriplex hortensis*		
45v	*Hyptis albida* (Lamiaceae)	Contellaria, Veronica? Camedris, Ehrenpreis?, beatrum (Petersen) Teucrium? (ELV)	*Lavandula angustifolia*	*Teucrium chamaedrys*	
46r		asclepias roots (Petersen) Asclepias? (ELV)	*Leucanthemum vulgare*	*Pulicaria dysenterica*	
46v		boragine, anchusa (Petersen)	*Inula conyza*	*Tanacetum balsamita*	
47r		Indian cucumber? (Petersen)	*Sempervivum tectorum*		
47v	*Cynoglossum grande* (Boraginaceae)	Doronicum, Pardalianchus (Holm) Gemswurz (Petersen) Anchusa? (ELV)	*Pulmonaria officinalis*	*Pulmonaria obscura*	
48r		(cyanus?) no! (Petersen) Hyasciamus (ELV)	*Adonis vernalis*		
48v		Geum (ELV)	*Ruta graveolens*		
49r	*Lithophragma affine* (Saxifragaceae)	Tamus communis (Holm) Scam[m]onia (Petersen) Convolvus (ELV)	*Nymphaea caerulea*		
49v		lunaria, (Holm): cyclamen. Alpenveilchen. (Petersen) Lunaria (ELV)			
50r		Artichoke, [illeg], species of sunflower? (Petersen)	*Astrantia major*		

(continued)

Table 16.4 (continued)

Voynich Codex folio	Tucker and Janick (Chap. 4)	Pre-1960s unedited tentative notations after Zandbergen (2017)	Sherwood and Sherwood (2008)	Velinska (2013a, b)	Bax (2014)
50v		Umbellifera? (coxcomb?) (O'Neill)	*Gentiana frigida*	*Phyteuma globularifolium*	
51r	*Fuchsia thymifolia* (Onagraceae)	Mandrake (Manley)	*Cakile maritima*		
51v	*Phacelia integrifolia* (Boraginaceae)	Cow's tongue, Licopsis, (Sedum majus, Hauswurz)? (Königskerze), Verbascum Thapsiforme, Mullein (Petersen) Lycopsis, Symphytum (ELV)	*Salvia officinalis*		
52r	*Anemone patens* (Ranunculaceae)	[illeg] (Petersen) Aristolochia (ELV)	*Anemone coronaria*		
52v		Acanthus? (ELV)	Fern	*Heliotropium europaeum*	
53r	*Ambrosia* sp., cf. *A. ambrosioides* (Asteraceae)	Elecampane (Petersen) Inula (ELV)	*Achillea ptarmica*	*Anchusa variegata*	
53v		hieracium (hawk weed) (Petersen)	*Hieracium aurantiacum*	*Epilobium alpestre*	
54r		thistle, leaves of boneset (Petersen) Carduus? (ELV)	*Cirsium oleraceum*		
54v			*Perovskia atriplicifolia*		
55r	*Diastema hispidum* (Gesneriaceae)		*Fumaria officinalis*	*Aquilegia atrata*	
55v		Elephants ear? (Petersen)	Cauliflower/ broccoli	*Allium ursinum*	
56r	*Phacelia campanularia* (Boraginaceae)	Boragine, echium, Dianthus flower (convulvulus?) (Petersen) Borago (ELV)	*Drosera* sp.		

56v		Aristolochia? (ELV)	*Cycas revoluta*		
57r	*Ipomoea nil* (Dioscoreaceae)	Geranium (ELV)	*Sherardia arvensis*		
65r	*Valeriana albonervata* (Valerianaceae)	Rue, dropwort, spiraea filipendula (Petersen) Geranium (ELV)	*Alchemilla vulgaris*	*Comarum palustre*	
65v		Chamomile. St. Jacobskraut (Petersen) Anthemis, some kind of Chamomila (ELV)	*Centaurea cyanus*	*Scandix pecten-veneris*	
66v		primula? (Petersen) Aristolochia? (ELV)	*Satureja montana*		
90v	*Caulanthus heterophyllus* (Brassicaceae)	not Löwenklau? Sp[h]ondylium? (Petersen)	*Eruca vesicaria*	*Ruta chalepensis*	
93r	*Helianthus annuus* (Asteraceae)	Sunflower (Helianthus), (like coxcomb?) (O'Neill)	*Cynara cardunculus*	*Helichrysum arenarium*, sunflower [sic]	
93v	*Manihot rubricaulis* (Euphorbiaceae)		Lupines		
94r	*Duranta erecta* (*D. repens*) (Verbenaceae)	Botriculum, (Lunaria? Mondfarn?) (Petersen)	*Botrychium lunaria*	*Botrychium lunaria*	
94v(1)			*Agrostemma githago*		
94v(2)			*Glycyrrhiza glabra*		
94v(3)			*Plantago lanceolata*	*Narcissus tazetta*	
95r	*Actaea rubra* f. *neglecta* (Ranunculaceae)	[illeg] (Petersen)	*Sambucus nigra*	*Actaea spicata*	
95v		Wermut, Artemisium, Absinthium	*Althaea rosea*	*Fumaria officinalis*	
96r		Dipsacus calendula (O'Neill)	*Angelica archangelica*	*Chenopodium*	
96v	*Dioscorea mexicana* (Dioscoreaceae)	Smilax tracheia or Chenopodium (bonus atruplex) good king Henry, atruplex or hastata (Petersen)	*Tamus communis*	*Fumaria officinalis*	

Petersen Theodore Petersen, *Holm* Theodor Holm, *ELV* Ethel Lilian Voynich, *O'Neill* Rev. Dr. Hugh O'Neill, *Manly* John M. Manly, *illeg Nomen illegitimum* (Latin for illegitimate name)

and a few by Bax (2014), a linguist. These are compared with the identifications of Tucker and Janick from Chap. 4. The notes of Petersen (Zandbergen 2016) were originally stored at the George C. Marshall Foundation (now missing) and also included identifications attributed to Ethel Voynich, Rev. Dr. Hugh O'Neill, and Theodor Holm, a Danish-American botanist who spent his last year at the Catholic University. Practically all the plants in Petersen's list are supposedly European, but he did mention O'Neill's' sunflower for folio 93r. The Sherwoods and Velinska have identified 130 plants in the Herbal section of the *Voynich Codex*, and some of the Velinska identifications are based on those of the Sherwoods. Many of their identifications have been accepted by other bloggers on the internet. We have commented on two of Bax's identifications in Chap. 11. Our conclusion is that their identifications of European species are incorrect. Three egregious misidentifications by the Sherwoods and Velinska are presented below and shown in Fig. 16.7.

Folio 93r The tenets of identifying phytomorphs have already have been discussed (Tucker and Janick 2017) and need not be repeated here except to note that nonbotanists usually start from the species, rather than the family, and produce egregious identifications. The phytomorph in folio 93r has been identified as the sunflower **Helianthus annuus**, Asteraceae, subfamily Heliantheae by Rev. Dr. Hugh O'Neill (1940), Tucker and Talbert (2013), and Tucker and Janick (2016), and has been confirmed by a number of *Helianthus* authorities, as discussed in Chap. 4. The Sherwoods first identified this plant as an artichoke (*Cynara cardunculus*, Carduoideae) with tuberous roots and blue thistle flowers that form at the end of a tall erect stalk, but later added that although the flowers resembled the flower of folio 93r (they did not) the leaves and roots did not and chose horse-heal (*Inula helenium,* subfamily Inuleae), a "sunflower-like" plant growing wild in Europe (0.91–1.52 m tall, with large leaves, and many yellow ray flowers, 5.08 cm in diameter, and thick, branching roots). However, the small flowers did not fit the identification. Velinska chose *Helichrysum arenarium,* subfamily Gnaphalieae, but the small flowers and multibranched stem did not fit either. We considered these two identifications to be correct regarding the family, but egregious on the subfamily and genus. The identification of folio 93r is critical to our interpretation of the *Voynich Codex* and we invite readers to examine Fig. 16.8 carefully.

Pelling's reluctance to accept the identification of the sunflower was equally problematic. He stated (2006: 126) that O'Neill "*controversially identified the plant in f33v (right in the centre of Quire E) as a sunflower*" and cited Jorge Stolfi, who stated that this image was only a moderate match to contemporary sunflowers, but also that it was quite different than the varieties initially brought across the Atlantic. He was incorrect. The image in folio 33v was identified by Tucker and Janick (2016) as a *Psacalium* or *Pippenalia* species. The sunflower image of O'Neill was in folio 93r (quire 17) and not folio 33v. This sunflower image matches, almost exactly, the first published European illustration by Rembert Dodoens in 1568, and engravings of Francisco Hernández based on drawings made by indigenous artists between 1570 and 1577.

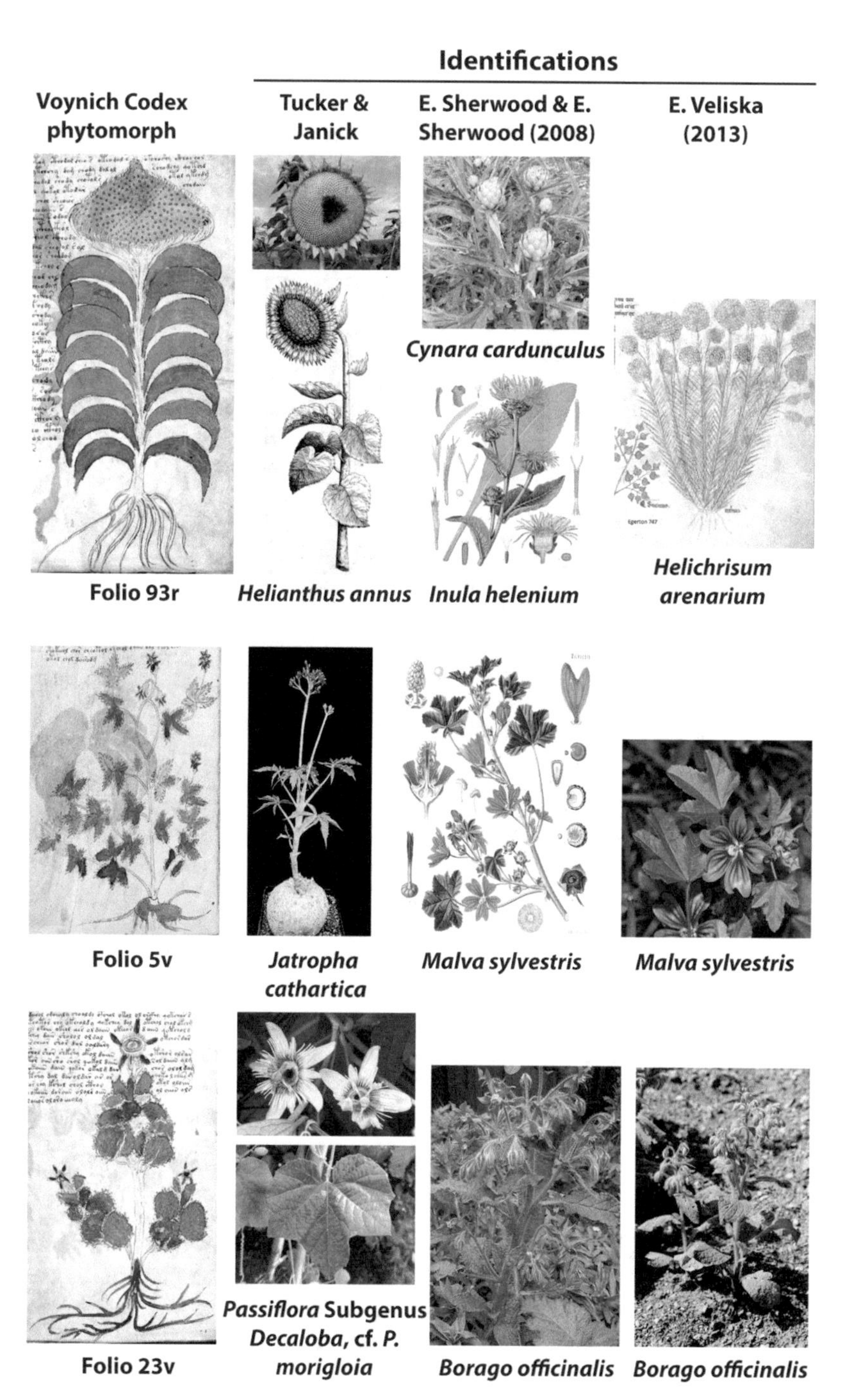

Fig. 16.7 Plant identifications of three phytomorphs in the *Voynich Codex* by Tucker and Janick (2016), Sherwood and Sherwood (2008), and E. Velinska (2013a)

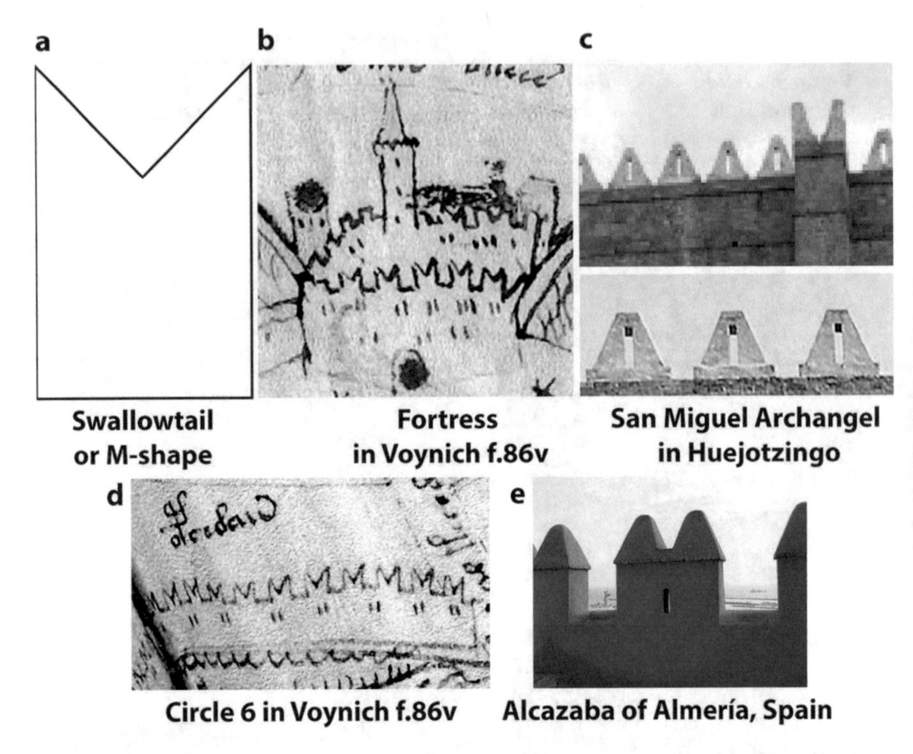

Fig. 16.8 Merlons. (**a**) "M" or swallowtail merlon shape; (**b**) image of monastery fortress in circle 2 of folio 86v of the *Voynich Codex* with swallowtail merlons; (**c**) pierced pyramidal merlons in the San Miguel Arcángel fortress at Huejotzingo, but note that the merlons on the stepped buttresses appear to be swallowtail; (**d**) swallowtail merlons in the passageway of circle 6 of folio 86v; (**e**) modified swallowtail-like merlons in the Alcazaba of Almería in Spain, tenth century

Folio 5v This phytomorph was identified as jicamilla (*Jatropha cathartica*, Euphorbiaceae) by Tucker and Talbert (2013) and characterized by palmate leaves, small red flowers, and large tuberous roots. The Sherwoods and Velinska identified it as mallow, *Malva sylvestris*, Malvaceae, but the lack of tuberous roots negates this identification.

Folio 23v There is no doubt that this species is a passion flower (*Passiflora*, subgenus *Decaloba,* cf. *P. morifolia*) and the distinctive flower is a giveaway. The phytomorph even shows the petiolar glands. In contrast, both Sherwoods classified this as borage (*Borago officinalis*) Boraginaceae, but the typical small blue flowers of borage are a very poor fit. We defy anyone to dispute this identification.

Swallowtail Merlons

The swallowtail (M-shaped) merlon (Fig. 16.8a) atop the fortress illustration of circle 2 in folio 86v (Fig. 16.8b) has been seized upon by many Voynich researchers to date and identify the author of the manuscript. We have associated this illustration with the monastery fortress at Huejotzingo that still exists (see Chap. 10, Figs. 10.8b, c). Swallowtail merlons are found in fifteenth century Italian castles and a similar style is found in Spain (Fig. 16.8e). Most of the merlons on the present monastery fortress at Huejotzingo are a different type, referred to as "aerated" or pierced pyramidal (Fig. 16.8c). The merlons atop the stepped buttresses have been described as "trapezoidal" by Angulo Iñiguez (1983), but "*swallow-tail battlements crown the stepped buttresses*," according to Perry (1992, page 97). Furthermore, this monastery was completed in the 1570s, before the *Voynich Codex* was composed. Thus, the artist clearly drew the fortress before it was finished and must have used the swallowtail merlons in the stepped buttresses as a guide. There is another drawing of swallowtail merlons in the *Voynich Codex* (Fig. 16.8d) that suggests that this might simply be the artist's style for merlons. We conclude that the swallowtail merlons illustrated in circle 2 of folio 86v are not an impediment to the conclusion that the image represents the current monastery fortress at Huejotzingo (Fig. 16.8c).

Archer Clothing

The clothing of the archer (sign of Sagittarius) in the zodiac (see Chap. 8, Table 8.1) has been used as evidence of a medieval date for the *Voynich Codex* by Velinska (2013b). However, the same medieval costume is found in the Apocalypse paintings of Juan Gerson (see Chap. 15, Fig. 15.6d) and we have proposed that he is the illustrator of the *Voynich Codex*. We conclude that the medieval costume cannot be used to date the *Voynich Codex*.

Boleite

The boleite that we identified (see Chap. 4, Fig. 4.58) is found in many locations of the world in trace quantities, in addition to the New World. However, we agree that the quality and quantity is unique to the southwest USA and Mexico. We realize that this infers Aztec exploration into this area or an exchange. We do not have evidence for this, but it is plausible in light of the wide array of plants in Aztec botanical gardens (Granziera 2005).

Astronomical Circles

The astronomical circles are typical of European astronomy. We agree that these circles are in general unexplainable and we need a better analysis of their meaning. The magic circle in folio 67v(1) (Fig. 16.9) is typical and appears to have Aztec allusions (Fernando Moreira, personal communication). It shows a circle at each corner (one clearly a T-O symbol for Earth), each with three or four faces in connected spheres; a central planet, presumably Venus; and both the rising and setting moon and sun. This bears certain resemblance to the famous Aztec sun stone (also called the Aztec calendar stone; Fig. 16.9) of about 1520, which has four squares, each containing three or four round circles (each assumed to represent different suns that were destroyed) surrounding a central image of the solar deity, *Tonatiuh* (with the tongue in the form of a sacrificial knife and holding a human heart in each of his outstretched, clawed hands). The elements highlighted in color show similarity to the Voynich image of the cosmos. The calendar stone may have been known by the illustrator of the *Voynich Codex*.

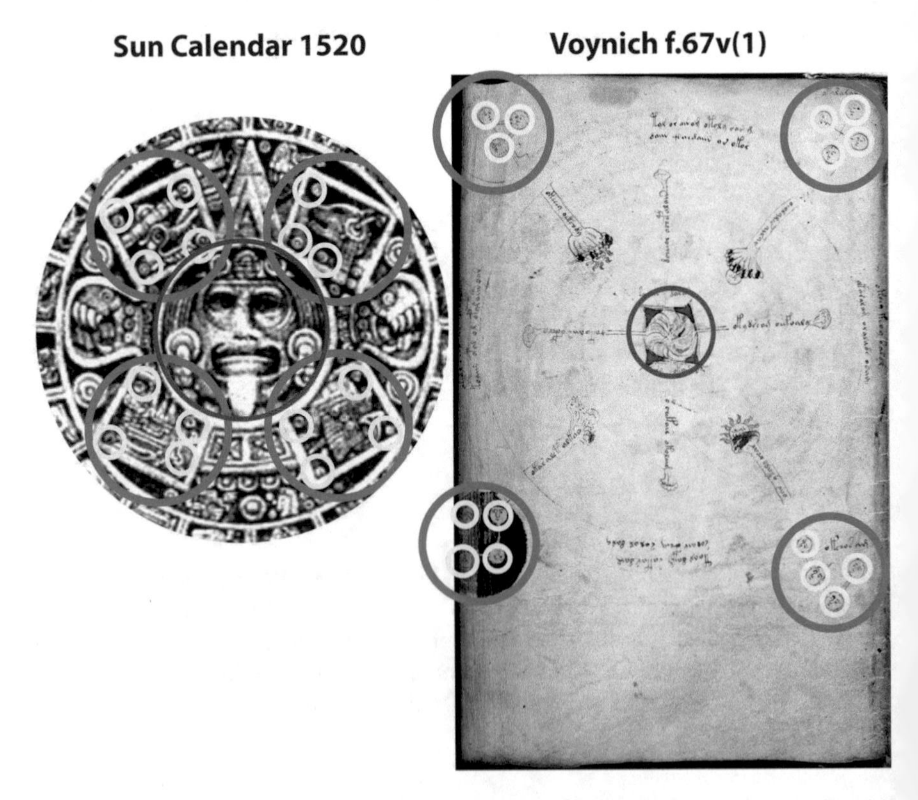

Fig. 16.9 A comparison of the Aztec calendar stone (about 1520) and folio 67v(1) of the *Voynich Codex* showing similar elements highlighted in blue, yellow, and red circles

Conclusion

We believe that the *Voynich Codex,* when fully deciphered, will prove to be one of the most important manuscripts in Aztec botany and herbal medicine of the early colonial period and will show new insights into the culture of colonial New Spain. It deserves to be fully deciphered and translated. We believe that the key has been found and that we have unraveled many of its mysteries. The identification of plants, animals, and a mineral determines that this was a Mesoamerican work. Internal evidence dates the manuscript to the sixteenth century in central Mexico. We surmise that the author and illustrator were probably two different people and, based on autographs embedded in the first botanical illustration, we surmise that the author was Gaspar de Torres, born in Santo Domingo of Spanish parents, one time master of students at the Colegio de Santa Cruz at Tlatelolco, Mexico, and governor of Cuba in 1580, whereas the illustrator was Juan Gerson, an indigenous Indian painter (*tlacuilo*) known for the ceiling paintings in the church at Tecamachalco. We estimate from internal evidence that the manuscript was created between 1565 and 1572. Various routes to it ending up in the court of Rudolf II have been proposed. Iconographic analysis indicates that the work fits into the genre of Spanish-*Tlacuilo* art. Progress has been made in decoding the alphabet, and a number of Nahuatl, Spanish, and Arabic words (all known to the elite of New Spain in the sixteenth century) have been deciphered. However, the main language of the *Voynich Codex* has not been deciphered and remains a mystery. We suspect that it is a synthetic (constructed) language, possibly made up of a lost dialect, lost language, or a Nahuatl lingua franca (Dakin 1981).

We conclude that the *Voynich Codex* was, in large part, a medicinal herbal and that the author and illustrator, both polymaths, expanded it to include features that were religious (ritual bathing, kabbalah), astrological (zodiac), and cosmological (astronomical). We present this analysis in the hope that it will encourage other researchers to join our quest to fully decipher the *Voynich Codex*, a document unfiltered through Spanish or Inquisitorial censors and of utmost historical importance.

Literature Cited

Angulo Iñiguez, D. 1983 [1932]. *Arquitectura Mudéjar Sevillana de los siglos XIII, XIV y XV*. Sevilla: Servicio de Publicaciones del Ayuntamiento de Sevilla. (originally published 1932 as "Discurso Inaugural" at the Universidad de Sevilla.)

Azpeitia, R.C. 1972. *Juan Gerson, Pintor indigena del siglo XVI—simbolo del mestizage—Tecamachalco Puebla*. Mexico: Fondo Editorial de la Plastica Mexicana.

Barlow, M. 1986. The Voynich manuscript—By Voynich? *Cryptologia* 10: 210–216.

Bax, S. 2014. A proposed partial decoding of the Voynich script. www.stephenbax.net. Version January 2014.

Brotherston, Gordon 1998. European scholasticism analyzed in Aztec terms: The case of the Codex Mexicanus. Boletim do CPA, 5/6 Campinas, Jan/Dec.169–181.

Codex Mexicanus.1571–1590. World Digital Library. https://www.wdl.org/en/item/1528t4/view/1/1

Dakin, K. 1981. The characteristics of a Nahuatl *lingua franca*. In *Nahuatl studies in memory of Fernando Horcasitas*, ed. Frances Karttunen, 55–67. Texas Linguistic Forum 18, University of Texas at Austin.

de Sahagún, B. 1951–1982. *Florentine Codex. General history of the things of New Spain*. [1540–1585] 12 vol. Trans. A.J.O. Anderson, and C.E. Dibble. Salt Lake City: Univ. Utah Press.

Emmart, E.W. 1940. *The Badianus manuscript*. Baltimore: Johns Hopkins Press.

Granziera, Patrizia. 2005. Huaxtepec: The sacred garden of an Aztec Emperor. *Landscape Research* 30 (1): 81–107.

Janick, J., and A.O. Tucker. 2017. Were Juan Gerson the illustrator and Gaspar de Torres the author of the Voynich Codex. *Notulae Botanicae Horti Agrobotanici Cluj-Napoca* 45 (2): 343–352 in press.

Knight, K. 2009. *The Voynich manuscript*. MIT. http://www.isi.edu/natural-language/people/voynich.pdf. 1 May 2014.

Lara, J. 2004. *City, temple, stage: Eschatological architecture and liturgical theatrics in New Spain*. Notre Dame: University of Notre Dame Press.

———. 2008. *Christian texts for Aztecs: Art and liturgy in Colonial Mexico*. Notre Dame: University of Notre Dame Press.

———. 2013. Temples of the Sun: Eschatological and biblical inspiration in Franciscan missionary architecture. Chapter 15. In *From La Florida to La California: Franciscan evangelization in the Spanish borderlands*, ed. T.J. Johnson and G. Melville. Berkely: The Academy of American Francisdan Hiustoryt.

Mackenzie, D. 1997. *A manual of manuscript transcription for the dictionary of the Old Spanish language*. Fifth ed. rev. R. Harris-Northall. Madison: Hispanic Seminary of Medieval Studies.

Montemorro, M.A., and D.H. Zanette. 2013. Keywords and co-occurrence patterns in the Voynich manuscript: An information-theoretic analysis. *PLoS One* 8 (6): e66344.

Morrill, P.C. 2014. *The Casa del Deán: New world imagery in a sixteenth century Mexican mural cycle*. Austin: Univ. Texas Press.

Pelling, N. 2006. *The curse of the Voynich: The secret history of the World's most mysterious manuscript*. Surbiton, Surrey: Compelling Press.

Perry, R. 1992. *Mexico's fortress monasteries*. Santa Barbara: Espadaña Press.

SantaColoma, R. 2009. Dating the Armadillo/The 1910 Vonich theory. https://photo57.wordpress.com/2009/05/27/dating-an-armadillo

Sherwood, E., and E. Sherwood. 2008. The Voynich botanical plants. http://www.edithsherwood.com/voynich_botanical_plants/

SilverMoon. 2007. The imperial college of Tlatelolco and the emergence of a new Nahua intellectual elite in New Spain (1500–1760). PhD thesis, Duke University, Durham, NC.

Tiltman, J.H. 1967. The Voynich manuscript: "The most mysterious manuscript in the world." National Security Agency/Central Security Service, Fort George G. Meade, Maryland. DOCID:631091. [released 23 April 2002]. www.nsa.gov/public_info/_files/tech_journals/voynich_manuscript_mysterious.pdf. 5 June 2017.

Tucker, A., and J. Janick. 2016. Identification of the phytomorphs of the Voynich Codex. *Horticultural Reviews* 44: 1–58.

———. 2017. Identification of plants in the 1584 murals of the Casa del Deán, Puebla, México. *Notulae Botanicae Horti Agrobotanici Cluj-Napoca* 45 (1): 1–8. https://doi.org/10.15835/nbha45110692.

Tucker, A.O., and R.H. Talbert. 2013. A preliminary analysis of the botany, zoology, and mineralogy of the Voynich Manuscript. *Herbalgram* 70(100). American Botanical Council. www.herbalgram.org

Varey, S., ed. 2000. *The Mexican treasury: The writings of Dr. Francisco Hernández*. Stanford: Stanford University Press.

Velinska, E. 2013a. The Voynich manuscript: Plant id list. ellievelinska.blogspot.com/2013/07/the-voynich-manuscript-plant-id-list.html. 5 June 2017.

Velinska E. 2013b. The Voynich manuscript: Hunting Tapestries. ellievelinska.blogspot.com/2013/09/the-voynich-manuscript-hunting.html. 5 June 2017.

Zandbergen, R. 2016. Pre-1960's tentative herb identifications. http://www.voynich.nu/extra/herb_oldid.html. 7 July, 2017.

Zandbergen, R. 2017. The history of the Voynich MS. www.voynich.nu/history.html. 5 June 2017.

Afterword: Conjectures on the Origins of the Voynich Codex

Jules Janick and Arthur O. Tucker

It is a riddle, wrapped in a mystery, inside an enigma; but perhaps there is a key.

Winston Churchill.

Churchill's 1939 aphorism refers to Russia, but it fits the *Voynich Codex* perfectly. At the end of this work, we propose a narrative of its origins based on key evidence gleaned from the *Codex* and informed conjecture. We perform this exercise with trepidation because we know that speculation is involved, and we are well aware that we are in danger of exposing ourselves to ridicule. But at this stage of our lives we take this opportunity to be simultaneously brave and foolish. We know that some of what we present may prove to be illusionary, but we aim to at least approximate the truth and hopefully lead others to join us in unraveling the *Voynich Codex*.

We take it as a given based on plant and animal identifications and the kabbalah map of central Mexico (folio 86v) that the *Voynich Codex* is a Mexican document composed between 1565 and 1572. We accept that the text seems to be in more than one hand, that the Voynichese symbol decoding is not completely accurate, and that the illustrations are a combination of botanical accuracy, fantasy, and humor, all at the same time. The document is complex and bizarre. Surely stranger than fiction is its circuitous route from its origins in Mexico in the sixteenth century, to the court of Rudolf II in Prague in the early seventeenth century, to its discovery by Polish book dealer Wilfrid Voynich at a Jesuit college in the twentieth century, and to its final resting spot as Manuscript 408 of the Beinecke Rare Book and Manuscript Library at Yale University in Connecticut, USA.

Can we conceive of a possible series of events that could explain the origins of this codex? Our conjecture evolves from the clue that the name of Gaspar de Torres and the ligated initials of Juan Gerson are embedded in the first botanical illustration in folio 1v. We propose that Gaspar de Torres is the instigator, author, and *force majeure* of the *Voynich Codex,* and that his motive was to organize a compendium

J. Janick, A. O. Tucker, *Unraveling the Voynich Codex*, Fascinating Life Sciences, https://doi.org/10.1007/978-3-319-77294-3

of indigenous knowledge current in New Spain. Juan Gerson, an indigenous painter, or *tlacuilo*, was signed on as the illustrator, and there are remarkable similarities among some of his paintings on the Tecamachalco chapel ceiling (completed in 1562), the murals at the Casa del Deán in Puebla (completed in 1584), and some of the Voynich illustrations. Early in our studies, we profiled the author/artist of the *Voynich Codex* based on the images and postulated that the person or team was a Nahua who attended El Colegio de Santa Cruz de Tlatelolco (in present-day Mexico City; also called Colegio de Santa Cruz), an institution organized by the Franciscans to train sons of the nobility to become priests (before this was prohibited). We turned out to be on the right track. Juan Gerson was indigenous, but trained at another school for sons of nobility, La Escuela de San José de los Naturales, and Gaspar de Torres, a *criollo* (a Spaniard born in New Spain), was associated with the Colegio de Santa Cruz as a staff member.

Gaspar de Torres was an intriguing person with an extraordinary career. Biographical information obtained from scattered sources provided a good fit to our supposition. He came from a wealthy, slave-holding family in Santo Domingo on the island of Espanola (present-day Hispaniola). He was highly educated, receiving a medical degree and a law degree from the Imperial and Pontifical University in Mexico City. From 1568 to 1572, he was employed as *maestre de nino*s (master of students) at the Colegio de Santa Cruz. He was also a practicing *licentiate* (lawyer) involved in Indian land rights. He had political aspirations and in 1580 briefly served as governor of Cuba, but was accused of misappropriation of funds by the royal treasurer and impeached. He was complex. He clearly had the resources to pursue his goal of providing a compendium of knowledge that included medicinal and botanical plants (Pharmaceutical and Herbal sections), Aztec culture (Balneological section), zodiac and cosmology (Astrological and Cosmological sections), and perhaps medical prescriptions, poetry, incantations, or folk wisdom (Recipe section). The map in folio 86v indicates that he was familiar with kabbalah and Puebla de Los Angeles, a city founded in 1530 by the Franciscan friar Toribio de Benavente, known as Motolinía, one of the famous Twelve Apostles of Mexico. Gaspar de Torres was clearly influenced by the former dean of the Colegio de Santa Cruz, Bernardino de Sahagún.

Sahagún (ca. 1499–1590), born Bernardino de Rivera in Sahagún, Spain, was a Franciscan friar, missionary priest, and ethnographer of the Aztecs. He arrived in New Spain in 1529, learned Nahuatl, and spent more than 50 years studying the culture and history of the Aztecs. He was one of the founders of the Colegio de Santa Cruz in 1536. He served as dean in 1560, but retired to the main convent of San Francisco in Mexico in 1563. Sahagún made the Colegio de Santa Cruz a center for research, translation, and literary production, and it became a focus for the intellectual elite of New Spain. In the 1540s, Motolinía charged Sahagún with the task of collecting information on the Aztecs, resulting in the seminal work: *Historia General de las Cosas de Nueva España* (*General History of the Things of New Spain)*, written in Nahuatl and Spanish. It is also known as the *Florentine Codex*. This enormous compilation, consisting of 12 books, 2400 pages of text, and 2468 illustrations by native artists, is now considered one of the great examples of ethnology and is a key resource for understanding Aztec culture. Indigenous students at the Colegio de Santa Cruz served as translators and mediators. The work was translated into Nahuatl in 1569, overlapping the tenure of Gaspar de Torres, and the final copy was completed in 1585. Three revisions exist.

Sahagún's masterpiece of ethnology was clearly an inspiration for Gaspar de Torres and we assume that many of the processes in its production must have been similar to those of the *Voynich Codex*, such as the use of collaborators and translators. Torres was likely in contact with Pedro de San Buenaventura, a former student, who instructed Sahagún on the calendar, and included medicine and animals in his *Himnos de los Dioses*. Torres must have been aware of the 1552 Aztec herbal (now known as the *Codex Cruz-Badianus*), written by Martinus de la Cruz, a Nahua staff physician, and translated into Latin by Juannes Badianus, a Nahua teacher of Latin, assuming that copies remained at the Colegio de Santa Cruz, as the original was shipped to Spain as a gift to Charles V shortly after its completion.

Gaspar de Torres chose Juan Gerson to organize the artwork and it is likely they were true collaborators. However, a number of different students could have been conscripted to do the tinting in the Herbal section because at least five images (folios 10r, 13v, 29r, 35r, and 41r) show a very sloppy application of color. An analysis of the illustrations indicates that Juan Gerson was the sole illustrator of the *Voynich Codex*. For example, it is clear that the illustrations in the Herbal and Pharmaceutical sections are by the same hand, based on a comparison of the same plant in both sections (*Ipomoea arborescens*, in folios 1v and 101v(2), and *I. pubescens* in folios 2v and 101v(3). The crude faces in the roots of the herbs are similar to the faces in the Cosmological section, indicating that all are by the same artist (see Chap. 9, Fig. 9.6). Similarly, the zodiac and balneological drawings are by the same hand, as the nymphs in both appear to be the same woman. It needs to be stressed that the *Voynich Codex* at present is known based on the illustrations of Juan Gerson because the text remains untranslated. Juan Gerson had fantastic and inventive skills. His drawings were made with practically no revision, and obviously rapidly executed, often showing wit and whimsy. Many of his drawings display complex variations on a theme. These include the 517 nymphs in various poses, all but 30 nude, and clearly appearing to be the same woman (a paramour?), the 42 apothecary jars in a dazzling display of forms, many sketches of animals in the marginalia and as zodiac signs, and an extraordinary display of virtuosity in displaying plant roots fantastically arranged into forms, such as animals and complex inventive patterns.

Gaspar de Torres chose to keep the text secret and wrote it in code. He had good cause, because the Inquisition would have considered much of the work to be heretical and the penalties were severe. The language of the *Voynich Codex* appears to be an invented synthetic language with many of the symbols selected from common scribal abbreviations (as shown in Table 5.5 of Chap. 5), and from the *Codex Osuna* of 1565. Based on the decipherment of Arthur O. Tucker and Rexford H. Talbert (2013), the Voynichese words display many cognates of Nahuatl, but the main language is not Classical Nahuatl and remains unknown to us.

We assume that the information on the plants in the Herbal section came from translators, who may have obtained identifications in different languages. Illiterate

Indian translators probably obtained information in the different languages they encountered, much as Francisco Hernández did from 1570 to 1577. Presumably, the synthetic language would include words in Nahuatl or some dialect, perhaps now extinct, a mixed Nahuatl lingua franca of commerce, or some other Mesoamerican languages, in addition to Spanish. We do not think that a cipher system was involved because the text is written in a fluid, rapid style without corrections. Furthermore, all attempts to discover a cipher system have failed gloriously and cryptological analyses have come back empty.

There are a number of characteristics of the *Voynich Codex* that point to Gaspar de Torres. The plants appear to be mainly medicinal plants and the most common plant of Mexico, maize, is conspicuously absent here as a food plant, although it may have been included in some of the missing pages. Medicinal plants would have been of interest to Gaspar de Torres, who was medically trained. The inclusion of a kabbalah image suggests that he might have had Jewish roots and genealogical investigations have confirmed that his grandparents were from converso families. This would help to explain the inclusion of a Hebrew letter *dalet* (ד) in the kabbalah map in the top sephiroth, representing *Ain Soph,* the infinite or God (see Chap. 10). The name Torres was common among *conversos* (converts) in New Spain. Luis de Torres, Christopher Columbus' interpreter and the first Spaniard to land in the New World, converted just before the voyage.

Sixteenth Century Works on Indigenous Medicine

Gaspar de Torres did not write in a vacuum. There were many books written in the sixteenth century that included information on native medicine practices and associated ceremonies in New Spain. Many of these works would be published decades or centuries later, often outside the New World, partly because of the Inquisition. They include:

1. Gonzalo Fernández de Oviedo y Valdés (1478–1557), *Historia General y Natural de las Indias*, Madrid, 1851–55.
2. Toribio de Benavente (Motolinía) (1482–1568), *Historia de los Indios de la Nueva España*, Mexico, 1858.
3. Bartolomene de Las Casas (1484–1566), *Historia de las Indias*, Madrid, 1875.

(4. and 5. Martinus de la Cruz & Juannes Badianus, translator, *Libellus de Medicinalibus Indorum Herbis* (*A Little Book of Indian Medicinal Herbs*), *1552, Codex Cruz-Badianus*, Baltimore, 1939.

6. Nicolás Monardes (1493–1588), *Historia Medicinal de las Cosas que se traen de Nuestras Indias Occidentales*, Seville, Spain, 1565.
7. Bernardino de Sahagún (ca. 1499–1590), *Historia General de las Cosas de la Nueva España* (*General History of the Things of New Spain* or *Florentine Codex*), Mexico, 1829–30.

8. Francisco Hernández de Toledo (1514–1587), *Resumen* de F. Ximénez, Mexico, 1615, and *Resumen* de A. N. Recchio, Rome, 1638, and *Edición* matritense, Madrid, 1790.
9. Gerónimo de Mendieta (1525–1604), *Historia eclesiástica Indiana*, Mexico, 1870.
10. Francisco Bravo (ca. 1525–ca. 1595), *Opera Medicinalia*, Mexico, 1570.
11. Agustín Farfan (ca. 1532–1604), *Tractado breve de anotomia y cirugia, y de algunas enfermedades* (*Brief Treatise on Anatomy and Surgery, and of some illnesses*), México, 1579, and *Tractado Brebe de Medicina*, (*A Brief Treatise on Medicine*), 1592.
12. Alonso López de Hinojosos (1535–1595), *Summa y Recopilación de Cirugía* (*Sum and Compilation of Surgery*), Mexico, 1578.
13. Juan Bautista de Pomar (1535–1601), *Relaciones Geográficas*, Mexico, 1891.
14. Diego Durán (1867), *Historia de las Indias de Nueva España e Islas de Tierra-Firme*, Mexico, 1867–1880.
15. José de Acosta (1540–1600), *Historia Natural y Moral de las Indias*, Seville, 1590.
16. Gregorio López (1542–1596), *Tesoro de Medicinas*, Madrid, 1708.
17. Antonio de Herrera y Tordesillas (1549–1625/26), *Décadas*, Madrid, 1601–1605.
18. Juan de Torquemada (ca. 1562–1624), *Monarquía Indiana*, Seville, 1615.
19. Juan de Barrios (1562–1645), *Valdadera Medicina, Cirugía y Astrología, en tres libros*, Mexico, 1607.
20. Juan de Cárdenas (1563–1609), *Primera Parte de los Problemas y Secretos Maravillosos de las Indias*, Mexico, 1591.

Bravo's *Opera Medicinalia* was dedicated to Gaspar de Torres' cousin, Luis de Villanueva, indicating interaction between these authors.

The transfer of the *Codex* from New Spain to Rudolf II will forever remain a mystery, but two paths were plausible. It could have been picked up by Francisco Hernández de Toledo, who was sent by King Philip II New Spain in 1570 as "*general protomedico of our Indies*" to analyze the medicinal plants of New Spain, returning in 1577. Francisco Hernández willed his books to Philip II and the *Codex* could have been passed off to his maternal nephew Rudolf II, who had lived with him as a child and was to become the Holy Roman emperor in Prague. Another more likely path was that the *Codex* was kept by Gaspar de Torres when he left Tlatelolco for Santo Domingo in 1572, and it could have remained in his home or have been brought there after he left in 1580. It was conceivable that the *Codex* was swept up in Sir Frances Drake's famous raid of Santo Domingo in 1586. Drake returned to London in December of that year, after raiding Cartagena (in present-day Colombia) and St. Augustine (in present-day Florida), and rescuing the colonists at Roanoke (in present-day North Carolina). The *Codex* would have belonged to the government and come into the possession of Sir Francis Walsingham, spymaster of Queen Elizabeth I. We favor this path, because the leader of the land raid was Francis Walsingham's stepson, Christopher Carleill. Francis Walsingham had had contact with John Dee while he was in Třeboň (in the present-day Czech Republic) through a courier named Francis Garland. According to a 1565 letter by Juannes Marcus Marci to Athanasius Kircher, Rudolf II paid 600 ducats to the bearer of the *Codex* and this sum was in John Dee's possession on 17 October 1586, based on an entry in his diary (see Chap. 15).

Clearly, all these suppositions await the final translation of the document, a feat that we had hoped to achieve, but have not fully accomplished. It is our fervent hope that this work will encourage other scholars to follow our lead. Success in the final decipherment will ultimately reveal the true origins of the *Voynich Codex*, which hopefully will no longer be the most mysterious book in the world.

Appendices

J. Janick, A. O. Tucker, *Unraveling the Voynich Codex*, Fascinating Life Sciences, https://doi.org/10.1007/978-3-319-77294-3

Appendix A

Table 1 Alphabetized list of Voynichese words in folio 86v and circles 5, 7, and 9 for city names compared with towns in the states of Mexico (District Federal, Guanajuato, Hidalgo, Mexico, Puebla, Tlaxcala, Veracruz, and Queretaro) that were obtained from the Directory of Cities and Towns in Mexico. (Falling Rain Genomics 2015)

Circle	Voynichese	Decoded Latin	Similar Mexican towns (states)
5		āhuāshnoll	
9		āhuoe	
9		āhuol	Ahualulca, Ahualulco (Guanajuato)
7		āhuollā	Ahualulca, Ahualulco (Guanajuato)
9		ātlā	Atla (Puebla)
7		ātlaachi	Atlaco (Hidalgo)
5		ātlācâoe	Atlaco (Hidalgo)
5		ātlachi	Atlaco (Hidalgo)
9		ātlachi	Atlaco (Hidalgo)
9		ātlāei	
9		ātlāei	
7		ātlahocâ	Atlatlahuaca, Atlatlahuca (Mexico)
5		ātlānml	Atlan (Puebla, Hidalgo)
7		ātlichi	Atlaxco, Atlixco (Puebla), Atlicos (Veracruz)
7		ātlmchi	
5		ātlmchi	
5		ātlmchoya	Atlmozoyahua (Veracruz)
7		ātlmi	
5		ātlshchi	
9		ch?ān	Chicayan (Veracruz)
9		chālli	Chalma (Puebla)
9		chāpoi	
7		mtlaachi	Manuel Avila Camache, Machanche (Queretaro)
9		ochāshni	
7		ohaahi	
7		ohuoshnoe	

Appendix B

Table 2 Alphabetized list of nymphs with locations in the zodiac folios. Duplicate words are in boldface

Voynichese words	Latin	Zodiac	Folio	Ring (outer ring = 1)	Order (12 o'clock = 1, clockwise)
[illegible]	?tloiâ	Gemini	71v(3)	5	2
[illegible]	āaaaān	Libra	72v	2	12
[illegible]	āaaācâ	Libra	72v	4	8
[illegible]	aaaāchi	Sagittarius	73v	1	4
[illegible]	āaaāchi	Libra	72v	4	6
[illegible]	āaaācholl	Cancer	71v(4)	2	8
[illegible]	āaaācholl	Virgo	72r(2)	2	18
[illegible]	āaaanoll	Cancer	71v(4)	4	2
[illegible]	āaaātli	Libra	72v	2	16
[illegible]	āaachai	Virgo	72r(2)	2	17
[illegible]	āaachi	Leo	72r(1)	4	5
[illegible]	āaachi	Virgo	72r(2)	4	2
[illegible]	āaachocâi	Virgo	72r(2)	4	6
[illegible]	aaah	Sagittarius	73v	5	1
[illegible]	āaai	Libra	72v	4	2
[illegible]	āaan	Libra	72v	2	2
[illegible]	āaānoll (unclear)	Leo	72r(1)	4	8
[illegible]	āaats	Libra	72v	2	20
[illegible]	āahutsi	Libra	72v	2	1
[illegible]	aazi	Sagittarius	73v	3	2
[illegible]	ācâatlocâ	Cancer	71v(4)	6	6
[illegible]	ācâhutsaāeocâ	Cancer	71v(4)	4	9
[illegible]	ācâocâni	Cancer	71v(4)	2	1
[illegible]	ācâoll ācâocuo	Cancer	71v(4)	4	8
[illegible]	ācâtlocâoll	Cancer	71v(4)	2	11
[illegible]	ācâtloe	Virgo	72r(2)	4	5
[illegible]	āchaan	Sagittarius	73v	3	1
[illegible]	āchmchi	Libra	72v	2	6
[illegible]	ācho?oz	Leo	72r(1)	4	6

(continued)

Table 2 (continued)

Voynichese words	Latin	Zodiac	Folio	Ring (outer ring = 1)	Order (12 o'clock = 1, clockwise)
	āchocâ	Libra	72v	2	4
	ācuā	Cancer	71v(4)	6	4
	ācuācâ	Virgo	72r(2)	4	1
	ācui	Libra	72v	2	17
	āeaaai	Leo	72r(1)	2	1
	āeācâoiâ	Gemini	71v(3)	3	13
	āeocâtloiâ	Cancer	71v(4)	4	10
	āeol (unclear)	Sagittarius	73v	5	3
	āeolhi	Gemini	71v(3)	3	6
	āeoll	Virgo	72r(2)	4	10
	āeolli	Sagittarius	73v	3	10
	āeolloiâ	Cancer	71v(4)	2	2
	āeoni	Gemini	71v(3)	3	11
	āhāhuoll	Leo	72r(1)	2	6
	āhuā??? on ocâoan (unclear)	Cancer	71v(4)	2	5
	āhuaaaani	Sagittarius	73v	3	9
	āhucâh	Virgo	72r(2)	4	3
	āhuinaiâ	Pisces	70r	2	7
	āhum?i	Virgo	72r(2)	4	8
	āhumai nocâ	Aries	70v	2	1
	āhumchinch	Virgo	72r(2)	2	9
	āhumchochi	Gemini	71v(3)	1	1
	āhuoahuāiâ	Taurus	71v(1)	4	5
	āhuocââeoe	Taurus	71v(1)	2	8
	āhuocâocâ	Cancer	71v(4)	6	7
	āhuocâocân	Leo	72r(1)	2	9
	āhuocâoeoiâchoz	Taurus	71v(1)	2	7
	āhuoeocâoe	Taurus	71v(2)	4	5
	āhuoll	Scorpio	73r	3	12
	āhuollocâ	Virgo	72r(2)	2	15
	āhutsācâchi	Cancer	71v(4)	2	9
	āitlaachi	Scorpio	73r	5	3

(continued)

Table 2 (continued)

Voynichese words	Latin	Zodiac	Folio	Ring (outer ring = 1)	Order (12 o'clock = 1, clockwise)
[illegible]	ālli	Libra	72v	4	1
[illegible]	āmācâ tsoeoiâ	Taurus	71v(2)	2	8
[illegible]	āmāchochi	Taurus	71v(2)	2	9
[illegible]	āmāi huo?? (unclear)	Cancer	71v(4)	4	3
[illegible]	ānoeo	Cancer	71v(4)	4	7
[illegible]	ānontsaaaā	Cancer	71v(4)	2	3
[illegible]	ānoshshnocâ	Cancer	71v(4)	6	2
[illegible]	āocâmah	Aries	70v	2	5
[illegible]	āoll oeoei	Taurus	71v(2)	2	1
[illegible]	ātktstschi	Aries	70v	2	8
[illegible]	ātl?chi	Gemini	71v(3)	3	10
[illegible]	ātlaā ocââcân oeocâi	Aries	70v	4	2
[illegible]	ātlaāaai ātlocâ ātlaocâoe	Aries	70v	4	3
[illegible]	ātlaaācâ	Aries	71r	4	3
[illegible]	ātlaaācâi	Libra	72v	2	3
[illegible]	ātlaaāch	Virgo	72r(2)	2	10
[illegible]	ātlāaāchi	Taurus	71v(1)	2	2
[illegible]	ātlaaāh	Virgo	72r(2)	4	4
[illegible]	ātlaaāni	Scorpio	73r	3	6
[illegible]	ātlaaaoei	Gemini	71v(3)	3	5
[illegible]	ātlaācâi	Pisces	70r	2	12
[illegible]	ātlaācâi	Aries	71r	2	3
[illegible]	ātlaācâi	Libra	72v	2	18
[illegible]	ātlaācâi	Libra	72v	2	19
[illegible]	ātlaācâoe	Aries	71r	2	2
[illegible]	ātlaāch	Sagittarius	73v	5	7
[illegible]	**ātlaachi**	Scorpio	73r	3	2
[illegible]	**ātlaāchi**	Virgo	72r(2)	2	4
[illegible]	**ātlaāchi**	Virgo	72r(2)	2	12
[illegible]	**ātlaāchi**	Scorpio	73r	3	16
[illegible]	**ātlaāchi**	Sagittarius	73v	1	2
[illegible]	**ātlaāchi**	Sagittarius	73v	1	3

(continued)

Table 2 (continued)

Voynichese words	Latin	Zodiac	Folio	Ring (outer ring = 1)	Order (12 o'clock = 1, clockwise)
[illegible]	**ātlaāchi**	Sagittarius	73v	3	6
[illegible]	**ātlaāchi**	Sagittarius	73v	3	15
[illegible]	**ātlāachi**	Scorpio	73r	3	4
[illegible]	ātlaachicâ	Scorpio	73r	3	1
[illegible]	ātlaachih	Scorpio	73r	3	8
[illegible]	ātlaāhui	Libra	72v	2	13
[illegible]	**ātlaai**	Libra	72v	4	10
[illegible]	**ātlaai**	Scorpio	73r	3	3
[illegible]	ātlaai chaaaicâ	Leo	72r(1)	2	12
[illegible]	**ātlaai** choll	Cancer	71v(4)	2	7
[illegible]	**ātlaai** oni	Gemini	71v(3)	3	4
[illegible]	ātlaaioni	Taurus	71v(2)	4	3
[illegible]	**ātlaan**	Pisces	70r	2	14
[illegible]	**ātlaān**	Scorpio	73r	3	13
[illegible]	**ātlaān**	Scorpio	73r	5	10
[illegible]	**ātlaān**	Sagittarius	73v	3	12
[illegible]	**ātlaān** oeoe	Aries	71r	2	9
[illegible]	ātlaānocâ	Pisces	70r	4	6
[illegible]	ātlaānoiâ	Virgo	72r(2)	2	1
[illegible]	ātlaānon	Virgo	72r(2)	4	11
[illegible]	ātlaāocâchi	Aries	71r	2	1
[illegible]	ātlaātl	Libra	72v	2	7
[illegible]	ātlaātlai noei	Aries	71r	2	7
[illegible]	ātlaatsch	Leo	72r(1)	4	9
[illegible]	ātlācâ ihutsonocâ	Aries	70v	2	9
[illegible]	ātlācââoeoiâ	Aries	71r	4	2
[illegible]	ātlacâi	Scorpio	73r	3	15
[illegible]	ātlācâi	Leo	72r(1)	2	8
[illegible]	ātlācâmch	Aries	71r	4	4
[illegible]	ātlācâmchi	Aries	71r	4	1
[illegible]	ātlācâocâ	Pisces	70r	6	1
[illegible]	ātlācâoe	Taurus	71v(1)	2	1
[illegible]	ātlācâtsi	Aries	70v	2	7

(continued)

Table 2 (continued)

Voynichese words	Latin	Zodiac	Folio	Ring (outer ring = 1)	Order (12 o'clock = 1, clockwise)
	ātlachi	Libra	72v	4	3
	ātlachi	Scorpio	73r	1	3
	ātlachi	Sagittarius	73v	5	10
	ātlāchi	Pisces	70r	2	3
	ātlāchi	Pisces	70r	2	6
	ātlachocâ	Scorpio	73r	3	9
	ātlai	Leo	72r(1)	2	3
	ātlai	Scorpio	73r	3	14
	ātlaocâ	Gemini	71v(3)	3	8
	ātlaocâ	Libra	72v	4	7
	ātlaocâchi	Virgo	72r(2)	2	13
	ātlaocâon	Gemini	71v(3)	1	5
	ātlāochon	Virgo	72r(2)	2	8
	ātlaon oeoichi	Aries	70v	4	4
	ātlcâcchi	Gemini	71v(3)	1	2
	ātlcâchoiâ	Aries	71r	2	10
	ātli	Pisces	70r	2	18
	ātli	Leo	72r(1)	4	2
	ātli	Scorpio	73r	5	4
	ātli	Sagittarius	73v	5	2
	ātli oe	Pisces	70r	2	1
	ātliaatsi	Leo	72r(1)	2	5
	ātliāchi	Pisces	70r	2	19
	ātlicâ	Libra	72v	2	10
	ātlicâi	Aries	70v	2	4
	ātlich	Gemini	71v(3)	3	12
	ātlichi	Pisces	70r	2	13
	ātlichi	Libra	72v	4	9
	ātlmaāchi	Leo	72r(1)	2	4
	ātlmāchi	Libra	72v	2	9
	ātlmāchicân	Aries	70v	2	6
	ātlmātsi	Taurus	71v(2)	4	1
	ātlmchocâ	Taurus	71v(2)	4	2

(continued)

Table 2 (continued)

Voynichese words	Latin	Zodiac	Folio	Ring (outer ring = 1)	Order (12 o'clock = 1, clockwise)
	ātlo?	Virgo	72r(2)	4	12
	ātlo?ocâ	Pisces	70r	4	1
	ātlo?oll	Cancer	71v(4)	2	4
	ātloāno	Scorpio	73r	5	2
	ātlocâ	Gemini	71v(3)	3	1
	ātlocâ	Gemini	71v(3)	3	3
	ātlocâ	Gemini	71v(3)	5	9
	ātlocâ	Scorpio	73r	5	6
	ātlocâ	Sagittarius	73v	1	1
	ātlocâ	Sagittarius	73v	5	4
	ātlocâ oeoe	Pisces	70r	2	10
	ātlocââchi	Taurus	71v(1)	4	1
	ātlocâahu on oloiâ	Taurus	71v(2)	2	7
	ātlocâatli	Aries	71r	2	4
	ātlocâch	Pisces	70r	2	4
	ātlocâchi	Pisces	70r	2	11
	ātlocâchi	Gemini	71v(3)	5	4
	ātlocâchocâ	Pisces	70r	4	8
	ātlocâchoe	Aries	71r	4	5
	ātlocâi	Pisces	70r	2	2
	ātlocâi	Pisces	70r	2	9
	ātlocâi	Aries	70v	4	5
	ātlocâi	Gemini	71v(3)	3	2
	ātlocâi	Leo	72r(1)	2	7
	ātlocâi	Scorpio	73r	1	1
	ātlocâmi tloeoiâchi	Aries	70v	2	10
	ātlocân	Sagittarius	73v	3	4
	ātlocânoe	Aries	71r	2	5
	ātlocâoc	Libra	72v	4	5
	ātlocâocâ	Aries	70v	2	3
	ātlocâocâh	Pisces	70r	2	15
	ātlocâocâi	Aries	71r	2	8
	ātlocâoe	Pisces	70r	4	2

(continued)

Table 2 (continued)

Voynichese words	Latin	Zodiac	Folio	Ring (outer ring = 1)	Order (12 o'clock = 1, clockwise)
	ātlocâoe	Cancer	71v(4)	4	6
	ātlocâoiâ	Pisces	70r	4	3
	ātlocâoiâ	Taurus	71v(2)	2	2
	ātlocâoll	Aries	70v	4	1
	ātlocâoll	Taurus	71v(1)	4	2
	ātlocâon	Gemini	71v(3)	3	15
	ātlocâtsi	Gemini	71v(3)	5	3
	ātlochi	Pisces	70r	2	16
	ātlocóchoe	Pisces	70r	2	5
	ātloe tsāe	Taurus	71v(1)	4	3
	ātloechi	Gemini	71v(3)	1	3
	ātloei	Scorpio	73r	5	1
	ātloemoiâ	Gemini	71v(3)	5	8
	ātloen	Gemini	71v(3)	3	9
	ātloeocâchi	Taurus	71v(2)	2	5
	ātloeoeocâi	Taurus	71v(1)	2	6
	ātloeoiâ	Pisces	70r	4	5
	ātloeoll	Taurus	71v(1)	2	5
	ātloh	Leo	72r(1)	2	2
	ātloiâ	Gemini	71v(3)	5	6
	ātloli	Taurus	71v(2)	4	4
	ātloll	Taurus	71v(1)	2	4
	ātloll ātlol	Taurus	71v(2)	2	6
	ātlon	Pisces	70r	2	17
	ātlon ocâchi	Gemini	71v(3)	3	14
	ātloni	Leo	72r(1)	4	7
	ātloshni	Gemini	71v(3)	1	4
	ātloshnmoh	Gemini	71v(3)	3	7
	ātlotlo?oz	Aries	70v	2	2
	ātsani	Virgo	72r(2)	2	3
	câocâi	Leo	72r(1)	4	4
	chācâoeoiâ	Pisces	70r	4	4
	chocâtsai	Scorpio	73r	3	11

(continued)

Table 2 (continued)

Voynichese words	Latin	Zodiac	Folio	Ring (outer ring = 1)	Order (12 o'clock = 1, clockwise)
[illegible]	choll māhuā	Leo	72r(1)	2	11
[illegible]???? [illegible]	hu???? aaaa?câi (unclear)	Cancer	71v(4)	4	4
[illegible]	iâon	Scorpio	73r	5	9
[illegible]	ihuo?	Libra	72v	4	4
[illegible]	ihuoaaae (unclear)	Sagittarius	73v	3	7
[illegible]	ihuoei	Cancer	71v(4)	6	1
[illegible]	ihuoll ocâācâi	Cancer	71v(4)	2	6
[illegible]	ihuolli	Sagittarius	73v	3	8
[illegible]	itlaaā	Sagittarius	73v	5	6
[illegible]	**itlaaāchi**	Scorpio	73r	1	4
[illegible]	**itlaaāchi**	Sagittarius	73v	3	11
[illegible]	itlaaāe	Virgo	72r(2)	2	14
[illegible]	itlaaāei	Scorpio	73r	3	5
[illegible]	**itlaāch**	Virgo	72r(2)	2	11
[illegible]	**itlaāch**	Libra	72v	2	8
[illegible]	**itlaachi**	Virgo	72r(2)	4	9
[illegible]	**itlaāchi**	Sagittarius	73v	3	5
[illegible]	itlaāe	Sagittarius	73v	3	3
[illegible]	itlaaoe	Sagittarius	73v	3	13
[illegible]	itlaatli	Libra	72v	2	15
[illegible]	itlācâoll	Pisces	70r	4	9
[illegible]	itlai	Sagittarius	73v	5	5
[illegible]	itlāoe tsāe	Cancer	71v(4)	6	3
[illegible]	itlocâi	Libra	72v	2	11
[illegible]	itlocâocācâ	Cancer	71v(4)	2	10
[illegible]	itlocâtschi	Taurus	71v(2)	2	3
[illegible]	itlococâ	Cancer	71v(4)	4	1
[illegible]	itlololl ācocâ	Cancer	71v(4)	4	5
[illegible]	maāahui	Libra	72v	2	14
[illegible]	maāatlai	Virgo	72r(2)	2	5
[illegible]	maācâchi	Virgo	72r(2)	2	6
[illegible]	maān?	Virgo	72r(2)	4	7

(continued)

Table 2 (continued)

Voynichese words	Latin	Zodiac	Folio	Ring (outer ring = 1)	Order (12 o'clock = 1, clockwise)
[illegible]	mācâchi	Virgo	72r(2)	2	16
[illegible]	mācui	Scorpio	73r	1	2
[illegible]	mahui	Scorpio	73r	5	5
[illegible]	mānoe	Gemini	71v(3)	5	5
[illegible]	maoei	Aries	71r	2	6
[illegible]	mchi	Scorpio	73r	3	10
[illegible]	mchollcholi	Taurus	71v(2)	2	10
[illegible]	**mcuai**	Pisces	70r	2	8
[illegible]	**mcuai**	Leo	72r(1)	4	3
[illegible]	mhuocâi	Taurus	71v(1)	2	10
[illegible]	mocâaai	Leo	72r(1)	4	1
[illegible]	moe ānāh	Taurus	71v(1)	2	9
[illegible]	moe ocâahu	Taurus	71v(2)	2	4
[illegible]	nocâācân	Pisces	70r	4	7
[illegible]	nocâocâ	Gemini	71v(3)	5	1
[illegible]	oaocâ	Cancer	71v(4)	6	5
[illegible]	ocâpi	Taurus	71v(1)	2	3
[illegible]	oco?	Virgo	72r(2)	2	2
[illegible]	ococâi	Gemini	71v(3)	5	7
[illegible]	oe	Scorpio	73r	5	8
[illegible]	ohuocâ	Sagittarius	73v	5	9
[illegible]	oiâ	Pisces	70r	4	10
[illegible]	ollāch	Libra	72v	2	5
[illegible]	quātlaācâi	Sagittarius	73v	3	14
[illegible]	tsaāatli	Virgo	72r(2)	2	7
[illegible]	tsaācâ	Sagittarius	73v	3	16
[illegible]	tsaaoch āpai (unclear)	Leo	72r(1)	2	10
[illegible]	tsācâtschi	Taurus	71v(1)	4	4
[illegible]	tsatl	Scorpio	73r	5	7
[illegible]	tsatlocâ	Scorpio	73r	3	7
[illegible]	utlai	Sagittarius	73v	5	8

? = no decipherment given

Appendix C

Table 3 Star names in Voynichese compared with Arabic names

Folio	Voynichese	Translation	Similar Arabic star name from Allen (1899)	Comments
68r(1)	[illegible]	ācââe		
68r(2)	[illegible]	ācâmaanai		
68r(1)	[illegible]	āchmacui	Acamar, page 219	Eridanus (constellation in Southern Hemisphere)
68r(2)	[illegible]	āchocmācâ		
68r(2)	[illegible]	ācholl		
68r(1)	[illegible]	ācuai		
68r(1)	[illegible]	ācui		
68r(3) seg 3	[illegible]	ācuich		
68r(1)	[illegible]	āecholl		
68r(2)	[illegible]	āhuāpāe	(Al) Agribah, page 130 (= the Ravens)	Canis Minor
68r(2)	[illegible]	āhumaāchi		
68r(2)	[illegible]	āmāei		
68r(3)	[illegible]	Āocâmaācâ (word next to Pleiades)		*aoccampa* (Karttunen, 1983: 11) = no longer anywhere, or disappears, suggesting that Pleiades can disappear
68r(1)	[illegible]	āpi		
68r(2)	[illegible]	ātlāaaā	Al Awwā[a] (barking dog)	Virgo
68r(2)	[illegible]	ātlaaaaāe	Al Awwā (barking dog)	Virgo
68r(2)	[illegible]	ātlaaaāchi	Al Awwā + *chi* (barking dog)	Virgo
68r(2)	[illegible]	ātlaaaāe	Al Awwā (barking dog)	Virgo
68r(2)	[illegible]	ātlaāācâ	Al Awwā + *ca* (barking dog)	Virgo, perhaps star Spica
68r(1)	[illegible]	ātlaaāchocâ	Al Awwā (barker or barking dog) + *choca* (howl in Nahuatl)	Virgo, perhaps star Spica
68r(1)	[illegible]	ātlaāe	Al Awwā (barking dog)	Virgo
68r(1)	[illegible]	ātlācâ	Alacel (page 467)	Virgo
68r(1)	[illegible]	ātlācâchi		Spear thrower = fisherman (Nahuatl)

(continued)

Table 3 (continued)

Folio	Voynichese	Translation	Similar Arabic star name from Allen (1899)	Comments
68r(3) seg 5	[Voynich script]	ātlācâmi		
68r(2)	[Voynich script]	ātlācâmmi	Alacast and Alcalst (ε Virginis, page 471)	Virgo
68r(1)	[Voynich script]	ātlāchocâi	Alasch'a (γ Scorpio, page 370)	
68r(1)	[Voynich script]	ātlāe	Al Awwā (barking dog)??	Virgo
68r(3) seg 7	[Voynich script]	ātlāei		
68r(1)	[Voynich script]	ātlāmachi	Alamac, Alamak, Alamech (γ Andromedae, page 36)	Andromeda
68r(3) seg 7	[Voynich script]	ātlān	Alanac, Alanak, Alioc (α Aurigae, pages 85, 87); Alanin (Draco), page 205; Alanac, Alanak, Alioc (α Aurigae, pages 85, 87); Alanin (Draco, page 205)	Auriga; Draco; Auriga
68r(1)	[Voynich script]	ātlāocâi		
68r(2)	[Voynich script]	ātlātsāi		
68r(3) seg 3	[Voynich script]	ātlichh		
68r(1)	[Voynich script]	ātlichi	Alasch'a (γ Scorpio, page 370)	Scorpio
68r(1)	[Voynich script]	ātlie		
68r(1)	[Voynich script]	ātlitlmn		
68r(2)	[Voynich script]	ātlmaāchoe		
68r(3) seg 3	[Voynich script]	ātlmāchi	Alamac, Alamak, Alamech (γ Andromedae, page 36)	Andromeda
68r(1)	[Voynich script]	ātlmāchi	Alamac, Alamak, Alamech (γ Andromedae, page 36)	Andromeda
68r(2)	[Voynich script]	ātlmāe		
68r(1)	[Voynich script]	ātlmchā	Alamac, Alamak, Alamech (γ Andromedae, page 36)	Andromeda
68r(1)	[Voynich script]	ātltsāe		
68r(1)	[Voynich script]	ātltsai		
68r(1)	[Voynich script]	chācâmachi		
68r(3) seg 1	[Voynich script]	chāoei	(Al) Thurayya	Pleiades (*Chow* [Egyptian])

(continued)

Table 3 (continued)

Folio	Voynichese	Translation	Similar Arabic star name from Allen (1899)	Comments
68r(2)	[illegible]	chmaācâchi		
68r(2)	[illegible]	chmācâ		
68r(3) seg 1	[illegible]	Chmācâchoi (word next to Pleiades)		*chamahu(a)* in Karttunen (1983: 450 = a swelling up (i.e., large star) = Aldebaran
68r(3) seg 3	[illegible]	choeocâcâ		Similar to *chaoei* + *caca* (frog) in Nahuatl (*ad(difta* = frog in Arabic
68r(1)	[illegible]	itlmāchi	Alamac, Alamak, Alamech (γ Andromedae, page 36)	Andromeda
68r(1)	[illegible]	itlmchi	Alamac, Alamak, Alamech (γ Andromedae, page 36)	Andromeda
68r(3) seg 5	[illegible]??[illegible]	m??pi		
68r(2)	[illegible]	maāeācâ		
68r(2)	[illegible]	mācâoe		
68r(2)	[illegible]	māchoe		
68r(1)	[illegible]	māpi		
68r(3) seg 5	[illegible]	mchi itli		
68r(1)	[illegible]	mtlmitlācâi (word under the image of the sun)	???x?	Sun = *Shams* (Arabic) *Tonaltzintli* (Nahuatl), *Mamalhuaztli* = fire sticks for the New Fire ceremony
68r(2)	[illegible]	oichmi		
68r(1)	[illegible]	olloe		
68r(1)	[illegible]	paaāe		
68r(1)	[illegible]	pācui		
68r(1)	[illegible]	tlāaaāchmi	Al Awwā (barking dog) + *chmi*	Virgo
68r(2)	[illegible]	tlāchoeoocâi		
68r(2)	[illegible]	tschoe		
68r(2)	[illegible]	tsmi		

? = no decipherment given
seg = segment
[a]Al Awwā (or Al'Awwa)

Appendix D

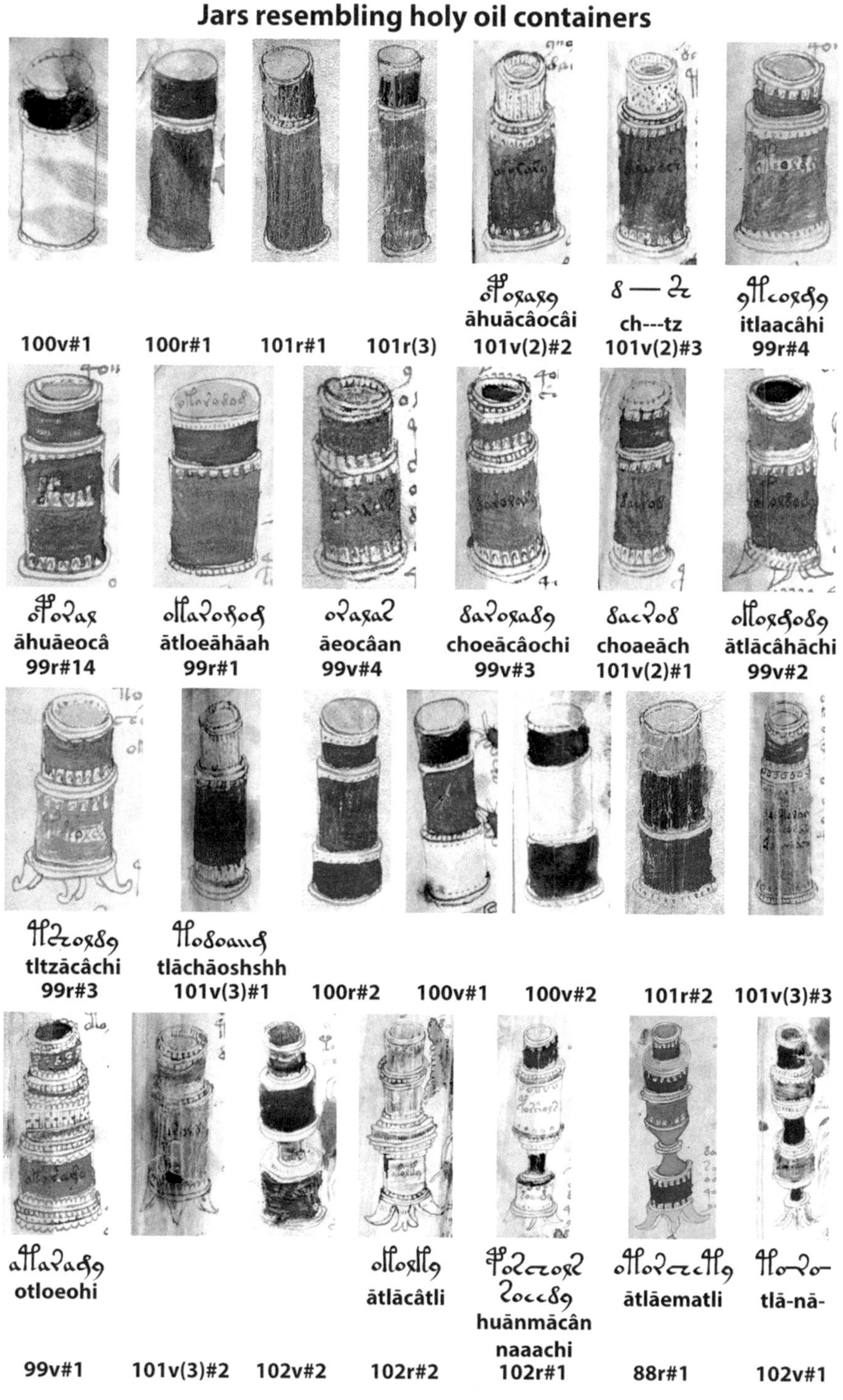

Jars resembling holy oil containers

100v#1 | 100r#1 | 101r#1 | 101r(3) | āhuācâocâi 101v(2)#2 | ch---tz 101v(2)#3 | itlaacâhi 99r#4

āhuāeocâ 99r#14 | ātloeāhāah 99r#1 | āeocâan 99v#4 | choeācâochi 99v#3 | choaeāch 101v(2)#1 | ātlācâhāchi 99v#2

tltzācâchi 99r#3 | tlāchāoshshh 101v(3)#1 | 100r#2 | 100v#1 | 100v#2 | 101r#2 | 101v(3)#3

otloeohi 99v#1 | 101v(3)#2 | 102v#2 | ātlācâtli 102r#2 | huānmācân naaachi 102r#1 | ātlāematli 88r#1 | tlā-nā- 102v#1

Fig. 1 Apothecary jars in the Pharmaceutical section rearranged by complexity

Jars resembling ciboria

āhuocâāshshei
89r#3

ātol iāeol
88v(3)#4

ātlācâi
88v(3)#2

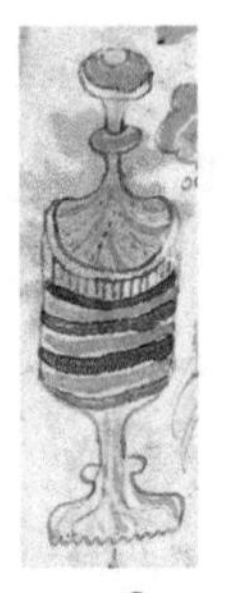
āchāei
88v(3)#1

itlich
88v(2)#2

ātloeācâi
89v#1

ātloei
89r#2

ātlocâchi
88r#2

ātlāeoha
88v(1)#2

itlāpi
88v(2)#3

ātlmtzi
88v(2)#1

choeohahocâ
88v(1)#3

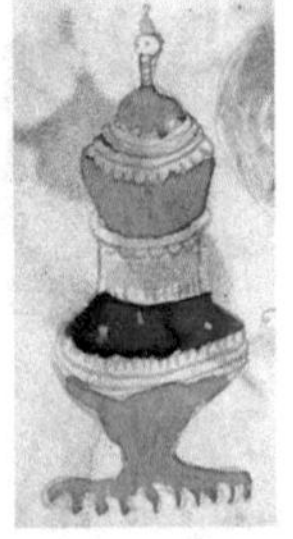
māaani
89r#1

tlāeoli
88v(3)#3

ātlācâhāchi
88r#13

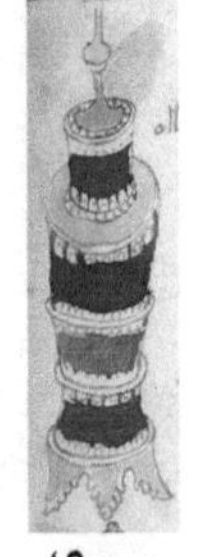
ātlācâi
88v(1)#1

tlāaaāeol
89v#2

Fig. 1 (continued)

Appendix E

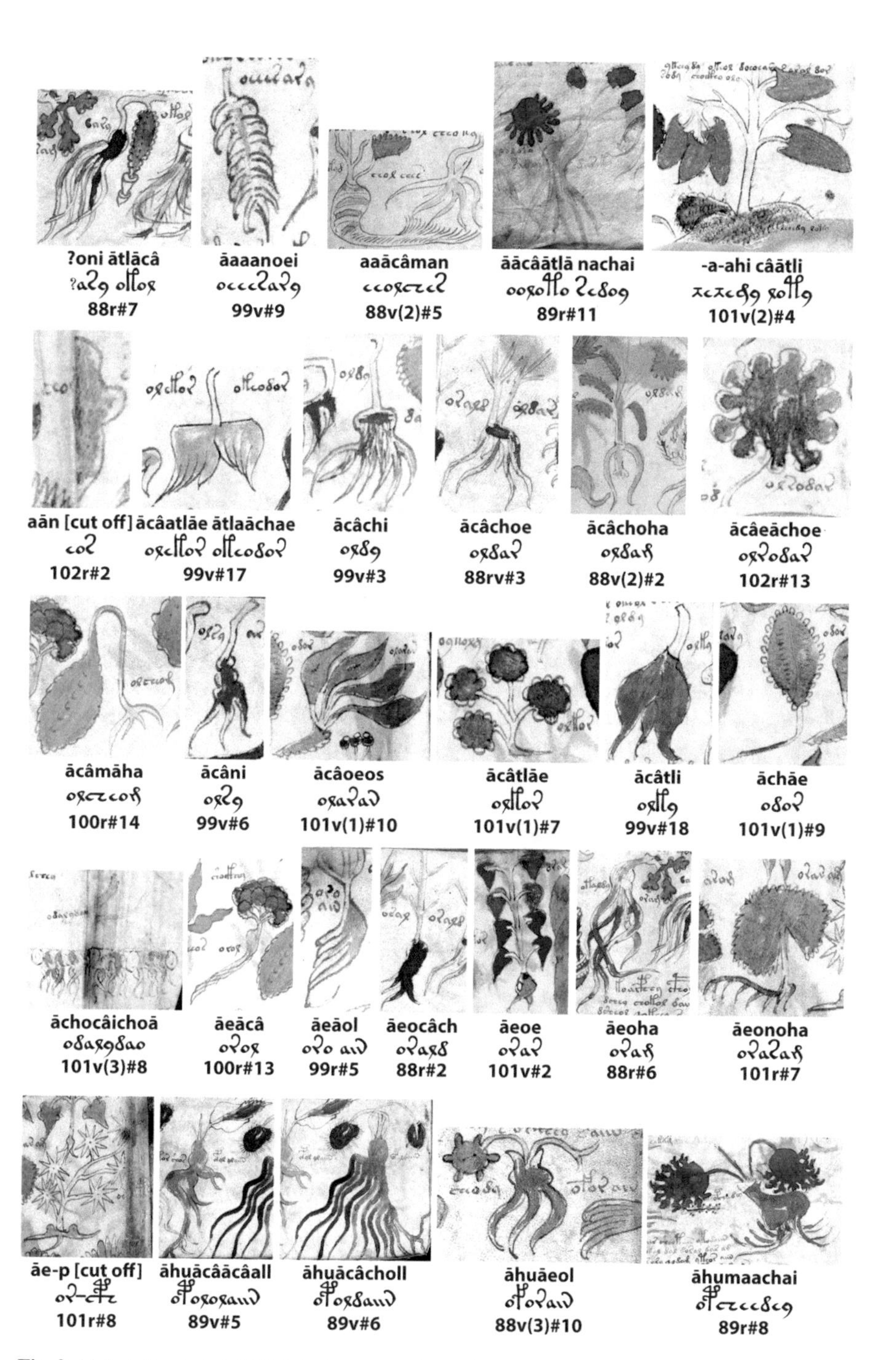

Fig. 2 Alphabetical list of 188 labeled phytomorphs in the Pharmaceutical section. The folio numbers are based on *The Voynich Manuscript: a facsimile of the complete work.* (Palatino Press, 2015)

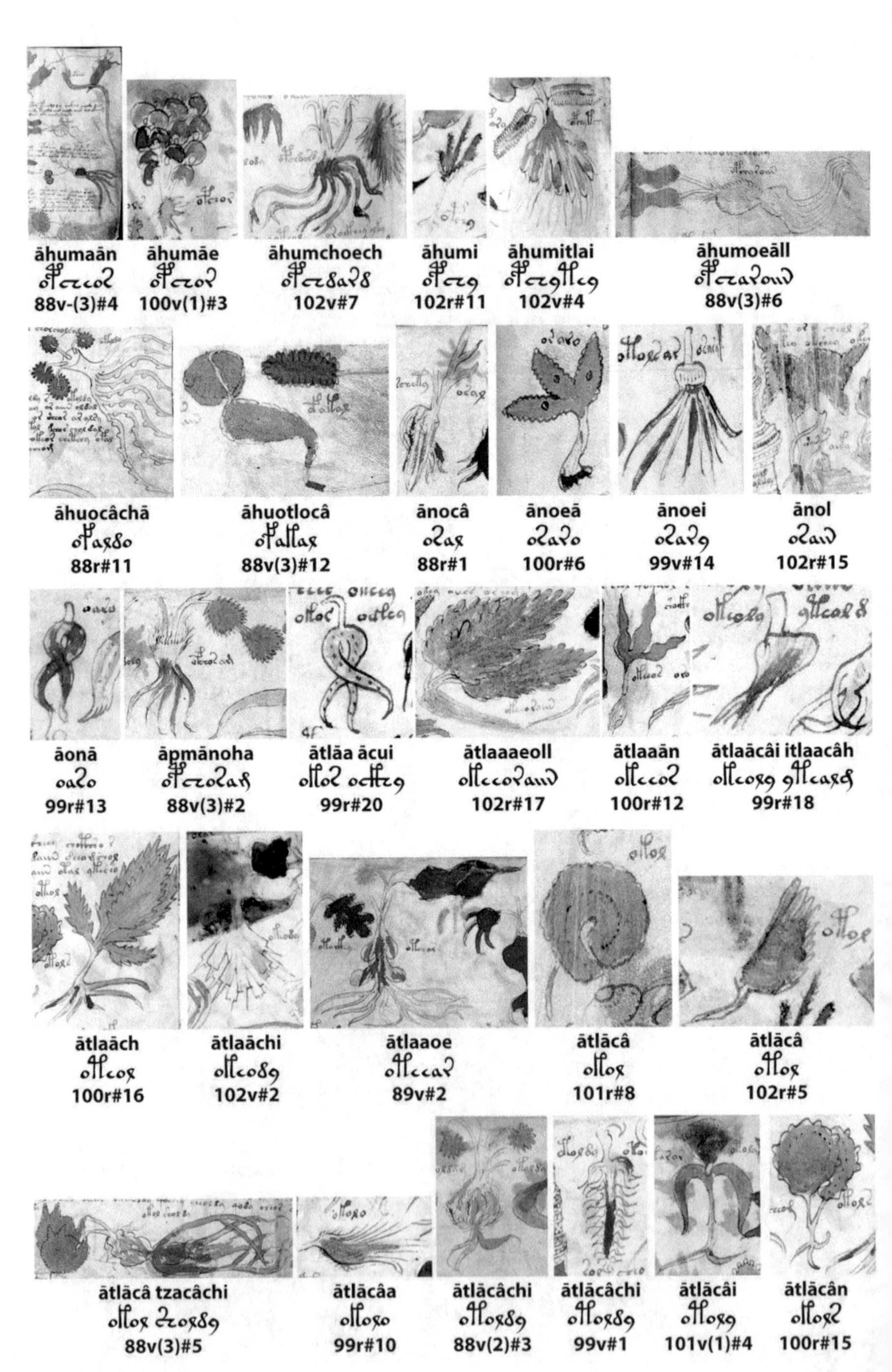

Fig. 2 (continued)

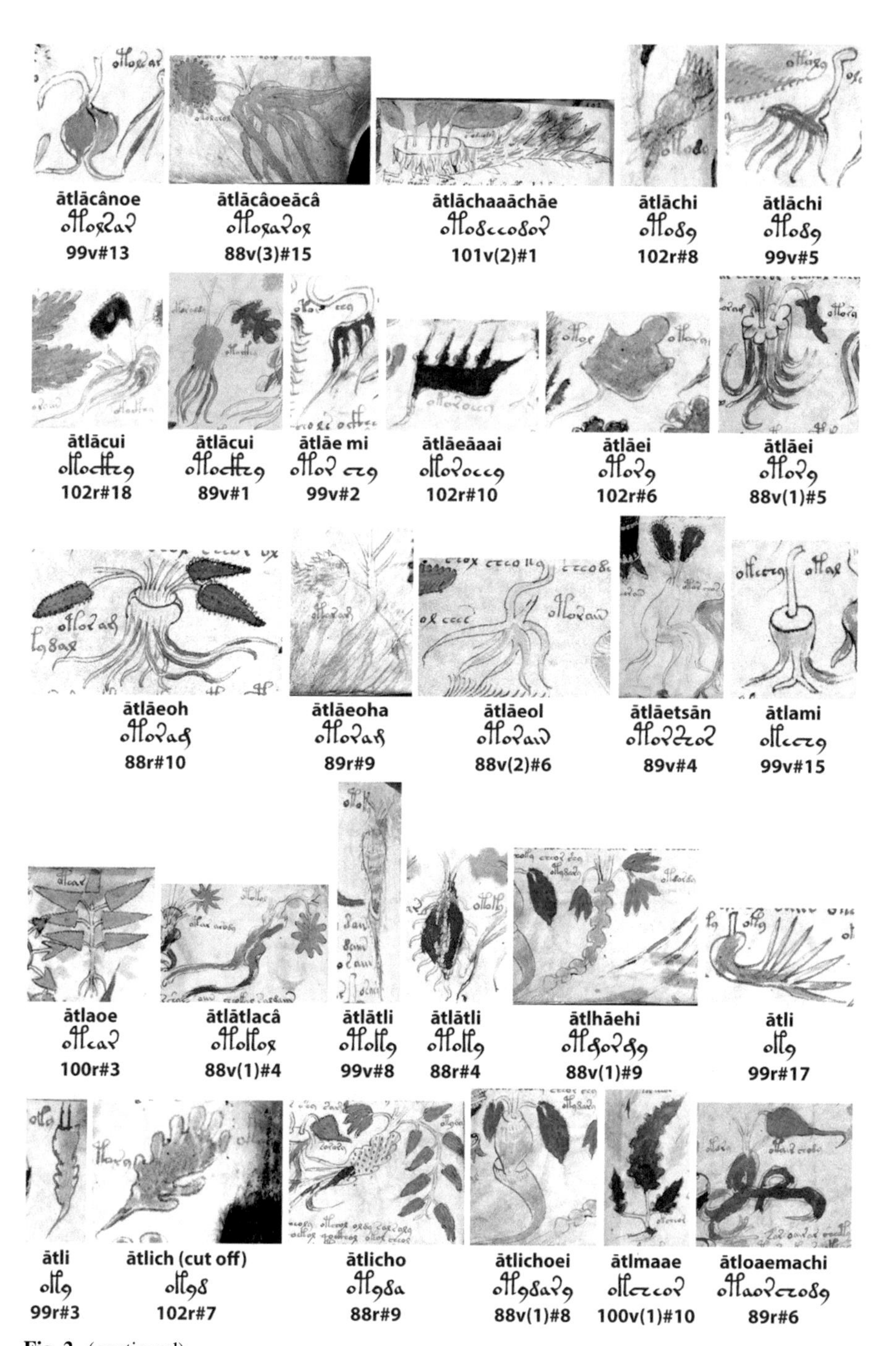

Fig. 2 (continued)

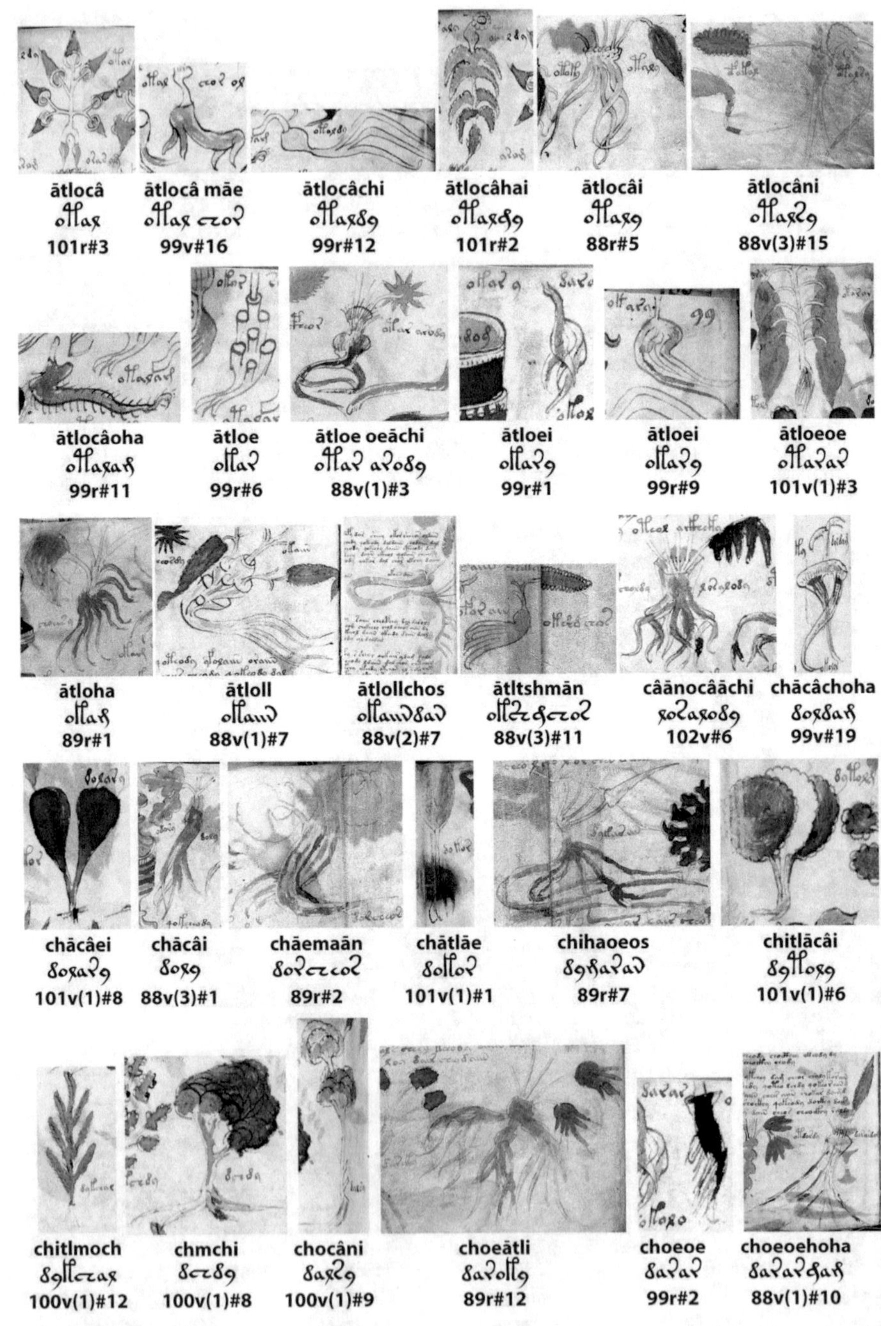

Fig. 2 (continued)

Fig. 2 (continued)

Fig. 2 (continued)

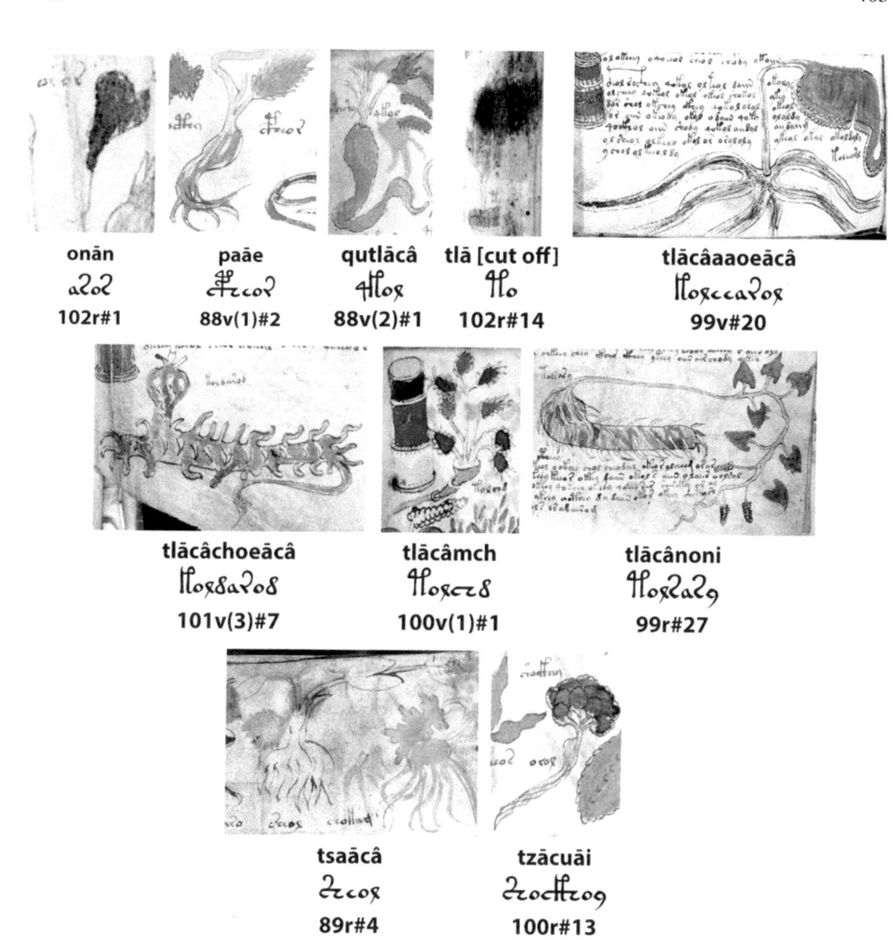

Fig. 2 (continued)

Index

J. Janick, A. O. Tucker, *Unraveling the Voynich Codex*, Fascinating Life Sciences, https://doi.org/10.1007/978-3-319-77294-3

Printed in the United States
By Bookmasters